Wind and Solar Power Systems

Mukund R. Patel, Ph.D., P.E.
U.S. Merchant Marine Academy
Kings Point, New York

Formerly
Principal Engineer, General Electric Company
Fellow Engineer, Westinghouse Reasearch Center

CRC Press
Boca Raton London New York Washington, D.C.

Contact Editor: Robert Stern
Project Editor: Ibrey Woodall
Marketing Manager: Barbara Glunn, Jane Lewis, Arline Massey, Jane Stark
Cover design: Dawn Boyd

Library of Congress Cataloging-in-Publication Data

Patel, Mukund R., 1942–
Wind and solar power systems / Mukund R. Patel.
p. cm.
Includes bibliographical references and index.
ISBN 0-8493-1605-7 (alk. paper)
1. Wind power plants. 2. Solar power plants. 3. Photovoltaic power systems. I. Title.
TK1541.P38 1999
621.31'2136--dc21 98-47934
CIP

Direct all inquiries to CRC Press LLC, 2000 Corporate Blvd., N.W., Boca Raton, Florida 33431.

Direct technical matters to Mukund R. Patel, 1199 Cobblestone Court, Yardley, Pa 19067.

International Standard Book Number 0-8493-1605-7
Library of Congress Card Number 98-47934
Printed in the United States of America 1 2 3 4 5 6 7 8 9 0
Printed on acid-free paper

...dedicated

to my mother, Shakariba,

who practiced ingenuity,

and

to my children, Ketan, Bina, and Vijul,

who flattered me by being engineers.

(Cover photo: Baix Ebre wind farm in Catalonia. With permission from Institut Catalia d'Energia, Spain.)

Preface

The total electricity demand in 1997 in the United States of America was three trillion kWh, with the market value of $210 billion. The worldwide demand was 12 trillion kWh in 1997, and is projected to reach 19 trillion kWh in 2015. This constitutes the worldwide average annual growth of 2.6 percent. The growth rate in the developing countries is projected to be approximately 5 percent, almost twice the world average.

Most of the present demand in the world is met by fossil and nuclear power plants. A small part is met by renewable energy technologies, such as the wind, solar, biomass, geothermal and the ocean. Among the renewable power sources, wind and solar have experienced a remarkably rapid growth in the past 10 years. Both are pollution free sources of abundant power. Additionally, they generate power near the load centers, hence eliminate the need of running high voltage transmission lines through rural and urban landscapes.

Since the early 1980s, the wind technology capital costs have declined by 80 percent, operation and maintenance costs have dropped by 80 percent and availability factors of grid-connected plants have risen to 95 percent. These factors have jointly contributed to the decline of the wind electricity cost by 70 percent to 5 to 7 cents per kWh. The grid-connected wind plant can generate electricity at cost under 5 cents per kWh. The goal of ongoing research programs funded by the U.S. Department of Energy and the National Renewable Energy Laboratory is to bring the wind power cost below 4 cents per kWh by the year 2000. This cost is highly competitive with the energy cost of the conventional power technologies. For these reasons, wind power plants are now supplying economical clean power in many parts of the world.

In the U.S.A., several research partners of the NREL are negotiating with U.S. electrical utilities to install additional 4,200 MW of wind capacity with capital investment of about $2 billion during the next several years. This amounts to the capital cost of $476 per kW, which is comparable with the conventional power plant costs. A recent study by the Electric Power Research Institute projected that by the year 2005, wind will produce the cheapest electricity available from any source. The EPRI estimates that the wind energy can grow from less than 1 percent in 1997 to as much as 10 percent of this country's electrical energy demand by 2020.

On the other hand, the cost of solar photovoltaic electricity is still high in the neighborhood of 15 to 25 cents per kWh. With the consumer cost of electrical utility power ranging from 10 to 15 cents per kWh nationwide, photovoltaics cannot economically compete directly with the utility power as yet, except in remote markets where the utility power is not available and the transmission line costs would be prohibitive. Many developing countries have large areas falling in this category. With ongoing research in the photovoltaic (pv) technologies around the world, the pv energy cost is expected to fall to 12 to 15 cents per kWh or less in the next several years as the learning curves and the economy of scale come into play. The research programs funded by DOE/NREL have the goal of bringing down the pv energy cost below 12 cents per kWh by 2000.

After the restructuring of the U.S. electrical utilities, as mandated by the Energy Policy Act (EPAct) of 1992, the industry leaders expect the power generation business, both conventional and renewable, to become more profitable in the long run. The reasoning is that the generation business will be stripped of regulated price and opened to competition among electricity producers and resellers. The transmission and distribution business, on the other hand, would still be regulated. The American experience indicates that the free business generates more profits than the regulated business. Such is the experience in the U.K. and Chile, where the electrical power industry had been structured similar to the EPAct of 1992 in the U.S.A.

As for the wind and pv electricity producers, they can now sell power freely to the end users through truly open access to the transmission lines. For this reason, they are likely to benefit as much as other producers of electricity. Another benefit in their favor is that the cost of the renewable energy would be falling as the technology advances, whereas the cost of the electricity from the conventional power plants would rise with inflation. The difference in their trends would make the wind and pv power even more advantageous in the future.

About the Author

Mukund R. Patel, Ph.D, P.E., is an experienced research engineer with 35 years of hands-on involvement in designing and developing state-of-the-art electrical power equipment and systems. He has served as principal power system engineer at the General Electric Company in Valley Forge, fellow engineer at the Westinghouse Research & Development Center in Pittsburgh, senior staff engineer at Lockheed Martin Corporation in Princeton, development manager at Bharat Bijlee Limited, Bombay, and 3M distinguished visiting professor of electrical power technologies at the University of Minnesota, Duluth. Presently he is a professor at the U.S. Merchant Marine Academy in Kings Point, New York.

Dr. Patel obtained his Ph.D. degree in electric power engineering from the Rensselaer Polytechnic Institute, Troy, New York; M.S. in engineering management from the University of Pittsburgh; M.E. in electrical machine design from Gujarat University and B.E.E. from Sardar University, India. He is a fellow of the Institution of Mechanical Engineers (U.K.), senior member of the IEEE, registered professional engineer in Pennsylvania, and a member of Eta Kappa Nu, Tau Beta Pi, Sigma Xi and Omega Rho.

Dr. Patel has presented and published over 30 papers at national and international conferences, holds several patents, and has earned NASA recognition for exceptional contribution to the photovoltaic power system design for UARS. He is active in consulting and teaching short courses to professional engineers in the electrical power industry.

About the Book

The book was conceived when I was invited to teach a course in the emerging electrical power technologies at the University of Minnesota in Duluth. The lecture notes and presentation charts I prepared for the course formed the first draft of the book. The subsequent teaching of a couple of short courses to professional engineers advanced the draft closer to the finished book. The book is designed and tested to serve as textbook for a semester course for university seniors in electrical and mechanical engineering fields. The practicing engineers will get detailed treatment of this rapidly growing segment of the power industry. The government policy makers would benefit by overview of the material covered in the book.

Chapters 1 through 3 cover the present status and the ongoing research programs in the renewable power around the world and in the U.S.A. Chapter 4 is a detailed coverage on the wind power fundamentals and the probability distributions of the wind speed and the annual energy potential of a site. It includes the wind speed and energy maps of several countries. Chapter 5 covers the wind power system operation and the control requirements. Since most wind plants use induction generators for converting the turbine power into electrical power, the theory of the induction machine performance and operation is reviewed in Chapter 6 without going into details. The details are left for the classical books on the subject. The electrical generator speed control for capturing the maximum energy under wind fluctuations over the year is presented in Chapter 7.

The power-generating characteristics of the photovoltaic cell, the array design, and the sun-tracking methods for the maximum power generation are discussed in Chapter 8. The basic features of the utility-scale solar thermal power plant using concentrating heliostats and molten salt steam turbine are presented in Chapter 9.

The stand-alone renewable power plant invariably needs energy storage for high load availability. Chapter 10 covers characteristics of various batteries, their design methods using the energy balance analysis, factors influencing their operation, and the battery management methods. The energy density and the life and operating cost per kWh delivered are presented for various batteries, such as lead-acid, nickel-cadmium, nickel-metal-hydride and lithium-ion. The energy storage by the flywheel, compressed air and the superconducting coil, and their advantages over the batteries are reviewed. The basic theory and operation of the power electronic converters and inverters used in the wind and solar power systems are presented in Chapter 11, leaving details for excellent books available on the subject.

The more than two billion people in the world not yet connected to the utility grid are the largest potential market of stand-alone power systems. Chapter 12 presents the design and operating methods of such power systems using wind and photovoltaic systems in hybrid with diesel generators. The newly developed fuel cell with potential of replacing diesel engine in urban areas is discussed. The grid-connected renewable power systems are covered in Chapter 13, with voltage and frequency control methods needed for synchronizing the generator with the grid. The theory and the operating characteristics of the interconnecting transmission line, the voltage regulation, the maximum power transfer capability, and the static and dynamic stability are covered.

Chapter 14 is about the overall electrical system design. The method of designing the system components to operate at their maximum possible efficiency is developed. The static and dynamic bus performance, the harmonics, and the increasingly important quality of power issues applicable to the renewable power systems are presented.

Chapter 15 discusses the total plant economy and the costing of energy delivered to the paying customers. It also shows the importance of a sensitivity analysis to raise confidence level of the investors. The profitability charts are presented for preliminary screening of potential sites. Finally, Chapter 16 discusses the past and present trends and the future of the green power. It presents the declining price model based on the learning curve, and the Fisher-Pry substitution model for predicting the market growth of the wind and pv power based on historical data on similar technologies. The effect of the utility restructuring, mandated by the EPAct of 1992, and its expected benefits on the renewable power producers are discussed.

At the end, the book gives numerous references for further reading, and name and addresses of government agencies, universities, and manufacturers active in the renewable power around the world.

Acknowledgment

The book of this nature on emerging technologies, such as the wind and photovoltaic power systems, cannot possibly be written without the help from many sources. I have been extremely fortunate to receive full support from many organizations and individuals in the field. They not only encouraged me to write the book on this timely subject, but also provided valuable suggestions and comments during the development of the book.

Dr. Nazmi Shehadeh, head of the Electrical and Computer Engineering Department at the University of Minnesota, Duluth, gave me the opportunity to develop and teach this subject to his students who were enthusiastic about learning new technologies. **Dr. Elliott Bayly**, president of the World Power Technologies in Duluth, shared with me and my students his long experience in the field. He helped me develop the course outline, which later became the book outline. **Dr. Jean Possbic** of Solarex Corporation in Frederick, Maryland and **Mr. Carl-Erik Olsen** of Nordtank Energy Group / NEG Micon, Denmark, kindly reviewed the draft and provided valuable suggestions for improvement. **Mr. Bernard Chabot** of ADEME, Valbonne, France, provided the profitability charts for screening the wind and photovoltaic power sites. **Mr. Ian Baring-Gould** of the National Renewable Energy Laboratory, Golden, Colorado, has been a source of useful information and the hybrid power plant simulation model.

Several institutions worldwide provided current data and reports on these rather rapidly developing technologies. They are the **American Wind Energy Association**, the **American Solar Energy Society**, the **European Wind Energy Association**, the **Risø National Laboratory**, Denmark, the **Tata Energy Research Institute**, India, and many corporations engaged in the wind and solar power technologies. Many individuals at these organizations gladly provided help I requested.

I gratefully acknowledge the generous support from all of you.

Mukund Patel
Yardley, Pennsylvania

Contents

1

Introduction

1.1 Industry Overview

The total annual primary energy consumption in 1997 was 390 quadrillion (10^{15}) BTUs worldwide[1] and over 90 quadrillion BTUs in the United States of America, distributed in segments shown in Figure 1-1. About 40 percent of the total primary energy is used in generating electricity. Nearly 70 percent of the energy used in our homes and offices is in the form of electricity. To meet this demand, 700 GW of electrical generating capacity is now installed in the U.S.A. For most of this century, the U.S. electric demand has increased with the gross national product (GNP). At that rate, the U.S. will need to install additional 200 GW capacity by the year 2010.

The new capacity installation decisions today are becoming complicated in many parts of the world because of difficulty in finding sites for new generation and transmission facilities of any kind. In the U.S.A., no nuclear power plants have been ordered since 1978[2] (Figure 1-2). Given the potential for cost overruns, safety related design changes during the construction, and local opposition to new plants, most utility executives suggest that none will be ordered in the foreseeable future. Assuming that no new nuclear plants are built, and that the existing plants are not relicensed at the expiration of their 40-year terms, the nuclear power output is expected to decline sharply after 2010. This decline must be replaced by other means. With gas prices expected to rise in the long run, utilities are projected to turn increasingly to coal for base load-power generation. The U.S.A. has enormous reserves of coal, equivalent to more than 250 years of use at current level. However, that will need clean coal burning technologies that are fully acceptable to the public.

An alternative to the nuclear and fossil fuel power is renewable energy technologies (hydro, wind, solar, biomass, geothermal, and ocean). Large-scale hydroelectric projects have become increasingly difficult to carry through in recent years because of competing use of land and water. Relicensing requirements of existing hydro plants may even lead to removal of some dams to protect or restore wildlife habitats. Among the other renewable

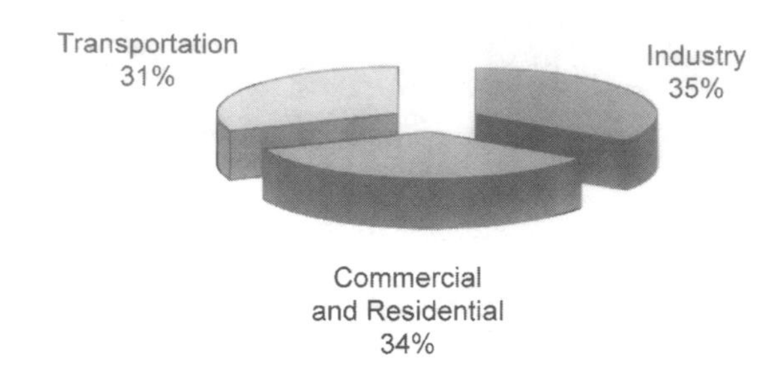

FIGURE 1-1
Primary energy consumption in the U.S.A. in three major sectors, total 90 quadrillion BTUs in 1997. (From U.S. Department of Energy, Office of the Integrated Analysis and Forecasting, Report No. DE-97005344, April 1997.)

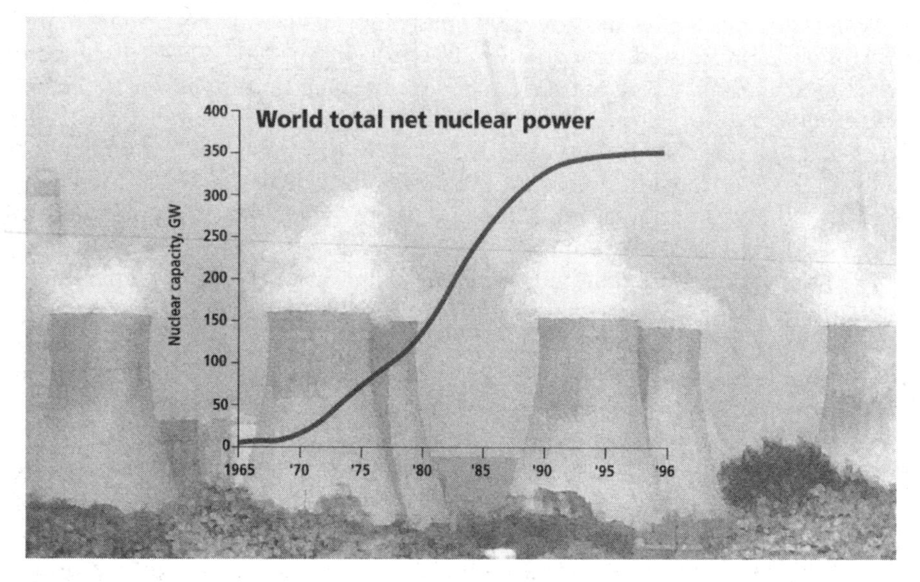

FIGURE 1-2
The stagnant nuclear power capacity worldwide. (From Felix, F., State of the nuclear economy, IEEE Spectrum, November 1997. ©1997 IEEE. With permission.)

power sources, wind and solar have recently experienced a rapid growth around the world. Having wide geographical spread, they can be generated near the load centers, thus simultaneously eliminating the need of high voltage transmission lines running through rural and urban landscapes.

The present status and benefits of the renewable power sources are compared with the conventional ones in Tables 1-1 and 1-2, respectively.

The renewables compare well with the conventionals in economy. Many energy scientists and economists believe that the renewables would get much more federal and state incentives if their social benefits were given full credit.

TABLE 1-1

Status of Conventional and Renewable Power Sources

Conventional	Renewables
Coal, nuclear, oil, and natural gas	Wind, solar, biomass geothermal, and ocean
Fully matured technologies	Rapidly developing technologies
Numerous tax and investment subsidies embedded in national economies	Some tax credits and grants available from some federal and/or state governments
Accepted in society under the 'grandfather clause' as necessary evil	Being accepted on its own merit, even with limited valuation of their environmental and other social benefits

TABLE 1-2

Benefits of Using Renewable Electricity

Traditional Benefits	Nontraditional Benefits Per Million kWh consumed
Monetary value of kWh consumed	Reduction in emission
U.S. average 12 cents/kWh	750–1000 tons of CO_2
U.K. average 7.5 pence/kWh	7.5–10 tons of SO_2
	3–5 tons of NOx
	50,000 kWh reduction in energy loss in power lines and equipment
	Life extension of utility power distribution equipment
	Lower capital cost as lower capacity equipment can be used (such as transformer capacity reduction of 50 kW per MW installed)

For example, the value of not generating one ton of CO_2, SO_2, and NOx, and the value of not building long high voltage transmission lines through rural and urban areas are not adequately reflected in the present evaluation of the renewables.

1.2 Incentives for Renewables

A great deal of renewable energy development in the U.S.A. occurred in the 1980s, and the prime stimulus for it was the passage in 1978 of the Public Utility Regulatory Policies Act (PURPA). It created a class of nonutility power generators known as the "qualified facilities (QFs)". The QFs were defined to be small power generators utilizing renewable energy sources and/or cogeneration systems utilizing waste energy. For the first time, PURPA required electric utilities to interconnect with QFs and to purchase QFs' power generation at "avoided cost", which the utility would have incurred by generating that power by itself. PURPA also exempted QFs from certain federal and state utility regulations. Furthermore, significant federal investment tax credit, research and development tax credit, and energy tax credit, liberally available up to the mid 1980s, created a wind rush in California, the state that also gave liberal state tax incentives. As of now, the financial incentives in the U.S.A. are reduced, but are still available under the Energy Policy Act of 1992, such as the energy tax credit of 1.5 cents per kWh. The potential impact of the 1992 act on renewable power producers is reviewed in Chapter 16.

Globally, many countries offer incentives and guaranteed price for the renewable power. Under such incentives, the growth rate of the wind power in Germany and India has been phenomenal.

1.3 Utility Perspective

Until the late 1980s, the interest in the renewables was confined primarily among private investors. However, as the considerations of fuel diversity, environmental concerns and market uncertainties are becoming important factors into today's electric utility resource planning, renewable energy technologies are beginning to find their place in the utility resource portfolio. Wind and solar power, in particular, have the following advantages to the electric utilities:

- Both are highly modular in that their capacity can be increased incrementally to match with gradual load growth.
- Their construction lead time is significantly shorter than those of the conventional plants, thus reducing the financial and regulatory risks.
- They bring diverse fuel sources that are free of cost and free of pollution.

Because of these benefits, many utilities and regulatory bodies are increasingly interested in acquiring hands on experience with renewable energy technologies in order to plan effectively for the future. The above benefits are discussed below in further details.

1.3.1 Modularity

The electricity demand in the U.S.A. grew at 6 to 7 percent until the late 1970s, tapering to just 2 percent in the 1990s as shown in Figure 1-3.

The 7 percent growth rate of the 1970s meant doubling the electrical energy demand and the installed capacity every 10 years. The decline in the growth rate since then has come partly from the improved efficiency in electricity utilization through programs funded by the U.S. Department of Energy. The small growth rate of the 1990s is expected to continue well into the next century.

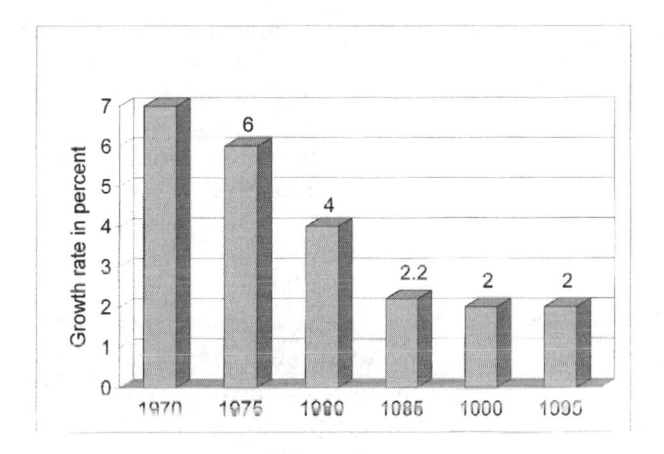

FIGURE 1-3
Growth of electricity demand in the U.S.A. (Source: U.S. Department of Energy and Electric Power Research Institute)

The economic size of the conventional power plant has been 500 MW to 1,000 MW capacity. These sizes could be justified in the past, as the entire power plant of that size, once built, would be fully loaded in just a few years. At a 2 percent growth rate, however, it could take decades before a 500 MW plant could be fully loaded after it is commissioned in service. Utilities are unwilling to take such long-term risks in making investment decisions. This has created a strong need of modularity in today's power generation industry.

Both the wind and the solar photovoltaic power are highly modular. They allow installations in stages as needed without losing the economy of size in the first installation. The photovoltaic (pv) is even more modular than the wind. It can be sized to any capacity, as the solar arrays are priced directly by the peak generating capacity in watts, and indirectly by square foot. The wind power is modular within the granularity of the turbine size. Standard wind turbines come in different sizes ranging from tens of kW to hundreds of kW. Prototypes of a few MW wind turbines are also tested and are being made commercially available in Europe. For utility scale installations, standard wind turbines in the recent past have been around 300 kW, but is now in the 500-1,000 kW range. A large plant consists of the required number and size of wind turbines for the initially needed capacity. More towers are added as needed in the future with no loss of economy.

For small grids, the modularity of the pv and wind systems is even more important. Increasing demand may be more economically added in smaller increments of the green power capacity. Expanding or building a new conventional power plant in such cases may be neither economical nor free from the market risk. Even when a small grid is linked by transmission line to the main network, installing a wind or pv plant to serve growing demand may be preferable to laying another transmission line. Local renewable power plants can also benefit small power systems by moving generation near the load, thus reducing voltage drop at the end of a long overloaded line.

In the developing countries like China and India, the demand has been rising at a 10 percent growth rate or more. This growth rate, when viewed with the large population base, makes these two countries rapidly growing electrical power markets for all sources of electrical energy, including the renewables.

1.3.2 Emission-Free

In 1995, the U.S.A. produced 3 trillion kWh of electricity, 70 percent of it (2 trillion kWh) from fossil fuels, a majority of that came from coal. The resulting emission is estimated to be 2 billion tons of CO_2, 15 million tons of SO_2 and 6 million tons of NOx. The health effects of these emissions are of significant concern to the U.S. public. The electromagnetic field emission around the high voltage transmission lines is another concern that has also recently become an environmental issue.

For these benefits, the renewable energy sources are expected to find importance in the energy planning in all countries around the world.

References

1. U.S. Department of Energy. 1997. "International Energy Outlook 1997 with Projections to 2015," *DOE Office of the Integrated Analysis and Forecasting, Report No. DE-97005344,* April 1997.
2. Felix, F. 1992. "State of the Nuclear Economy," *IEEE Spectrum,* November 1997, p. 29-32.

2

Wind Power

The first use of wind power was to sail ships in the Nile some 5000 years ago. The Europeans used it to grind grains and pump water in the 1700s and 1800s. The first windmill to generate electricity in the rural U.S.A. was installed in 1890. Today, large wind-power plants are competing with electric utilities in supplying economical clean power in many parts of the world.

The average turbine size of the wind installations has been 300 kW until the recent past. The newer machines of 500 to 1,000 kW capacity have been developed and are being installed. Prototypes of a few MW wind turbines are under test operations in several countries, including the U.S.A. Figure 2-1 is a conceptual layout of modern multimegawatt wind tower suitable for utility scale applications.[1]

Improved turbine designs and plant utilization have contributed to a decline in large scale wind energy generation costs from 35 cents per kWh in 1980 to less than 5 cents per kWh in 1997 in favorable locations (Figure 2-2). At this price, wind energy has become one of the least-cost power sources. Major factors that have accelerated the wind-power technology development are as follows:

- high-strength fiber composites for constructing large low-cost blades.
- falling prices of the power electronics.
- variable-speed operation of electrical generators to capture maximum energy.
- improved plant operation, pushing the availability up to 95 percent.
- economy of scale, as the turbines and plants are getting larger in size.
- accumulated field experience (the learning curve effect) improving the capacity factor.

2.1 Wind in the World

The wind energy stands out to be one of the most promising new sources of electrical power in the near term. Many countries promote the wind-power

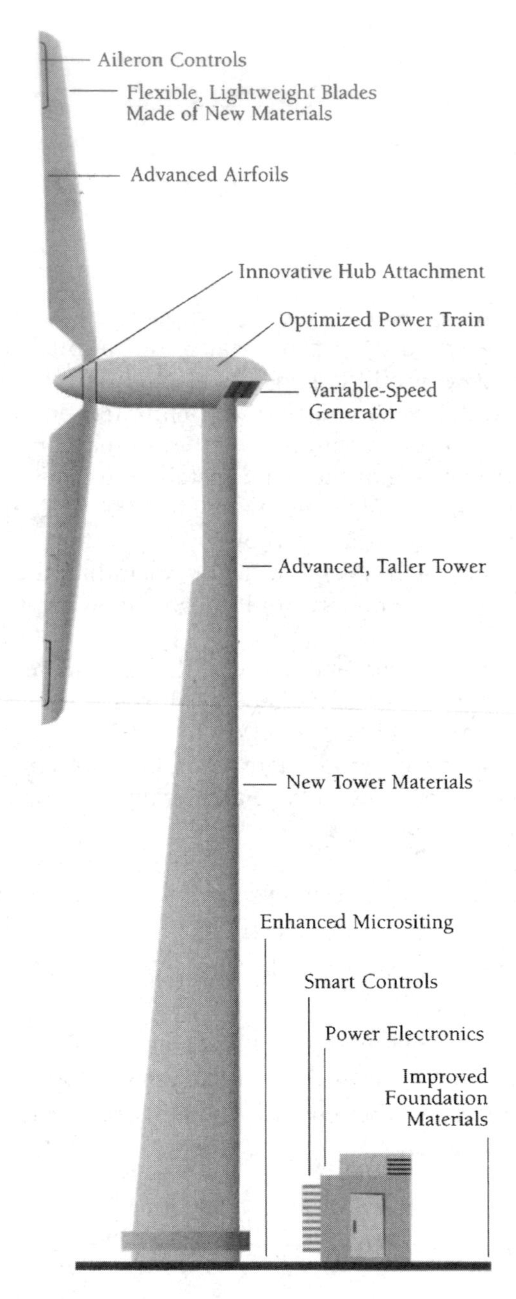

FIGURE 2-1
Modern wind turbine for utility scale power generation.

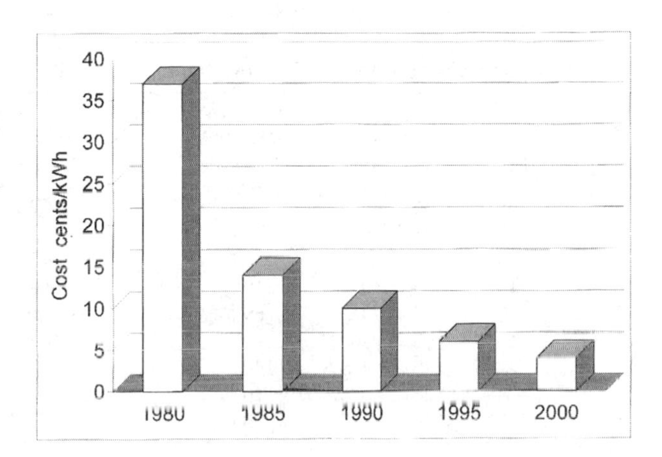

FIGURE 2-2
Declining cost of wind-generated electricity. (Source: AWEA/DOE/IEA.)

technology by national programs and market incentives. The International Energy Agency (IEA), with funding from 14 countries, supports joint research projects and information exchange on wind-power development.[2] These countries are Austria, Canada, Denmark, Finland, Germany, Italy, Japan, the Netherlands, New Zealand, Norway, Spain, Sweden, the United Kingdom, and the United States of America. By the beginning of 1995, more than 25,000 grid-connected wind turbines were operating in the IEA-member countries, amounting to a rated power capacity of about 3,500 MW. Collectively, these turbines are producing more than 6 million MWh of energy every year. The annual rate of capacity increase presently is about 600 MW.

According to the AWEA and the IEA, the 1994, 1995, and 1997 installed capacity in countries listed in Table 2-1 were 3,552 MW, 4,776 MW, and 7,308 MW, respectively. The 1995 sales of new plants set a record of 1224 MW ($1.5 billion), boosting the global installed capacity by 35 percent to nearly 4,776 MW. The most explosive growth occurred in Germany installing 500 MW and India adding 383 MW to their wind capacities. New wind plants installed in 1996-97 added another 2,532 MW to that total with an annual growth rate of 24 percent.

Much of the new development around the world can be attributed to government policies to promote the renewables energy sources. For example, the United Kingdom's nonfossil fuel obligation program will add 500 MW of wind power to the UK's power grid within this decade.

A 1994 study, commissioned by the American Wind Energy Association (AWEA) with Arthur D. Little Inc., concluded that in the 10 overseas wind farm markets, between 1,935 MW and 3,525 MW of wind capacity would be added by the year 2000 (Table 2-2). This translates to between $2 to $3.5 billion in sales.

TABLE 2-1

Installed Wind Capacity in Selected Countries, 1994, 1995 and 1997

Country	1994 MW	1995 MW	1997 MW	Growth 1994-1995 Percent	Annual Growth Rate 1995-97 Percent
Germany	643	1136	2079	76.7	35.2
United States	1785	1828	2000	2.4	4.7
Denmark	540	614	1141	13.7	36.3
India	182	565	1000	210	33.0
Netherlands	153	259	325	69	12.0
United Kingdom	147	193	308	31	26.3
Spain	72	145	455	100	77.1
China	30	36	—	20	—
TOTAL	3552	4776	7308	35.4	23.7

(Source: United States data from Energy Information Administration, Annual Energy Outlook 1997, DOE/EIA Report No. 0383-97, Table A17, Washington, D.C., December 1996. Other countries data from the Amercian Wind Energy Association, Satus Report of International Wind Projects, Washington, D.C., March 1996.)

TABLE 2-2

Projected Wind Capacity Addition in Megawatts Between 1994 and 2000

Country	Addition planned Megawatts
United Kingdom	100–300
Spain	150–250
Germany	200–350
India	700–1200
China	350–600
Mexico	150–300
Argentina	100–150
Chile	100–200
Australia	50–75
New Zealand	50–100
Total	1950–3525

(Source: American Wind Energy Association/Arthur D. Little, Inc.)

2.2 The U.S.A.

Large-scale wind-power development in the U.S.A. has been going on since the late 1970s. In 1979, a 2 MW experimental machine was installed jointly by the Department of Energy and NASA on the Howard Knob Mountain near Boone, North Carolina. The system was designed and built by the

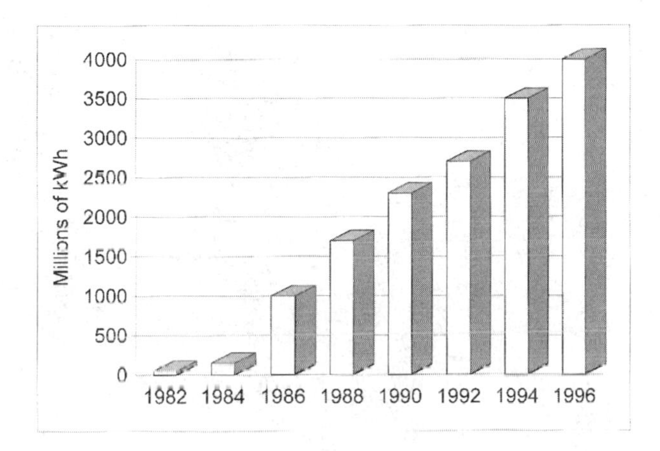

FIGURE 2-3
Electricity generated by U.S. wind-power plants since 1982. (Source: AWEA. With permission.)

General Electric Company using two 61-meter diameter rotor blades from Boeing Aerospace Corporation. It was connected to the local utility grid and operated successfully.

Beginning 1984, the electrical energy derived from the wind and delivered to the paying users is increasing rapidly (Figure 2-3). Until the late 1980s, most wind power plants in the U.S.A. were owned and operated by private investors or cooperatives. In California alone, there was more than 1500 MW of wind generating capacity in operation by the end of 1991. The Southern California Edison Company had over 1,120 MW of wind turbines under contract, with over 900 MW installed and connected to their grid. Major benefits to the Southern California Edison are the elimination of building new generating plant and transmission lines.

The technology development and the resulting price decline have caught the interest of a number of electric utilities outside California that are now actively developing wind energy as one element of the balanced resource mix.[3] Projects are being built in Alaska, California, Minnesota, Texas, Vermont, Washington, and Wyoming. During 1994, several new wind-energy projects started, particularly in the Midwest. A 73-turbine 25 MW plant was completed in Minnesota. Iowa and Wyoming producers and utilities plan to add significant wind-power capacity in the near future.[4] In 1995, a 35 MW utility-scale wind plant came on-line in Texas, which is expandable to 250 MW capacity, enough to supply 100,000 homes. This plant has a 25-year fixed-price contract to supply Austin City at 5 cents per kWh for generation and 1 cent per kWh for transmission. Nearly 700 MW of additional wind capacity was brought on-line in the U.S.A. by the end of 1996. The current and near-term wind-power capacity plans in the U.S.A. are shown in Figure 2-4.

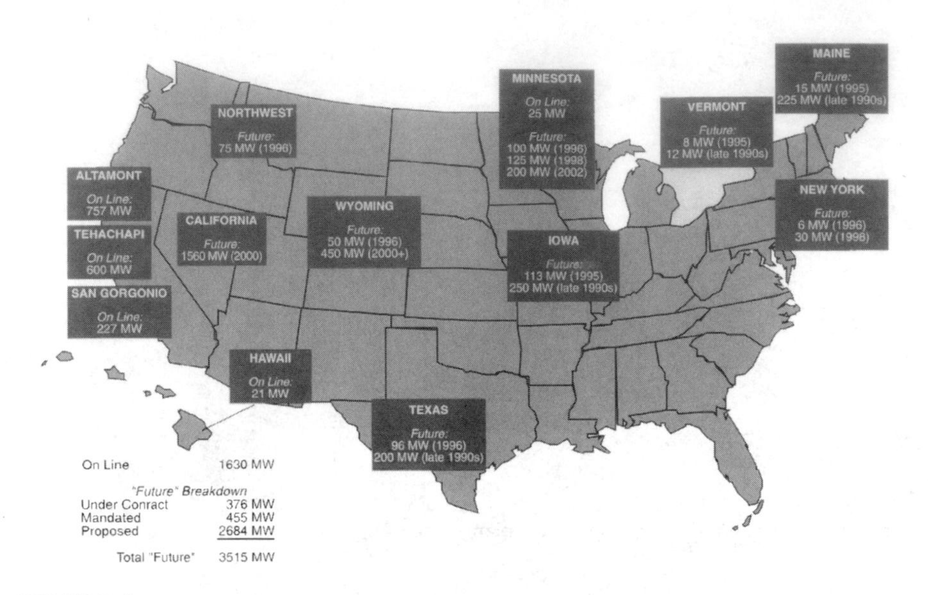

FIGURE 2-4
Wind-power capacity plans in the U.S.A., current and near future. (Source: NREL/IEA Wind Energy Annual Report, 1995.)

About 90 percent of the usable wind resource in the U.S.A. lies in the Great Plains. In Minnesota, Northern States Power Company started installing 100 MW wind-power capacity with a plan to expand to at least 425 MW by 2020. Several contractors working with the National Renewable Energy Laboratory are negotiating power purchase agreements with utilities for 4,200 MW of wind capacity at an estimated $2 billion capital investment.

The Energy Information Administration estimates that the U.S. wind capacity will reach 12,000 MW by 2015. Out of this capacity, utilities and wind-power developers have announced plans for more than 4,200 MW of new capacity in 15 states by 2006. The 1.5 cents per kWh federal tax credit that took effect January 1, 1994, is certainly helping the renewables.

State legislators in Minnesota have encouraged the wind power development by mandating that Northern States Power Company acquire 425 MW of wind generation by 2002. After commissioning a 25 MW wind plant in the Buffalo Ridge Lake Benton area in southwestern Minnesota near Holland, the Northern States Power is now committed to develop up to 100 MW of wind capacity over the next few years in the same area. This area has good steady wind and is accessible to the transmission lines. To support this program, the state has funded the "Wind Smith" education program at junior colleges to properly train the work force with required skills in installing, operating, and repairing the wind power plants.

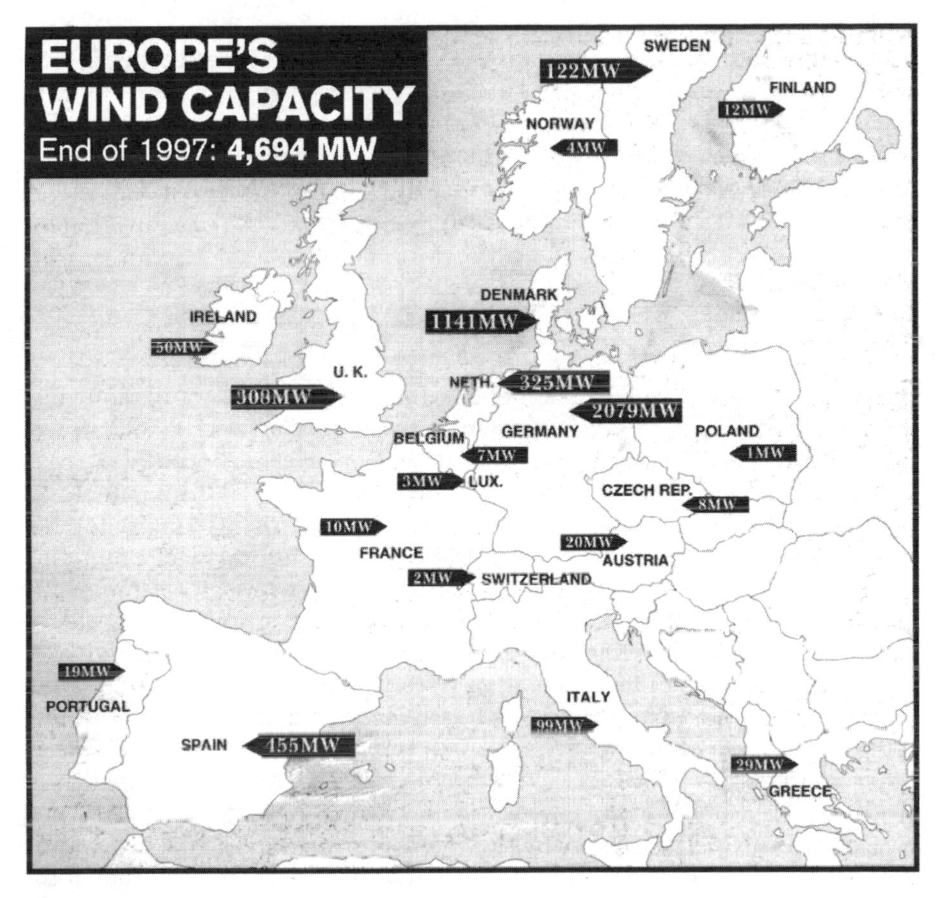

FIGURE 2-5
Installed wind capacity in European countries as of December 1997. (Source: Wind Directions, Magazine of the European Wind Energy Association, London, January 1998. With permission.)

2.3 Europe

The wind-power picture in Europe is rapidly growing. The 1995 projections on the expected wind capacity in 2000 have been met in 1997, in approximately one-half of the time. Figure 2-5 depicts the wind capacity installed in the European countries at the end of 1997. The total capacity installed was 4,694 MW. The new targets adopted by the European Wind Energy Association are 40,000 MW capacity by 2010 and 100,000 MW by 2020. These targets form part of a series of policy objectives agreed by the association in November

1997. Germany and Denmark lead Europe in the wind power. Both have achieved phenomenal growth through guaranteed tariff based on the domestic electricity prices. Germany has a 35-fold increase between 1990 and 1996. With 2,079 MW installed capacity, Germany is now the world leader. The former global leader, the U.S.A., has seen only a small increase during this period, from 1,500 MW in 1990 to approximately 2,000 MW in 1997.

2.4 India

India has 9 million square kilometers land area with a population over 900 million, of which 75 percent live in agrarian rural areas. The total power generating capacity has grown from 1,300 MW in 1950 to about 100,000 MW in 1998 at an annual growth rate of about nine percent. At this rate, India needs to add 10,000 MW capacity every year. The electricity network reaches over 500,000 villages and powers 11 million agricultural water-pumping stations. Coal is the primary source of energy. However, coal mines are concentrated in certain areas, and transporting coal to other parts of the country is not easy. One-third of the total electricity is used in the rural areas, where three-fourths of the population lives. The transmission and distribution loss in the electrical network is relatively high at 25 percent. The environment in a heavily-populated area is more of a concern in India than in other countries. For these reasons, the distributed power system, such as wind plants near the load centers, are of great interest to the state-owned electricity boards. The country has adopted aggressive plans for developing these renewables. As a result, India today has the largest growth rate of the wind capacity and is one of the largest producers of wind energy in the world.[5] In 1995, it had 565 MW of wind capacity, and some 1,800 MW additional capacity is in various stages of planning. The government has identified 77 sites for economically feasible wind-power generation, with a generating capacity of 4,000 MW of grid-quality power.

It is estimated that India has about 20,000 MW of wind power potential, out of which 1,000 MW has been installed as of 1997. With this, India now ranks in the first five countries in the world in wind-power generation, and provides attractive incentives to local and foreign investors. The Tata Energy Research Institute's office in Washington, D.C., provides a link between the investors in India and in the U.S.A.

2.5 Mexico

Mexico has over a decade of experience with renewable power systems. The two federally-owned utilities provide power to 95 percent of Mexico's population. However, there are still 90,000 hard-to-access villages with fewer than 1,000 inhabitants without electricity. These villages are being powered by renewable systems with deep cycle lead-acid batteries for energy storage. The wind resource has been thoroughly mapped in collaboration with the U.S. National Renewable Energy Laboratory.[6]

2.6 Ongoing Research and Development

The total government research and development funding in the International Energy Agency member countries in 1995 was about $200 million. The U.S. Department of Energy funded about $50 million worth of research and development in 1995. The goal of these programs is to further reduce the wind electricity-generation cost to less than 4 cents per kWh by the year 2000. The Department of Energy and the U.S. national laboratories also have a number of programs to promote the wind-hybrid power technologies throughout the developing world, with particular emphasis on Latin America and the Pacific Rim countries.[7] These activities include feasibility studies and pilot projects, project financing and supporting renewable energy education efforts.

References

1. U.S. Department of Energy. 1995. "Wind Energy Programs Overview," *NREL Report No. DE-95000288,* March 1995.
2. International Energy Agency. 1995. "Wind Energy Annual Report," *International Energy Agency Report by NREL,* March 1995.
3. Utility Wind Interest Group. 1995. "Utilities Move Wind Technology Across America," *1995 Report.* November 1995.
4. Anson, S., Sinclair, K., and Swezey, B. 1994. "Profiles in Renewables Energy," *Case studies of successful utility-sector projects, DOE/NREL Report No. DE-930000081,* National Renewable Energy Laboratory, Golden, Colorado, August 1994.
5. Gupta, A. K. 1997. "Power Generation from Renewables in India," *Ministry of Non-Conventional Energy Sources,* New Delhi, India, 1997.
6. Schwartz, M. N. and Elliott, D. L. 1995. "Mexico's Wind Resources Assessment Project," *DOE/NREL Report No. DE-AC36-83CH10093,* National Renewable Energy Laboratory, Golden, Colorado, May 1995.
7. Hammons, T. J., Ramakumar, R., Fraser, M., Conners, S. R., Davies, M., Holt, E. A., Ellis, M., Boyers, J., and Markard, J. 1997. "Renewable Energy Technology Alternatives for Developing Countries," *IEEE Power Engineering Review,* December 1997, p. 10-21.

3

Photovoltaic Power

The photovoltaic (pv) power technology uses semiconductor cells (wafers), generally several square centimeters in size. From the solid-state physics point of view, the cell is basically a large area p-n diode with the junction positioned close to the top surface. The cell converts the sunlight into direct current electricity. Numerous cells are assembled in a module to generate required power (Figure 3-1). Unlike the dynamic wind turbine, the pv installation is static, does not need strong tall towers, produces no vibration or noise, and needs no cooling. Because much of the current pv technology uses crystalline semiconductor material similar to integrated circuit chips, the production costs have been high. However, between 1980 and 1996, the capital cost of pv modules per watt of power capacity has declined from more than $20 per watt to less than $5 per watts (Figure 3-2). During the same period, the cost of pv electricity has declined from almost $1 to about $0.20 per kWh, and is expected to decline to $0.15 per kWh by the year 2000 (Figure 3-3). The installed capacity in the U.S. has risen from nearly zero in 1980 to approximately 200 MW in 1996 (Figure 3-4). The world capacity of pv systems was about 350 MW in 1996, which could increase to almost 1,000 MW by the end of this century (Figure 3-5).

The pv cell manufacturing process is energy intensive. Every square centimeter cell area consumes a few kWh before it faces the sun and produces the first kWh of energy. However, the manufacturing energy consumption is steadily declining with continuous implementation of new production processes (Figure 3-6).

The present pv energy cost is still higher than the price the utility customers pay in most countries. For that reason, the pv applications have been limited to remote locations not connected to the utility lines. With the declining prices, the market of new modules has been growing at more than a 15 percent annual rate during the last five years. The United States, the United Kingdom, Japan, China, India, and other countries have established new programs or have expanded the existing ones. It has been estimated that the potential pv market, with new programs coming in, could be as great as 1,600 MW by 2010. This is a significant growth projection, largely attributed to new manufacturing plants installed in the late 1990s to manufacture low cost pv cells and modules to meet the growing demand.

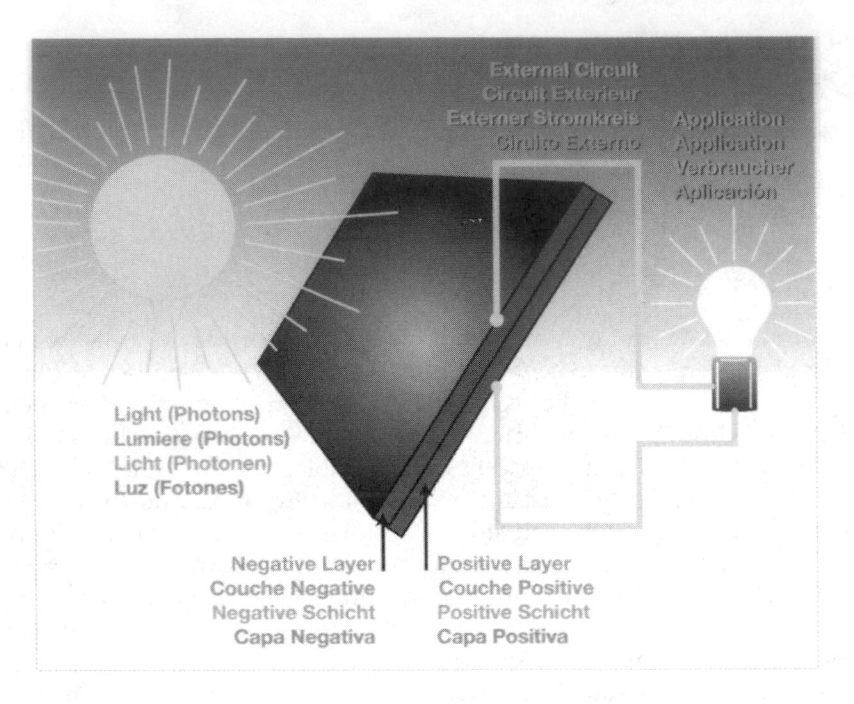

FIGURE 3-1
Photovoltaic module in sunlight generates direct current electricity. (Source: Solarex Corporation, Frederick, Md. With permission.)

Major advantages of the photovoltaic power are as follows:

- short lead time to design, install, and start up a new plant.
- highly modular, hence, the plant economy is not a strong function of size.
- power output matches very well with peak load demands.
- static structure, no moving parts, hence, no noise.
- high power capability per unit of weight.
- longer life with little maintenance because of no moving parts.
- highly mobile and portable because of light weight.

Almost 40 percent of the pv modules installed in the world are produced in the United States of America. Approximately 40 MW modules were produced in the U.S.A. in 1995, out of which 19 MW were produced by Siemens Solar Industries and 9.5 MW by Solarex Corporation.

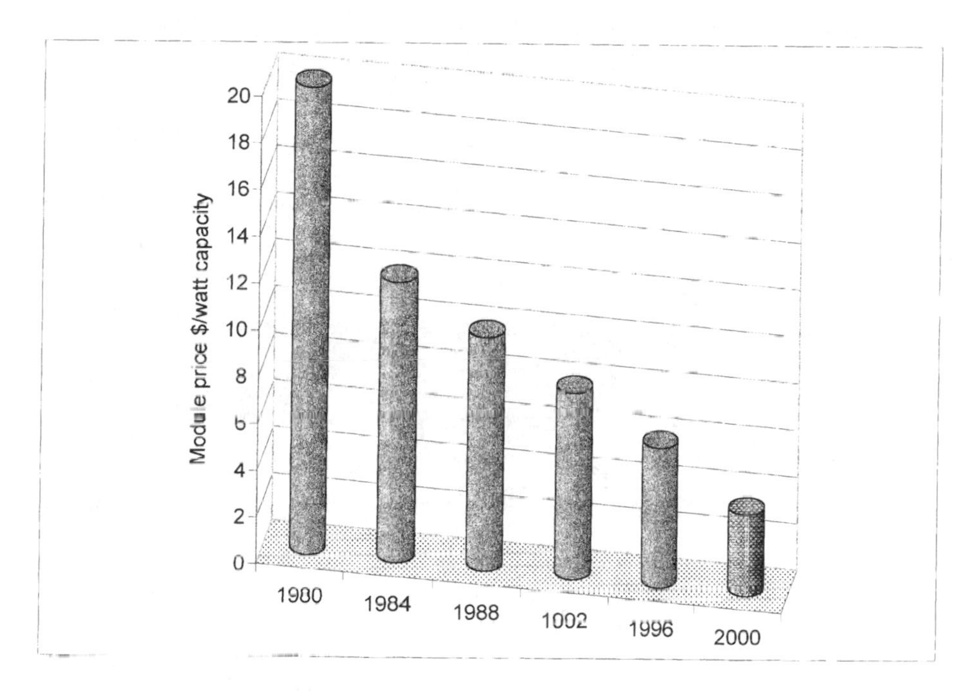

FIGURE 3-2
Photovoltaic module price trend.

3.1 Present Status

At present, pv power is extensively used in stand-alone power systems in remote villages, particularly in hybrid with diesel power generators. It is expected that this application will continue to find expanding markets in many countries. The driving force is the energy need in developing countries, and the environmental concern in developed countries. In the United States, city planners are recognizing the favorable overall economics of the pv power for urban applications. Tens of thousands of private, federal, state and commercial pv systems have been installed over the last 20 years. More than 65 cities in 24 states have installed such systems for a variety of needed services. These cities, shown in Figure 3-7, are located in all regions of the country, dispensing the myth that pv systems require a sunbelt climate to work effectively and efficiently.

The U.S. utilities have started programs to develop power plants using the newly available low-cost pv modules. Idaho Power has a pilot program to supply power to selected customers not yet connected to the grid. Other

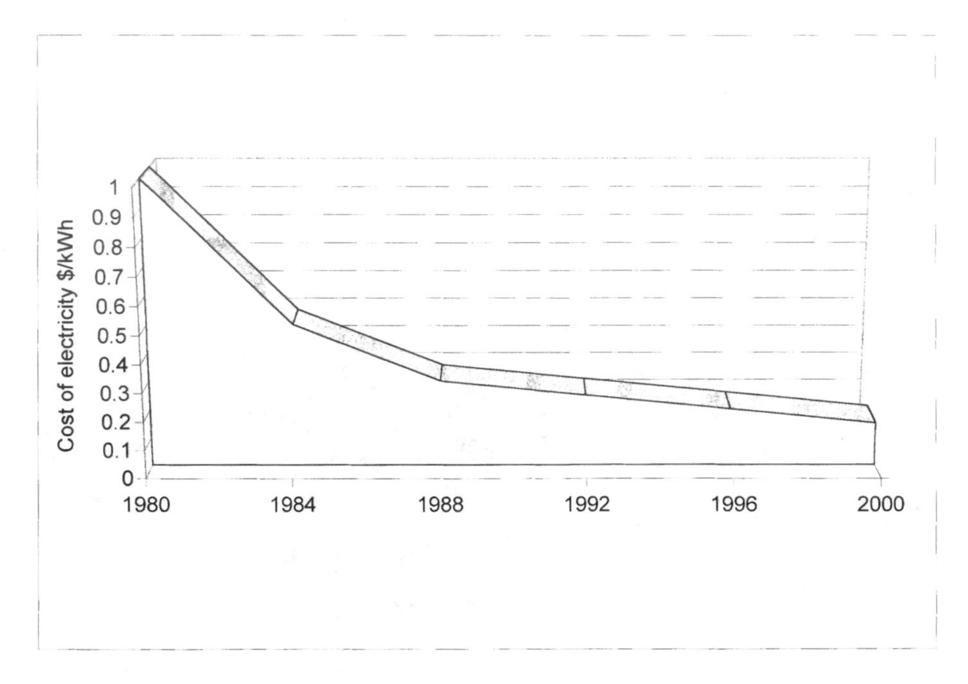

FIGURE 3-3
Photovoltaic energy price trend.

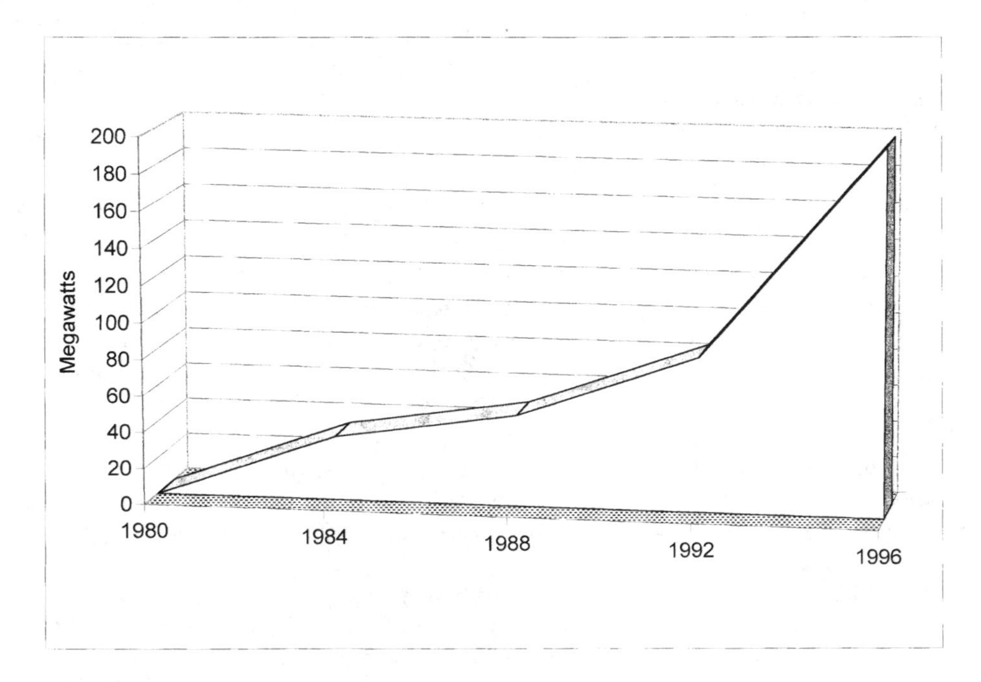

FIGURE 3-4
Cumulative capacity of pv installations in the U.S.A.

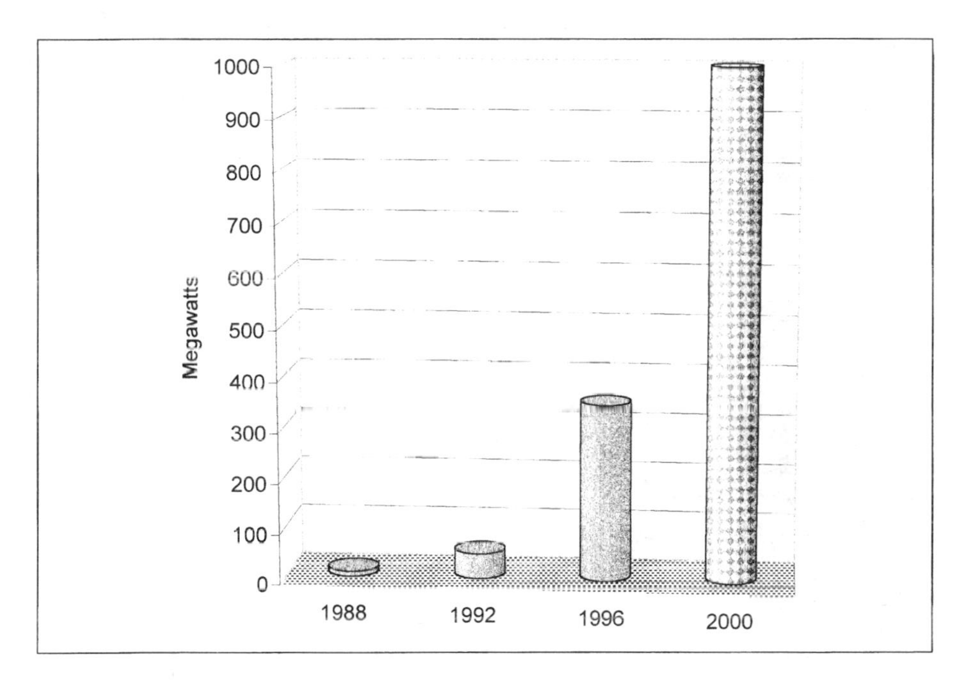

FIGURE 3-5
Cumulative capacity of pv installations in the world.

utilities such as Southern California Edison, the municipal utility of Austin, Delmarva Power and Light, and New York Power Authority are installing such systems to meet peak demands. The Pacific Gas and Electric's utility-scale 500 kW plant at Kerman is designed to deliver power during the local peak demand. It generates 1.1 MWh of energy annually.

The roof of the Aquatic Center in Atlanta (Figure 3-8), venue of the 1996 Olympic swimming competition, is one of the largest grid-connected power plants. It generates 345 kW of electric power, and is tied into the Georgia Power grid lines. Its capacity is enough to power 70 homes connected to the network. It saves 330 tons of CO_2, 3.3 tons of SO_2 and 1.2 tons of NOx yearly. Installations are under way to install a similar 500 kW grid-connected pv system to power the Olympics Games of 2000 in Sydney, Australia.

In 14 member countries of the International Energy Agency, the pv installations are being added at an average annual growth rate of 27 percent. The total installed capacity increased from 54 MW in 1990 to 176 MW in 1995. The IEA estimates that by the end of 2000, 550 MW of additional capacity will be installed, 64 percent off-grid and 36 percent grid-connected (Figure 3-9).

India is implementing perhaps the most number of pv systems in the world for remote villages. About 30 MW capacity has already been installed, with more being added every year. The country has a total production capacity of 8.5 MW modules per year. The remaining need is met by imports. A

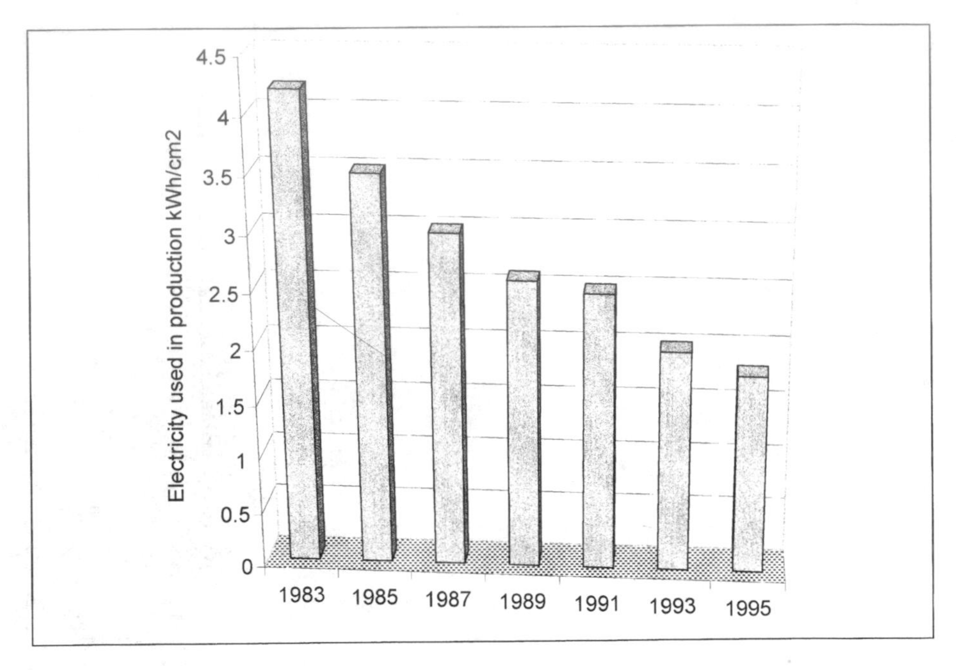

FIGURE 3-6
Energy consumption per cm^2 of pv cell manufactured. (Source: U.S. Department of Commerce and Dataquest, Inc.)

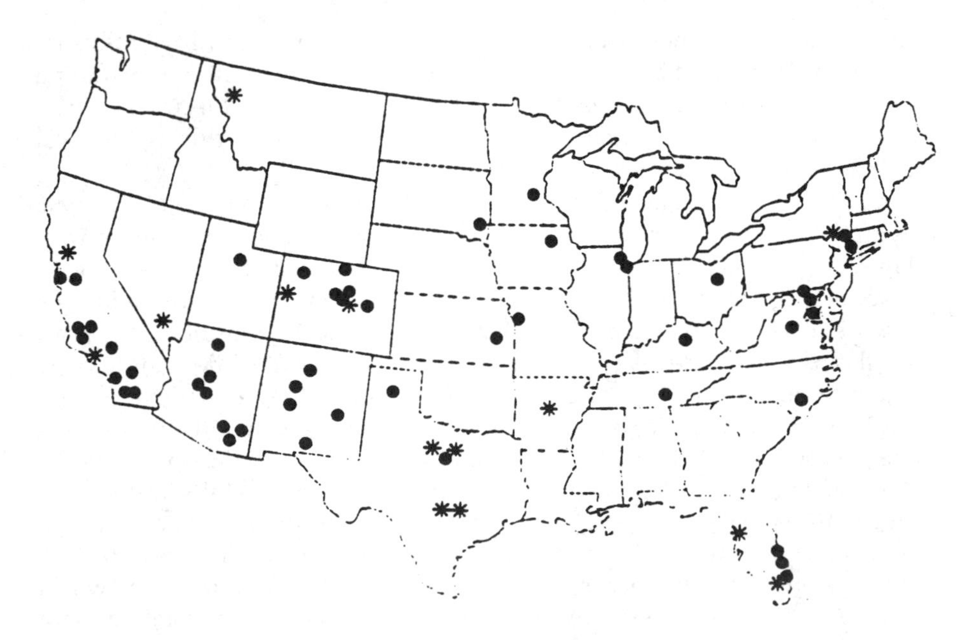

FIGURE 3-7
U.S. cities with pv installations. (Source: Department of Energy, 1995.)

FIGURE 3-8
Roof (pv) of Atlanta's Aquatic Center with 345 kW grid-connected power system. (Source: Georgia Institute of Technology. With permission.)

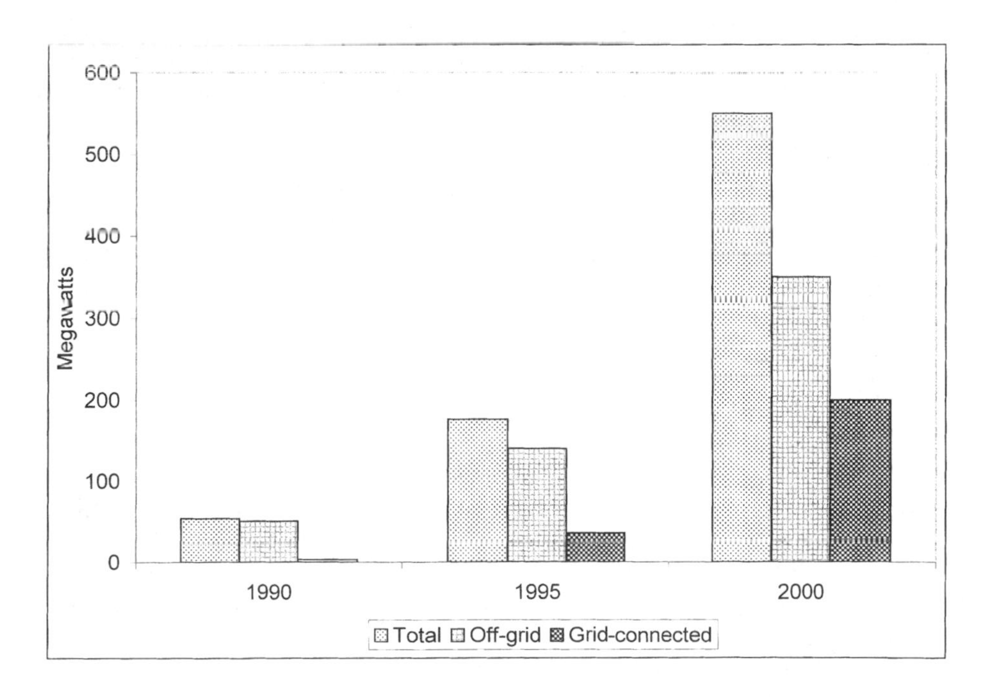

FIGURE 3-9
Cumulative pv installations in the IEA countries in 1990, 1995, and projected to 2000. (Source: International Energy Agency.)

700 kW grid-connected pv plant has been commissioned, and a 425 kW capacity is under installation in Madhya Pradesh. The state of West Bengal has decided to convert the Sagar Island into a pv island. The island has 150,000 inhabitants in 16 villages spread out in an area of about 300 square kilometers. The main source of electricity at present is diesel, which is expensive and is causing severe environmental problems on the island.

The state of Rajasthan has initialed a policy to purchase pv electricity at an attractive rate of $0.08 per kWh. In response, a consortium of Enron and Amoco has proposed installing a 50 MW plant using thin film cells. When completed, this will be the largest pv power plant in the world.

The studies at the Arid Zone Research Institute, Jodhpur, indicate significant solar energy reaching the earth surface in India. About 30 percent of the electrical energy used in India is for agricultural needs. Since the availability of solar power for agricultural need is not time critical (within a few days), India is expected to lead the world in pv installations in near future.

3.2 Building Integrated pv Systems

In new markets, the near-term potentially large application of the pv technology is for cladding buildings to power air-conditioning and lighting loads. One of the attractive features of the pv system is that its power output matches very well with the peak load demand. It produces more power on a sunny summer day when the air-conditioning load strains the grid lines (Figure 3-10). The use of pv installations in buildings has risen from a mere 3 MW in 1984 to 16 MW in 1994, at a rate of 18 percent per year.

In the mid 1990s, the DOE launched a 5-year cost-sharing program with Solarex Corporation of Maryland to develop and manufacture low cost, easy to install, pre-engineered Building Integrated Photovoltaic (BIPV) modules. Such modules made in shingles and panels can replace traditional roofs and walls. The building owners have to pay only the incremental cost of these components. The land is paid for, the support structure is already in there, the building is already wired, and developers may finance the BIPV as part of their overall project. The major advantage of the BIPV system is that it produces power at the point of consumption. The BIPV, therefore, offers the first potentially widespread commercial implementation of the pv technology in the industrialized countries. The existing programs in the U.S.A., Europe, and Japan could add 200 MW of BIPV installations by the year 2010. Worldwide, the Netherlands plans to install 250 MW by 2010, and Japan has plans to add 185 MW between 1993 and 2000.

Figure 3-11 shows a building-integrated and grid-connected pv power system recently installed and operating in Germany.

In August 1997, The DOE announced that it will lead an effort to place one million solar-power systems on home and building roofs across the

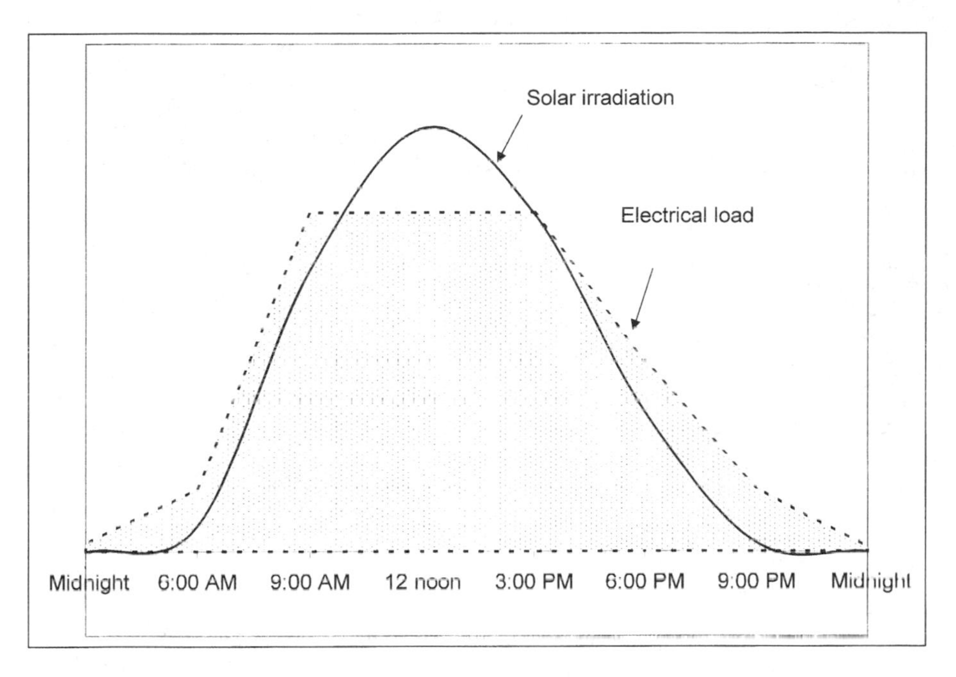

FIGURE 3-10
Power usage in commercial building on a typical summer day.

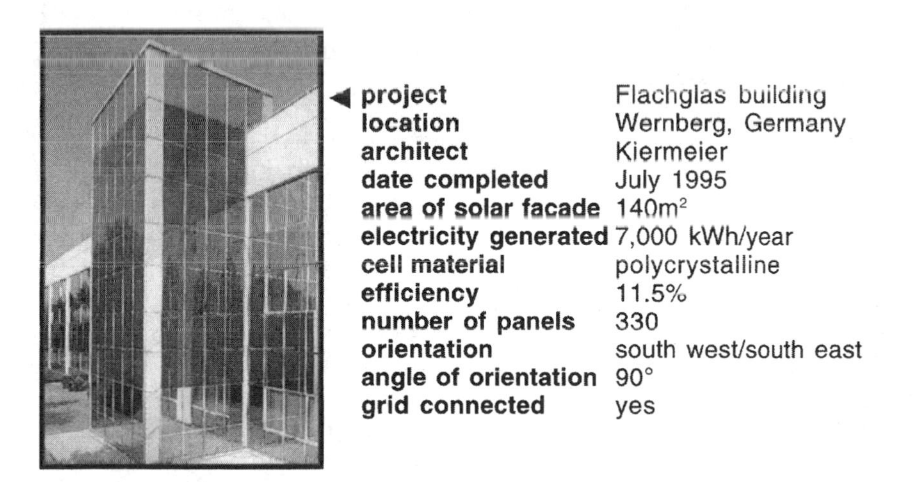

FIGURE 3-11
Building-integrated pv systems in Germany. (Source: Professional Engineering, Publication of the Institution of Mechanical Engineers, April 1997, U.K.. With permission.)

U.S.A. by the year 2010. This "Million Solar Roof Initiative" is expected to increase the momentum for more widespread use of solar power, lowering the cost of photovoltaic technologies.

3.3 pv Cell Technologies

In making comparisons between alternative power technologies, the most important measure is the energy cost per kWh delivered. In pv power, this cost primarily depends on two parameters, the photovoltaic energy conversion efficiency, and the capital cost per watt capacity. Together, these two parameters indicate the economic competitiveness of the pv electricity.

The conversion efficiency of the photovoltaic cell is defined as follows:

$$\eta = \frac{\textit{electrical power output}}{\textit{solar power impinging the cell}}$$

The continuing development efforts to produce more efficient low cost cells have resulted in various types of pv technologies available in the market today, in terms of the conversion efficiency and the module cost. The major types are discussed in the following sections:[1]

3.3.1 Single-Crystalline Silicon

The single crystal silicon is the widely available cell material, and has been the workhorse of the industry. In the most common method of producing this material, the silicon raw material is first melted and purified in a crucible. A seed crystal is then placed in the liquid silicon and drawn at a slow constant rate. This results in a solid, single-crystal cylindrical ingot (Figure 3-12). The manufacturing process is slow and energy intensive, resulting in high raw material cost presently at $25 to $30 per pound. The ingot is sliced using a diamond saw into 200 to 400 μm (0.005 to 0.010 inch) thick wafers. The wafers are further cut into rectangular cells to maximize the number of cells that can be mounted together on a rectangular panel. Unfortunately, almost half of the expensive silicon ingot is wasted in slicing ingot and forming square cells. The material waste can be minimized by making the full size round cells from round ingots (Figure 3-13). Using such cells would be economical where the panel space is not at a premium. Another way to minimize the waste is to grow crystals on ribbons. Some U.S. companies have set up plants to draw pv ribbons, which are then cut by laser to reduce waste.

3.3.2 Polycrystalline and Semicrystalline

This is relatively a fast and low cost process to manufacture thick crystalline cells. Instead of drawing single crystals using seeds, the molten silicon is cast into ingots. In the process, it forms multiple crystals. The conversion efficiency is lower, but the cost is much lower, giving a net reduction in cost per watt of power.

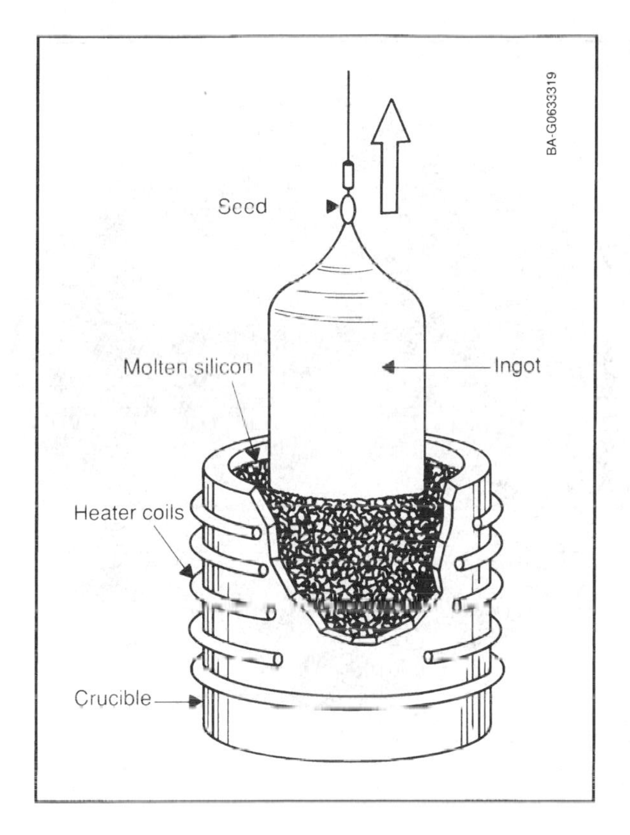

FIGURE 3-12
Single crystal ingot making by Czochralski process. (Source: Cook, G., Photovoltaic Fundamental, DOE/NREL Report DE91015001, February 1995.)

3.3.3 Thin Films

These are new types of photovoltaics entering the market. Copper Indium Diselenide, Cadmium Telluride, and Gallium Arsenide are all thin film materials, typically a few μm or less in thickness, directly deposited on glass, stainless steel, ceramic or other compatible substrate materials. This technology uses much less material per square area of the cell, hence, is less expensive per watt of power generated.

3.3.4 Amorphous Silicon

In this technology, amorphous silicon vapor is deposited on a couple of μm-thick amorphous (glassy) films on stainless steel rolls, typically 2,000-feet long and 13-inches wide. Compared to the crystalline silicon, this technology uses only 1 percent of the material. Its efficiency is about one-half of the crystalline silicon at present, but the cost per watt generated is projected to

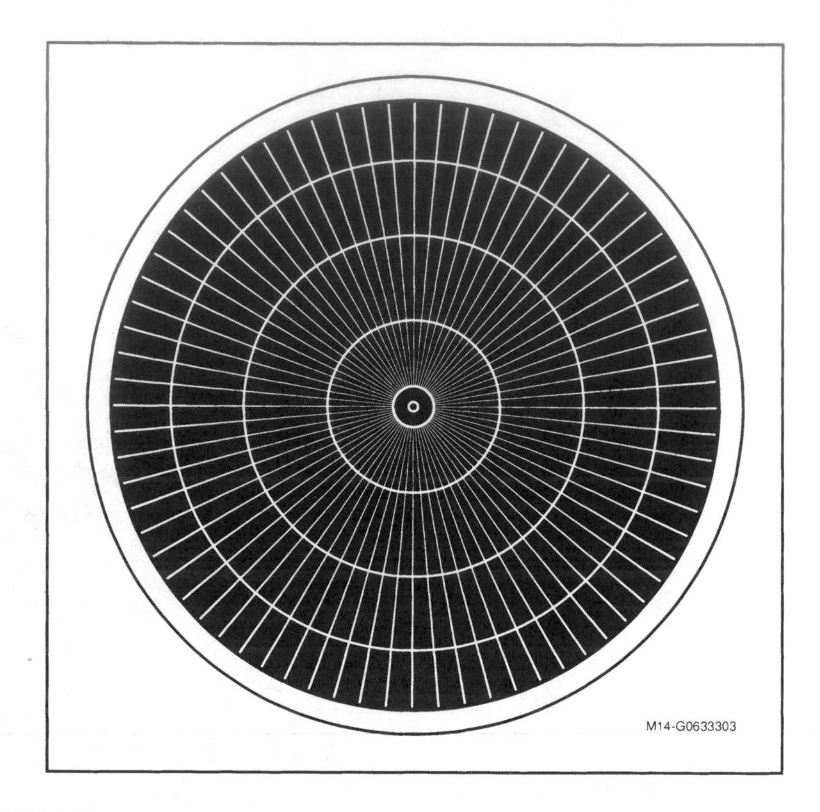

FIGURE 3-13
Round-shape pv cell reduces material waste typically found in rectangular cell. (Depiction based on cell used by Applied Solar Energy Corporation, City of Industry, California.)

be significantly lower. On this premise, two large plants to manufacture amorphous silicon panels started in the U.S.A. in 1996.

3.3.5 Spheral

This is yet another technology that is being explored in the laboratories. The raw material is low-grade silicon crystalline beads, presently costing about $1 per pound. The beads are applied on typically 4-inch squares of thin perforated aluminum foil. In the process, the impurities are pushed to the surface, from where they are etched away. Since each sphere works independently, the individual sphere failure has negligible impact on the average performance of the bulk surface. According to a Southern California Edison Company's estimate, 100 square feet of spheral panels can generate 2,000 kWh per year in an average southern California climate.

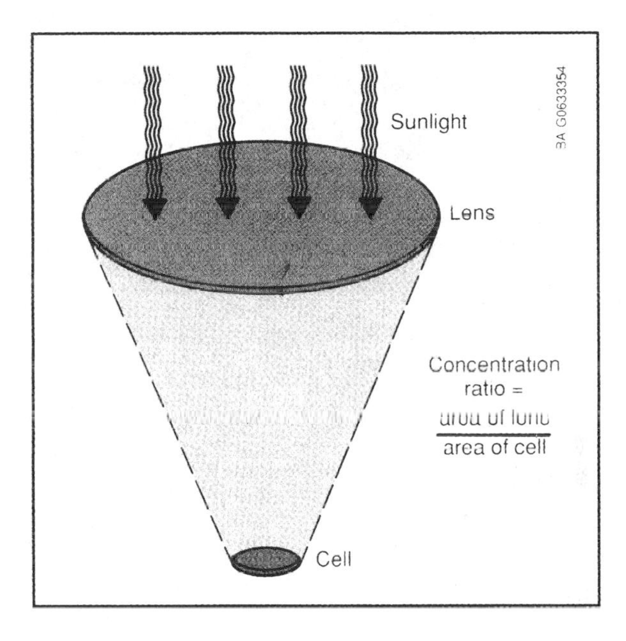

FIGURE 3-14
Lens concentrating the sunlight on small area reduces the need of active cell material. (Source: Photovoltaic Fundamental, DOE/NREL Report DE91015001, February 1995.)

3.3.6 Concentrated Cells

In an attempt to improve the conversion efficiency, the sunlight is concentrated into tens or hundreds of times the normal sun intensity by focusing on a small area using low cost lenses (Figure 3-14). The primary advantage is that such cells require a small fraction of area compared to the standard cells, thus significantly reducing the pv material requirement. However, the total module area remains the same to collect the required sun power. Besides increasing the power and reducing the size or number of cells, such cells have additional advantage that the cell efficiency increases under concentrated light up to a point. Another advantage is that they can use small area cells. It is easier to produce high efficiency cells of small areas than to produce large area cells with comparable efficiency. On the other hand, the major disadvantage of the concentrator cells is that they require focusing optics adding into the cost.

The annual production of various pv cells in 1995 is shown in Table 3-1. Almost all production has been in the crystalline silicon and the amorphous silicon cells, with other types being in the development stage. The present status of the crystalline silicon and the amorphous silicon technologies is shown in Table 3-2. The former is dominant in the market at present and the latter is expected to be dominant in the near future.

TABLE 3-1

Production Capacities of Various pv Technologies in 1995

PV Technology	1995 Production
Crystalline Silicon	55 MW
Amorphous Silicon	9 MW
Ribbon Si, GaAs, CdTe	1 MW
TOTAL	65 MW

(Source: Carlson, D. E., Recent Advances in Photovoltaics, 1995. Proceedings of the Intersociety Engineering Conference on Energy Conversion, 1995.)

TABLE 3-2

Comparison of Crystalline and Amorphous Silicon Technologies

	Crystalline Silicon	Amorphous Silicon
Present Status	Workhorse of terrestrial and space applications	New rapidly developing technology, tens of MW yearly production facilities have been built in 1996 to produce low cost cells
Thickness	200-400 μm (.004-.008 inch)	2 μm (less then 1 percent of that in crystalline silicon)
Raw Material Cost	High	About 3 percent of that in crystalline silicon
Conversion Efficiency	16-18 percent	8-9 percent
Module Costs (1995)	\$6–8 per watt, expected to fall slowly due to the matured nature of this technology	\$6-8 per watt, expected to fall rapidly to \$2 per watt in 2000 due to heavy DOE funding to fully develop this new technology

3.4 pv Energy Maps

The solar energy impinging the surface of the earth is depicted in Figure 3-15, where the white areas get more solar radiation per year. The yearly 24-hour average solar flux reaching the horizontal surface of the earth is shown in Figure 3-16, whereas Figure 3-17 depicts that in the month of December. Notice that the 24-hour average decreases in December. This is due to the shorter days and clouds, and not due to the cold temperature. As will be seen later, the pv cell actually converts more solar energy into electricity at low temperatures.

It is the total yearly energy capture potential of the site that determines the economical viability of installing a power plant. Figure 3-18 is useful in this regard, as it gives the annual average solar energy per day impinging on the surface always facing the sun at right angle. Modules mounted on a sun-tracking structure receive this energy. The electric energy produced per day is obtained by multiplying the map number by the photoconversion efficiency of the modules installed at the site.

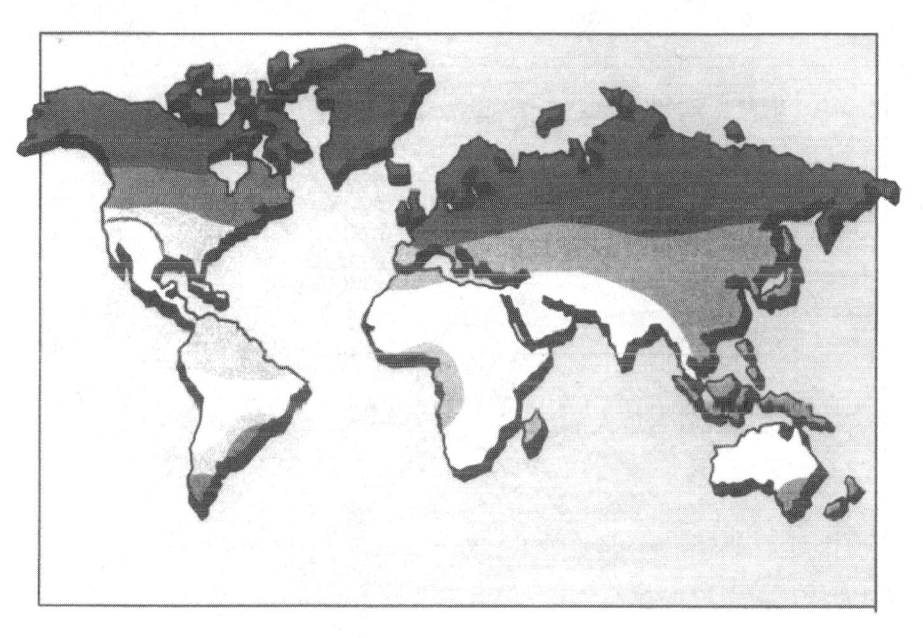

FIGURE 3-15
Solar radiation by regions of the world (higher energy potential in the white areas).

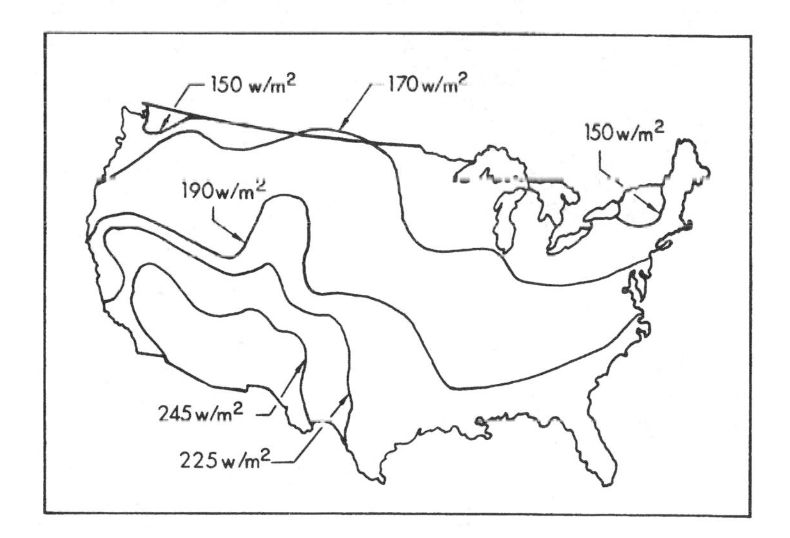

FIGURE 3-16
Yearly 24-hour average solar radiation in watts/m^2 reaching the horizontal surface of the earth. (Source: Profiles in Renewable Energy, DOE/NREL Report No. DE-930000081, August 1994.)

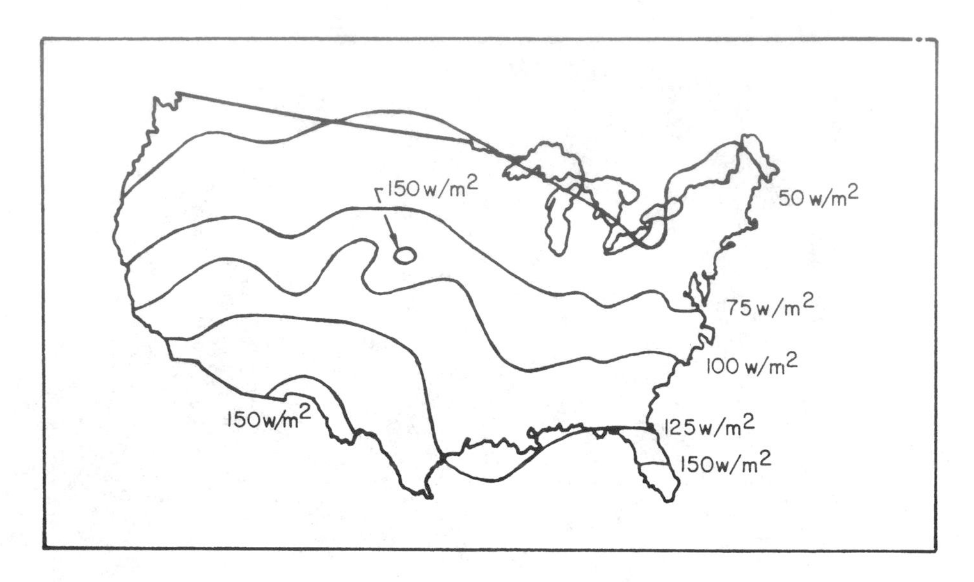

FIGURE 3-17
December 24-hour average solar radiation in watts/m^2 reaching the horizontal surface of the earth. (Source: Profiles in Renewable Energy, DOE/NREL Report No. DE-930000081, August 1994.)

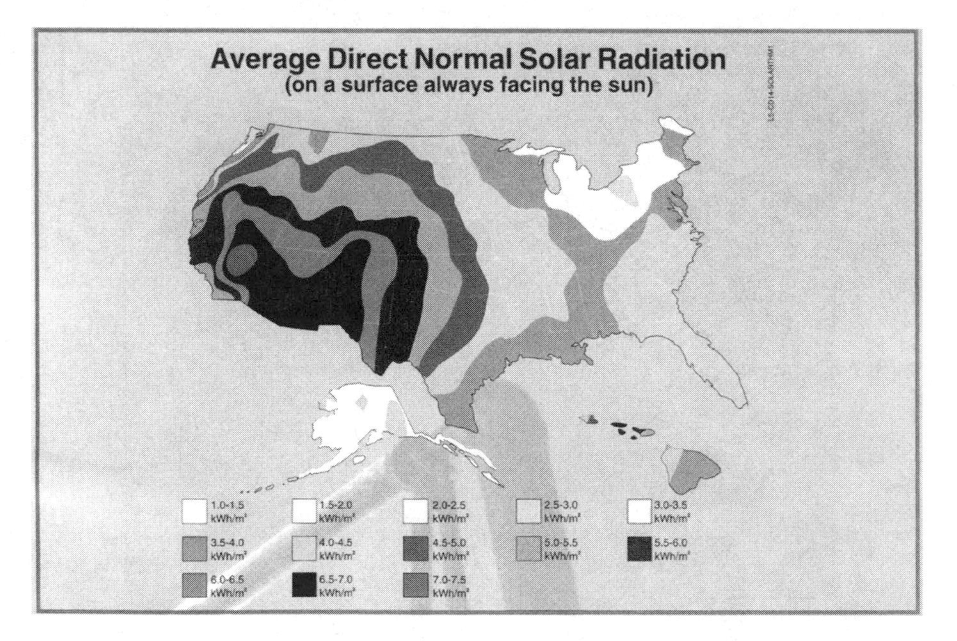

FIGURE 3-18
Yearly 24-hour average solar energy in kWh/m^2 reaching surface always facing the sun at right angel. (Source: A. Anson, Profiles in Renewable Energy, DOE/NREL Report No. DE-930000081, August 1994.)

References

1. Carlson, D. E. 1995. "Recent Advances in Photovoltaics," *1995 Proceedings of the Intersociety Engineering Conference on Energy Conversion.* 1995, p. 621-626.

4

Wind Speed and Energy Distributions

The wind turbine captures the wind's kinetic energy in a rotor consisting of two or more blades mechanically coupled to an electrical generator. The turbine is mounted on a tall tower to enhance the energy capture. Numerous wind turbines are installed at one site to build a wind farm of the desired power production capacity. Obviously, sites with steady high wind produce more energy over the year.

Two distinctly different configurations are available for the turbine design, the horizontal axis configuration (Figure 4-1) and the vertical axis configuration (Figure 4-2). The vertical axis machine has the shape of an egg beater, and is often called the Darrieus rotor after its inventor. It has been used in the past because of specific structural advantage. However, most modern wind turbines use horizontal-axis design. Except for the rotor, all other components are the same in both designs, with some difference in their placement.

4.1 Speed and Power Relations

The kinetic energy in air of mass "m" moving with speed V is given by the following in SI units:

$$\textit{Kinetic Energy} = \frac{1}{2} \cdot m \cdot V^2 \text{ joules.} \tag{4-1}$$

The power in moving air is the flow rate of kinetic energy per second. Therefore:

$$\textit{Power} = \frac{1}{2} \cdot (\textit{mass flow rate per second}) \cdot V^2 \tag{4-2}$$

If we let P = mechanical power in the moving air
ρ = air density, kg/m^3
A = area swept by the rotor blades, m^2
V = velocity of the air, m/s

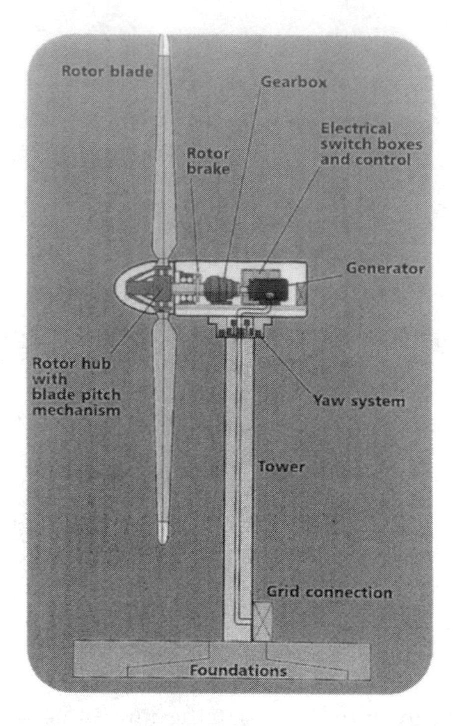

FIGURE 4-1
Horizontal axis wind turbine showing major components. (Courtesy: Energy Technology Support Unit, DTI, U.K.. With permission.)

FIGURE 4-2
Vertical axis 34-meter diameter wind turbine built and tested by DOE/Sandia National Laboratory during 1994 in Bushland, Texas.

then, the volumetric flow rate is A·V, the mass flow rate of the air in kilograms per second is ρ·A·V, and the power is given by the following:

$$P = \frac{1}{2}(\rho A V) \cdot V^2 = \frac{1}{2}\rho A V^3 \ \textit{watts}. \tag{4-3}$$

Two potential wind sites are compared in terms of the specific wind power expressed in watts per square meter of area swept by the rotating blades. It is also referred to as the power density of the site, and is given by the following expression:

$$\textit{Specific Power of the site} = \frac{1}{2}\rho \cdot V^3 \ \textit{watts per } m^2 \textit{ of the rotor swept area.} \tag{4-4}$$

This is the power in the upstream wind. It varies linearly with the density of the air sweeping the blades, and with the cube of the wind speed. All of the upstream wind power cannot be extracted by the blades, as some power is left in the downstream air which continues to move with reduced speed.

4.2 Power Extracted from the Wind

The actual power extracted by the rotor blades is the difference between the upstream and the downstream wind powers. That is, using Equation 4-2:

$$P_o = \frac{1}{2} \ \textit{mass flow rate per second} \cdot \left\{V^2 = V_o^2\right\} \tag{4-5}$$

where P_o = mechanical power extracted by the rotor, i.e., the turbine output power
V = upstream wind velocity at the entrance of the rotor blades
V_o = downstream wind velocity at the exit of the rotor blades.

The air velocity is discontinuous from V to V_o at the "plane" of the rotor blades in the macroscopic sense (we leave the aerodynamics of the blades for many excellent books available on the subjects). The mass flow rate of air through the rotating blades is, therefore, derived by multiplying the density with the average velocity. That is:

$$\textit{mass flow rate} = \rho \cdot A \cdot \frac{V + V_o}{2} \tag{4-6}$$

The mechanical power extracted by the rotor, which is driving the electrical generator, is therefore:

$$P_o = \frac{1}{2}\left[\rho \cdot A \cdot \frac{(V + V_o)}{2}\right] \cdot \left(V^2 - V_o^2\right) \tag{4-7}$$

The above expression can be algebraically rearranged:

$$P_o = \frac{1}{2}\rho \cdot A \cdot V^3 \frac{\left(1+\frac{V_o}{V}\right)\left[1-\left(\frac{V_o}{V}\right)^2\right]}{2} \tag{4-8}$$

The power extracted by the blades is customarily expressed as a fraction of the upstream wind power as follows:

$$P_o = \frac{1}{2}\rho \cdot A \cdot V^3 \cdot C_p \tag{4-9}$$

where

$$C_p = \frac{\left(1+\frac{V_o}{V}\right)\left[1-\left(\frac{V_o}{V}\right)^2\right]}{2} \tag{4-10}$$

The C_p is the fraction of the upstream wind power, which is captured by the rotor blades. The remaining power is discharged or wasted in the downstream wind. The factor C_p is called the power coefficient of the rotor or the rotor efficiency.

For a given upstream wind speed, the value of C_p depends on the ratio of the down stream to the upstream wind speeds, that is (V_o/V). The plot of power coefficient versus (V_o/V) shows that C_p is a single, maximum-value function (Figure 4-3). It has the maximum value of 0.59 when the (V_o/V) is one-third. The maximum power is extracted from the wind at that speed ratio, when the downstream wind speed equals one-third of the upstream speed. Under this condition:

$$P_{max} = \frac{1}{2}\rho \cdot A \cdot V^3 \cdot 0.59 \tag{4-11}$$

The theoretical maximum value of C_p is 0.59. In practical designs, the maximum achievable C_p is below 0.5 for high-speed, two-blade turbines, and between 0.2 and 0.4 for slow speed turbines with more blades (Figure 4-4). If we take 0.5 as the practical maximum rotor efficiency, the maximum power output of the wind turbine becomes a simple expression:

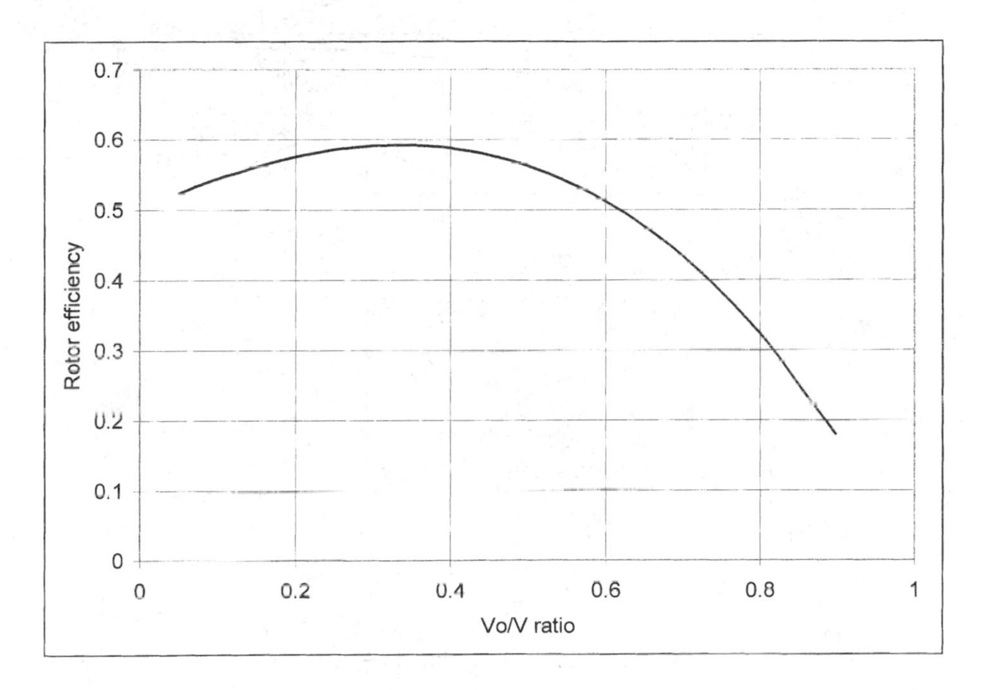

FIGURE 4-3
Rotor efficiency versus V_o/V ratio has single maximum. Rotor efficiency is the fraction of available wind power extracted by the rotor and fed to the electrical generator.

$$P_{max} = \frac{1}{4} \cdot \rho \cdot V^3 \text{ watts per m}^2 \text{ of swept area.} \quad (4\text{-}12)$$

4.3 Rotor Swept Area

As seen in the power equation, the output power of the wind turbine varies linearly with the rotor swept area. For the horizontal axis turbine, the rotor swept area is given by:

$$A = \frac{\pi}{4} D^2 \quad \text{where D is the rotor diameter.} \quad (4\text{-}13)$$

For the Darrieus vertical axis machine, determination of the swept area is complex, as it involves elliptical integrals. However, approximating the blade shape as a parabola leads to the following simple expression for the swept area:

$$A = \frac{2}{3} \cdot (\textit{Maximum rotor width at the center}) \cdot (\textit{Height of the rotor}). \quad (4\text{-}14)$$

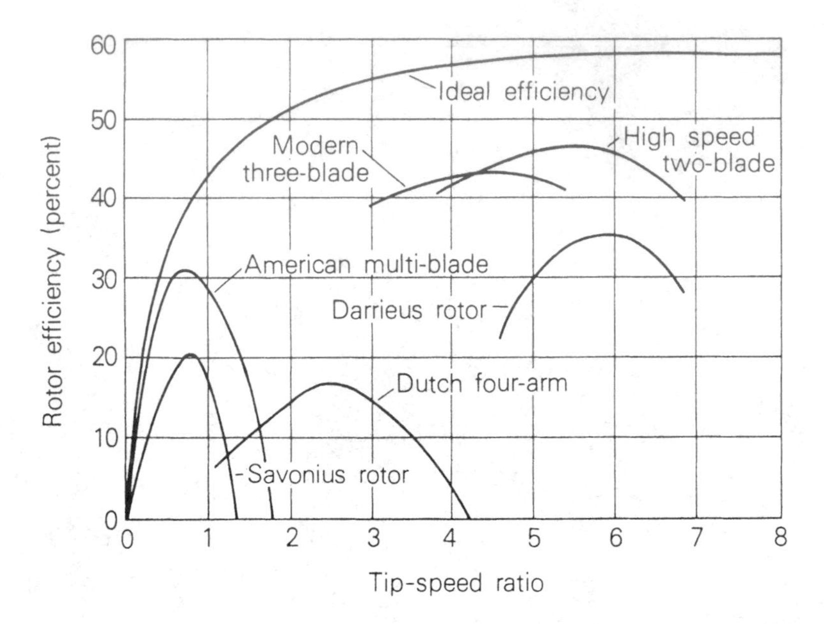

FIGURE 4-4
Rotor efficiency versus tip speed ratio for rotors with different numbers of blades. Two-blade rotors have the highest efficiency. (Source: Eldridge, F.R., Wind Machines, Energy Research and Development Administration, Washington, DC, Report AER-75-12937, p. 55, 1975.)

The wind turbine efficiently intercepts the wind energy flowing through the entire swept area even though it has only two or three thin blades with solidity between 5 to 10 percent. The solidity is defined as the ratio of the solid area to the swept area of the blades. The modern 2-blade turbine has low solidity ratio. Hence, it requires little blade material to sweep large areas.

4.4 Air Density

The wind power varies linearly with the air density sweeping the blades. The air density ρ varies with pressure and temperature in accordance with the gas law:

$$\rho = \frac{p}{R \cdot T} \tag{4-15}$$

where p = air pressure
T = temperature on the absolute scale
R = gas constant.

The air density at sea level, one atmospheric pressure (14.7 psi) and 60°F is 1.225 kg/m^3. Using this as the reference, ρ is corrected for the site specific temperature and pressure. The temperature and the pressure both in turn vary with the altitude. Their combined effect on the air density is given by the following equation, which is valid up to 6,000 meters (20,000 feet) of site elevation above the sea level:

$$\rho = \rho_o \cdot e^{-\left\{\frac{0.297\, H_m}{3048}\right\}} \tag{4-16}$$

where H_m is the site elevation in meters.

Equation 4-16 is often written in a simple form:

$$\rho = \rho_o - 1.194 \cdot 10^{-4} \cdot H_m \tag{4-17}$$

The air density correction at high elevations can be significant. For example, the air density at 2,000-meter elevation would be 0.986 kg/m³, 20 percent lower than the 1.225 kg/m³ value at sea level.

For ready reference, the temperature varies with the elevation:

$$T = 15.5 - \frac{19.83\, H_m}{3048}\ °C \tag{4-18}$$

4.5 Global Wind Patterns

The global wind patterns are created by uneven heating and the spinning of the earth. The warm air rises near the equator, and the surface air moves in to replace the rising air. As a result, two major belts of the global wind patterns are created. The wind between the equator and about 30° north and south latitudes move east to west. These are called the trade winds because of their use in sailing ships for trades. There is little wind near the equator, as the air slowly rises upward, rather than moving westward. The prevailing winds move from west to east in two belts between latitudes 30° and 60° north and south of the equator. This motion is caused by circulation of the trade winds in a closed loop.

In many countries where the weather systems come from the west, the wind speed in the west is generally higher than in the east.

Two features of the wind, its speed, and the direction, are used in describing and forecasting weather (Figure 4-5). The speed is measured with an instrument called anemometer, which comes in several types. The most common type has three or four cups attached to spokes on a rotating shaft. The wind turns the cups and the shaft. The angular speed of the spinning shaft is calibrated in terms of the linear speed of the wind. In the United States, the wind speed is reported in miles per hour or in nautical miles per

FIGURE 4-5
Old multiblade windmill and modern three-blade wind turbine. (Source: Vestas Wind Systems, Denmark.)

hour (knots). In other countries, it is reported in kilometers per hour or some times in meters per second.

The wind direction is measured with an instrument called the weather vane. It has a broad, flat blade attached to a spoke pivoted at one end. The wind impinging on the blade turns the spoke and lines up the blades in the wind direction. The wind direction is indicated by an arrow fastened to the spoke, or by an electric meter remotely controlled by the weather vane. The wind direction is often indicated in terms of a 360 degrees circular scale. On such scale, 0° indicates the north, 90° indicates the east, 180° indicates the south, and 270° indicates the west directions.

An optical sensor developed at the Georgia Institute of Technology may soon replace the conventional anemometer and improve the measurement accuracy. The mechanical anemometer can register readings at a single location where it is actually placed. A complex array of traditional anemometers is needed to monitor the wind speed over a large area such as in a wind farm. The new optical sensor is able to measure average crosswind speeds and directions over a long distance and is more accurate than the mechanical anemometer. Figure 4-6 depicts the sensor's layout and the working principle. The sensor is mounted on a large telescope and a helium neon laser of about 50 millimeters diameter. It projects a beam of light on to a target about 100 meters away. The target is made of the type of retroflective material used on road signs. The sensor uses a laser beam degradation phenomenon known as the residual turbulent scintillation effect. The telescope collects laser light reflected from the target and sends it through a unique optical path in the instrument.

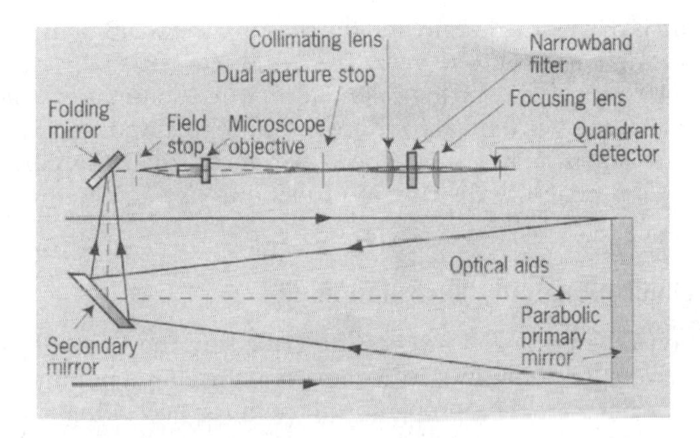

FIGURE 4-6
Optical wind speed sensor construction. (Source: Georgia Institute of Technology, Atlanta, GA. With permission.)

Each of the two tiny detectors monitors a spot on the target inside the laser beam and picks up shadowy waves or fringes moving across the laser beam. The waves are visible on the target itself. The fringes look much like the shadows of waves seen at the bottom of a swimming pool on a sunny day. Something akin to these shadows can be seen if a turbulent wind is viewed with a laser beam. Each detector registers the moment when a dark fringe passes its view. By digitizing the detected points, a computer can measure time and separation, and, therefore, the average wind speed.

The major advantage of the optical sensor is that it can measure the wind speed over a wide range from a faint wind to a wild wind over a large area.

4.6 Wind Speed Distribution

Having the cubic relation with the power, the wind speed is the most critical data needed to appraise the power potential of a candidate site. The wind is never steady at any site. It is influenced by the weather system, the local land terrain, and the height above the ground surface. The wind speed varies by the minute, hour, day, season, and year. Therefore, the annual mean speed needs to be averaged over 10 or more years. Such a long term average raises the confidence in assessing the energy-capture potential of a site. However, long-term measurements are expensive, and most projects cannot wait that long. In such situations, the short term, say one year, data is compared with a nearby site having a long term data to predict the long term annual wind speed at the site under consideration. This is known as the "measure, correlate and predict (mcp)" technique.

Since wind is driven by the sun and the seasons, the wind pattern generally repeats over the period of one year. The wind site is usually described by the speed data averaged over the calendar months. Sometimes, the monthly data is aggregated over the year for brevity in reporting the overall "windiness" of various sites. The wind-speed variations over the period can be described by a probability distribution function.

4.6.1 Weibull Probability Distribution

The variation in wind speed are best described by the Weibull probability distribution function 'h' with two parameters, the shape parameter 'k', and the scale parameter 'c'. The probability of wind speed being v during any time interval is given by the following:

$$h(v) = \left(\frac{k}{c}\right)\left(\frac{v}{c}\right)^{(k-1)} e^{-\left(\frac{v}{c}\right)^{k}} \quad \text{for } 0 < v < \infty \tag{4-19}$$

In the probability distribution chart, h is plotted against v over a chosen time period, where:

$$h = \frac{\text{fraction of time wind speed is between } v \text{ and } (v + \Delta v)}{\Delta v} \tag{4-20}$$

By definition of the probability function, probability that the wind speed will be between zero and infinity during that period is unity, i.e.:

$$\int_0^\infty h \cdot dv = 1 \tag{4-21}$$

If we choose the time period of one year, then express the probability function in terms of the number of hours in the year, such that:

$$h = \frac{\text{number of hours the wind is between } v \text{ and } (v + \Delta v)}{\Delta v} \tag{4-22}$$

The unit of 'h' is hours per year per meter/second, and the integral (4-21) becomes 8,760 (the total number of hours in the year) instead of unity.

Figure 4-7 is the plot of h versus v for three different values of k. The curve on the left with $k = 1$ has a heavy bias to the left, where most days are windless ($v=0$). The curve on the right with $k = 3$ looks more like a normal bell shape distribution, where some days have high wind and equal number

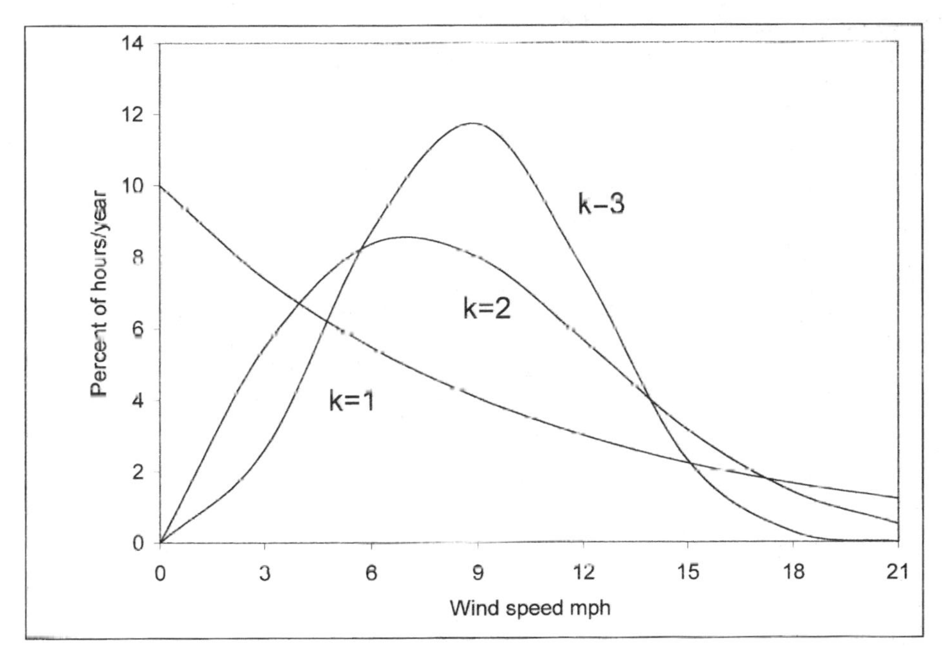

FIGURE 4-7
Weibull probability distribution function with scale parameter c=10 and shape parameters k = 1, 2 and 3.

of days have low wind. The curve in the middle with k = 2 is a typical wind distribution found at most sites. In this distribution, more days have lower than the mean speed, while few days have high wind. The value of k determines the shape of the curve, hence is called the 'shape parameter'.

The Weibull distribution with k = 1 is called the exponential distribution which is generally used in the reliability studies. For k>3, it approaches the normal distribution, often called the Gaussian or the bell-shape distribution.

Figure 4-8 shows the distribution curves corresponding to k = 2 with different values of c ranging from 8 to 16 mph (1 mph = 0.446 m/s). For greater values of c, the curves shift right to the higher wind speeds. That is, the higher the c, the more number of days have high winds. Since this shifts the distribution of hours at a higher speed scale, the c is called the scale parameter.

At most sites the wind speed has the Weibull distribution with k = 2, which is specifically known as the Rayleigh distribution. The actual measurement data taken at most sites compare well with the Rayleigh distribution, as seen in Figure 4-9. The Rayleigh distribution is then a simple and accurate enough representation of the wind speed with just one parameter, the scale parameter "c".

Summarizing the characteristics of the Weibull probability distribution function:

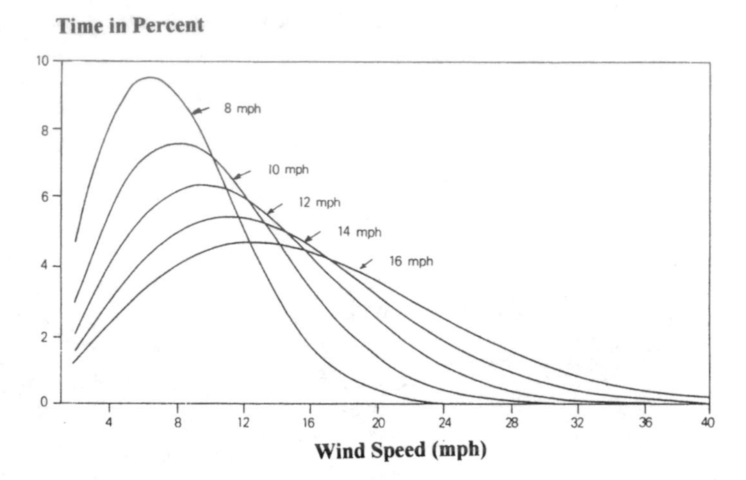

FIGURE 4-8
Weibull probability distribution with shape parameter k = 2 and the scale parameters ranging from 8 to 16 miles per hour (mph).

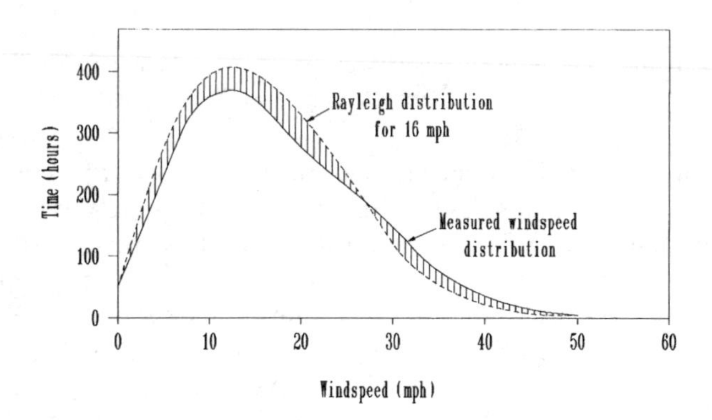

FIGURE 4-9
Rayleigh distribution of hours/year compared with measured wind-speed distribution at St. Ann's Head, England.

k = 1 makes it the exponential distribution, $h = \lambda \cdot e^{-\lambda V}$ where $\lambda = 1/c$

k = 2 makes it the Rayleigh distribution, $h = 2\lambda^2 \cdot V \cdot e^{-(\lambda V)^2}$, and (4-23)

k > 3 makes it approach a normal bell-shape distribution.

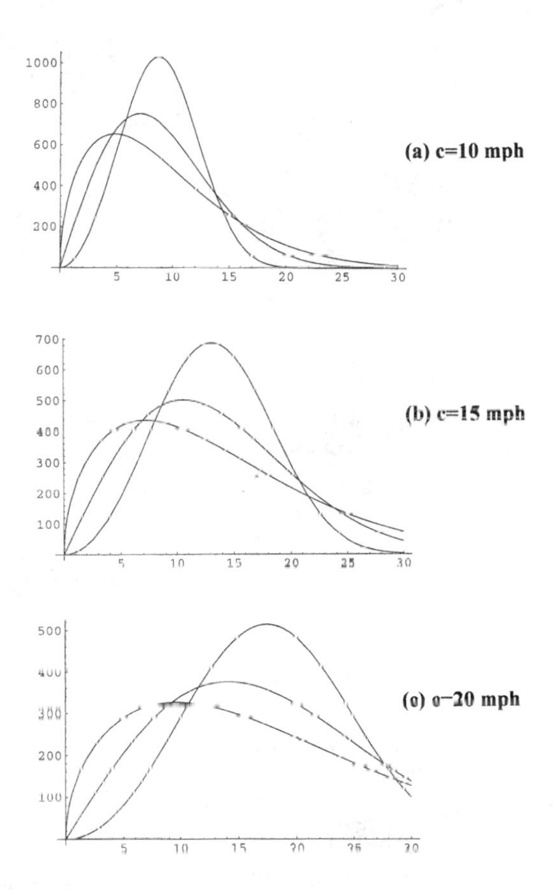

FIGURE 4-10
Weibull distributions of hours/year with three different shape parameters k = 1.5, 2, and 3.

Since most wind sites would have the scale parameter ranging from 10 to 20 miles per hour (about 5 to 10 m/s), and the shape parameter ranging from 1.5 to 2.5 (rarely 3.0), our discussion in the following sections will center around those ranges of c and k.

Figure 4-10 displays the number of hours on the vertical axis versus the wind speed on the horizontal axis with distributions of different scale parameters c = 10, 15, and 20 mph and shape parameters k = 1.5, 2, and 3. The values of h in all three sets of curves are the number of hours in a year in the speed interval v + Δv divided by Δv. Figure 4-11 depicts the same plots in the three-dimensional h-v-k space. It shows the effect of k in shifting the shape from the bell shape in the front right hand side (k = 3) to the Rayleigh and flatter shapes as the value of k decreases from 3.0 to 1.5. It is also observed from these plots that as c increases, the distribution shifts to the higher speed values.

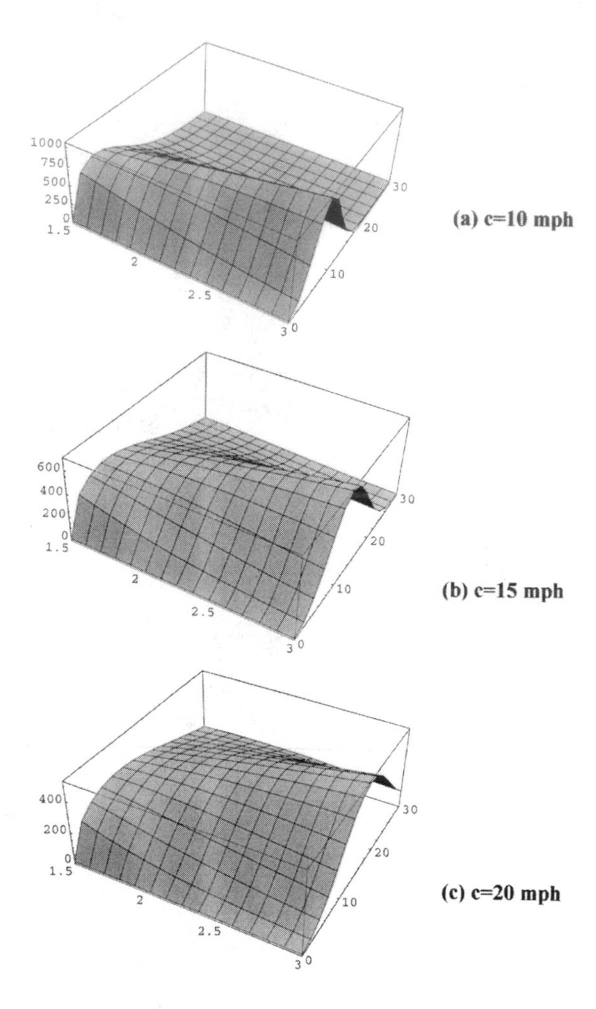

FIGURE 4-11
Three-dimensional h-v-k plots with c ranging from 10 to 20 mph and k ranging from 1.5 to 3.0.

4.6.2 Mode and Mean Speeds

We now define the following terms applicable to the wind speed:

Mode speed is defined as the speed corresponding to the hump in the distribution function. This is the speed the wind blows most of the time.

Mean speed over the period is defined as the total area under the h-v curve integrated from $v = 0$ to ∞, divided by the total number of hours in the period (8,760 if the period is one year). The annual mean speed is therefore the weighted average speed and is as follows:

$$V_{mean} = \frac{1}{8760}\int_0^{\infty} h \cdot v \cdot dv \tag{4-24}$$

For c and k values in the range found at most sites, the integral expression can be approximated to the Gamma function:

$$V_{mean} = c\,\Gamma\left(1 + \frac{1}{k}\right) \tag{4-25}$$

For the Rayleigh distribution with k = 2, the Gamma function can be further approximated to the following:

$$V_{mean} = 0.90 \cdot c \tag{4-26}$$

This is a very simple relation between the scale parameter c and V_{mean}, which can be used with reasonable accuracy. For example, most sites are reported in terms of their mean wind speeds. The c parameter in the corresponding Rayleigh distribution is then $c = V_{mean}/0.9$. The k parameter is of course 2.0 for the Rayleigh parameters. Thus, we have the Rayleigh distribution of the site using the generally reported mean speed as follows:

$$h(v) = \frac{2v}{c^2} e^{-\left(\frac{v}{c}\right)^2} = \frac{2v}{\left(V_{mean}\right)^2} e^{-\left(\frac{v}{V_{mean}}\right)^2} \tag{4-27}$$

4.6.3 Root Mean Cube Speed

The wind power is proportional to the speed cube, and the energy collected over the year is the integral of $h \cdot v^3 \cdot dv$. We, therefore, define the "root mean cube" or the "rmc" speed in the manner similar to the root mean square (rms) value in the alternating current circuits:

$$V_{rmc} = \sqrt[3]{\frac{1}{8760}\int_0^{\infty} h \cdot v^3 \cdot dv} \tag{4-28}$$

The rmc speed is useful in quickly estimating the annual energy potential of the site. Using V_{rmc} in Equation 4-12 gives the annual average power:

$$P_{rmc} = \frac{1}{4}\rho \cdot V_{rmc}^{\ 3} \text{ watts/m}^2 \tag{4-29}$$

TABLE 4-1

Comparison of Three Wind Farm Sites with the Same Mean Wind Speed but Significantly Different Specific Power Density

Site	Annual mean wind speed Meters/second	Annual average specific power Watts/m^2
Culebra, Puerto Rico	6.3	220
Tiana beach, New York	6.3	285
San Gorgonio, California	6.3	365

Then, multiplying the P_{rmc} value by the total number of the hours in the year gives the annual energy production potential of the site.

The importance of the rmc speed is highlighted in Table 4-1. It compares the wind power density at three sites with the same annual average wind speed of 6.3 m/s. The San Gorgonio site in California has 66 percent greater power density than the Culebra site in Puerto Rico. The difference comes from having the different shape factors k, and, hence, the different rmc speeds, although all have the same annual mean speed.

4.6.4 Mode, Mean, and rmc Speeds Compared

The important difference between the mode, mean, and the rmc speeds is illustrated in Table 4-2. The values of the three speeds are compiled for four shape parameters (k = 1.5, 2.0, 2.5, and 3.0) and three scale parameters (c = 10, 15, and 20 mph). The wind power densities are calculated using the respective speeds in the wind power equation $P = 1/2\,\rho \cdot V^3$ watts/m^2 using the air mass density of 1.225 kg/m^3.

We observe the following from the c=15 rows:

1. For k = 1.5, the power density using the mode speed is 230 versus the correct value of 4,134 watts/m^2 using the rmc speed. The ratio of the incorrect to correct value of the power density is 1 to 18 — a huge difference.
2. For k = 2, the power densities using the mode and rmc speeds are 731 and 2,748 watts/m^2, respectively, in the ratio of 1 to 3.76. The corresponding power densities with the mean and the rmc speeds are 1,439 and 2,748 watts/m^2 in the ratio of 1 to 1.91.
3. For k = 3, the power densities using the mode and rmc speeds are 1,377 and 2,067 watts/m^2, respectively, in the ratio of 1 to 1.50. The corresponding power densities with the mean and the rmc speeds are 1,472 and 2,067 watts/m^2, in the ratio of 1 to 1.40.

Thus, regardless of the shape and the scale parameters, use of the mode or the mean speed in the power density equation would introduce a significant error in the annual energy estimate, sometimes of several folds, making the estimates completely useless.

TABLE 4-2

Influence of the Shape and Scale Parameters on the Mode, Mean and RMC Speeds and the Energy Density

c	k	Mode Speed	Mean Speed	RMC Speed	Pmode W/m²	Pmean W/m²	Prmc W/m²	Ermc KWh/yr
10	1.5	4.81	9.03	12.60	68	451	1225	5366
	2.0	7.07	8.86	11.00	216	426	814	3565
	2.5	8.15	8.87	10.33	331	428	675	2957
	3.0	8.74	8.93	10.00	409	436	613	2685
15	1.5	7.21	13.54	18.90	230	1521	4134	18107
	2.0	10.61	13.29	16.49	731	1439	2748	12036
	2.5	12.23	13.31	15.49	1120	1444	2278	9978
	3.0	13.10	13.39	15.00	1377	1472	2067	9053
20	1.5	9.61	18.05	25.19	544	3604	9790	42880
	2.0	14.14	17.72	22.00	1731	3410	6514	28531
	2.5	16.30	17.75	20.66	2652	3423	5399	23648
	3.0	17.47	17.86	20.00	3266	3489	4900	21462

P = wind power density in watts per square meter of the blade swept area = $0.5\ \rho\ V^3$ where ρ = 1.225 kg/m³

The last column is the energy potential of the site in kWh per year per m² of the blade area, assuming the rotor efficiency Cp of 50 percent (i.e., the maximum power that can be converted into electrical power is $0.25\ \rho\ V^3$).

The last column in Table 4-2 is the yearly energy potentials of the corresponding sites in kWh per year per square meter of the blade area for the given k and c values. These values are calculated for the rotor efficiency C_p of 50 percent, which is the maximum that can be practically achieved.

4.6.5 Energy Distribution

If we define the energy distribution function:

$$e = \frac{kWh\ contribution\ in\ the\ year\ by\ the\ wind\ between\ v\ and\ (v + \Delta v)}{\Delta v} \tag{4-30}$$

Then, it would look like that in Figure 4-12, which is for the Rayleigh speed distribution. The wind speed curve has the mode at 5.5 m/s and the mean at 6.35 m/s. However, because of the cubic relation with the speed, the maximum energy contribution comes from the wind speed at 9.45 m/s. Above this speed, although V^3 continues to increase in cubic manner, the number of hours at those speeds decreases faster than V^3. The result is an overall decrease in the yearly energy contribution. For this reason, it is advantageous to design the wind power to operate at variable speeds in order to capture the maximum energy available during high wind periods.

Figure 4-13 is a similar chart showing the speed and energy distribution functions for shape parameter 1.5 and scale parameter 15 mph. The mode speed is 10.6 mph, the mean speed is 13.3 and the rmc speed is 16.5 mph.

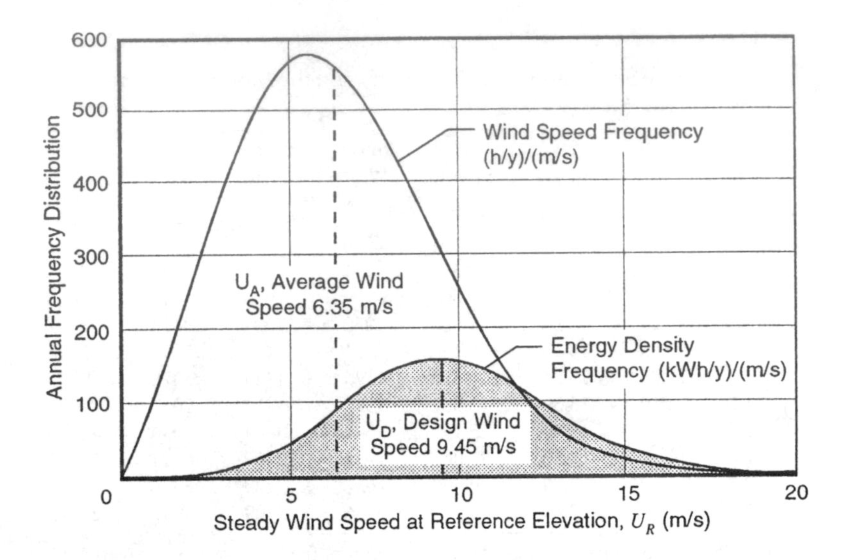

FIGURE 4-12
Rayleigh distributions of hours and energy per year versus wind speed with c = 10 and k = 2.

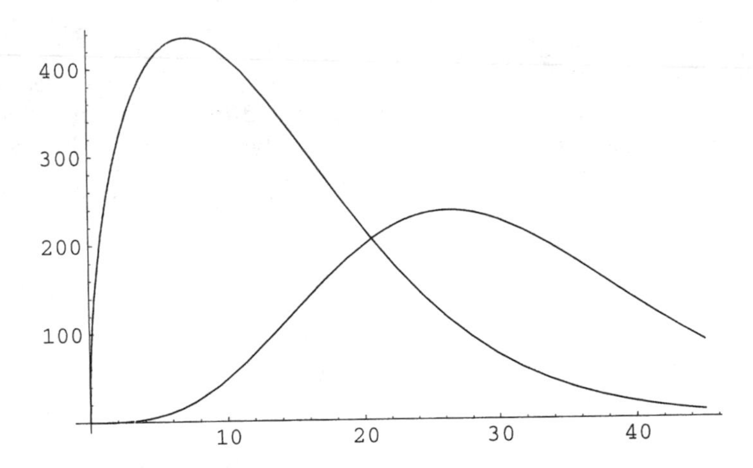

FIGURE 4-13
Rayleigh distributions of hours and energy per year versus wind speed with c = 15 and k = 1. 5.

The energy distribution function has the mode at 28.5 mph. That is, the most energy is captured at 28.5 mph wind speed, although the probability of wind blowing at that speed is low.

The comparison of Figures 4-12 and 4-13 shows that as the shape parameter value decreases from 2.0 to 1.5, the speed and the energy modes move farther apart. On the other hand, as the speed distribution approaches the bell shape for k > 3, the speed and the energy modes get closer to each other.

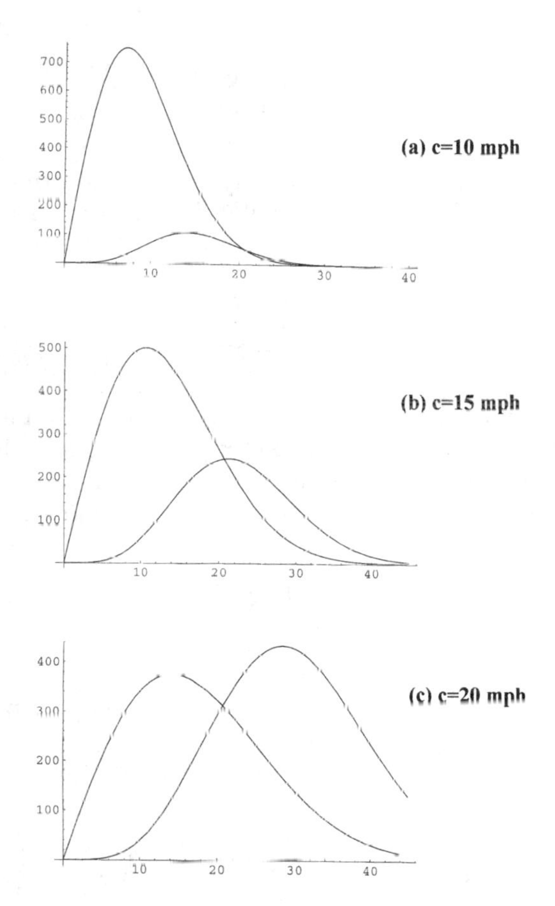

FIGURE 4-14
Rayleigh distributions of hours and energy per year with k = 2 and c = 10, 15, and 20 mph.

Figure 4-14 compares the speed and the energy distributions with k = 2 (Rayleigh) and c = 10, 15, and 20 mph. As seen here, the relative spread between the speed mode and the energy mode remains about the same, although both shift to the right as c increases.

4.6.6 Digital Data Loggers

The mean wind speed over a period of time is obtained by adding numerous readings taken over the period and dividing the sum by the number of readings. Many digital data loggers installed over the last few decades collected average wind speed data primarily for the meteorological purpose, as opposed to assessing the wind power. They logged the speed every hour, and then averaged over the day, which in turn, was averaged over the month and over the year. The averaging was done as follows:

$$V_{avg} = \frac{1}{n}\sum_{i=1}^{n} V_i \tag{4-31}$$

As seen earlier, for assessing the wind power, the rmc speed is what matters. The rmc equivalent of the digital data logging is as follows:

$$V_{rmc} = \sqrt[3]{\frac{1}{n}\sum_{i=1}^{n} V_i^3} \tag{4-32}$$

The above equation does not take into account the variation in the air mass density, which is also a parameter (although of second order) in the wind power density. Therefore, a better method of collecting the wind power data is to digitize the yearly average power density as follows:

$$P_{rmc} = \frac{1}{2n}\sum_{i=1}^{n} \rho_i \cdot V_i^3 \tag{4-33}$$

where n = number of observations in the averaging period
ρ_i = air density (kg/m3), and
V_i = wind speed (m/s) at the ith observation time.

4.6.7 Effect of Height

The wind shear at ground surface causes the the wind speed increase with height in accordance with the expression

$$2 = V_1 \cdot \left(\frac{h_2}{h_1}\right)^{\alpha} \tag{4-34}$$

where V_1 = wind speed measured at the reference height h_1
V_2 = wind speed estimated at height h_2, and
α = ground surface friction coefficient.

The friction coefficient is low for smooth terrain and high for rough ones (Figure 4-15). The values of α for typical terrain classes are given in Table 4-3.

The wind speed does not increase with height indefinitely. The data collected at Merida airport in Mexico show that typically the wind speed increases with height up to about 450 meter height, and then decreases (Figure 4-16)[1]. The wind speed at 450 meters height can be four to five times greater than that near the ground surface.

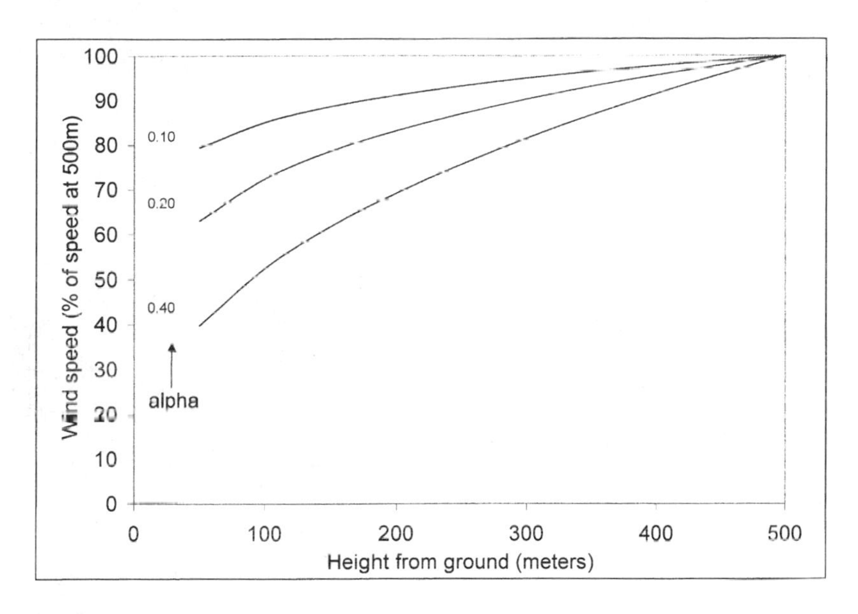

FIGURE 4-15
Wind speeds variation with height over different terrain. Smooth terrain has lower friction, developing a thin layer above.

TABLE 4-3
Friction Coefficient of Various Terrain

Terrain Type	Friction Coefficient α
Lake, ocean and smooth hard ground	0.10
Foot high grass on level ground	0.15
Tall crops, hedges, and shrubs	0.20
Wooded country with many trees	0.25
Small town with some trees and shrubs	0.30
City area with tall buildings	0.40

4.6.8 Importance of Reliable Data

Some of the old wind-speed data around the world may have been primarily collected for meteorological use with rough instruments and relatively poor exposure to the wind. This is highlighted by the recent wind resource study of Mexico.[1] Significant differences in the old and the new data have been found, as listed in Table 4-4. The 1983 OLADE Atlas of Mexico indicates very low energy potential, whereas the 1995 NREL data reports several times more energy potential. The values from the OLADE Atlas are from a few urban locations where anemometers could be poorly exposed to the prevailing

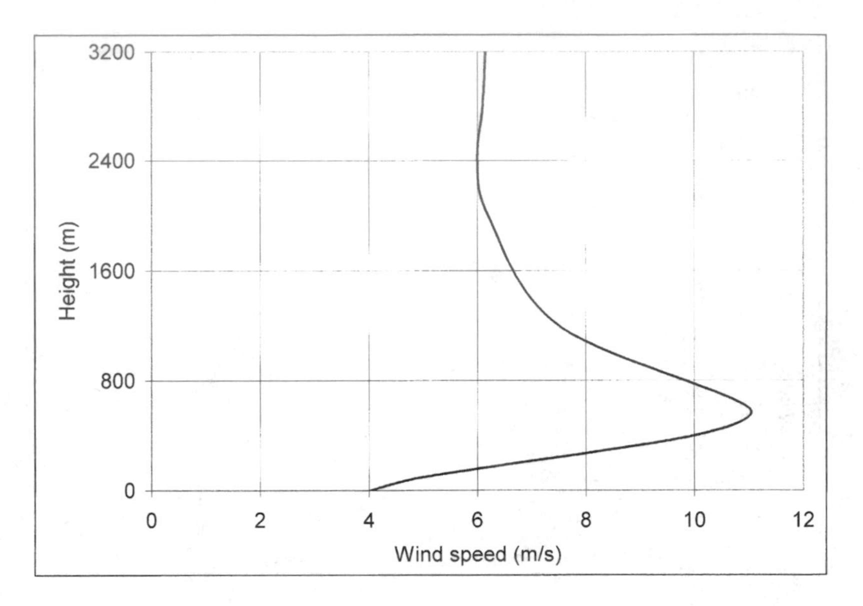

FIGURE 4-16
Wind-speed variations with height measured at Merida airport in Mexico. (Source: Schwartz and Elliott, DOE/NREL Report DE95009202, May 1995.)

TABLE 4-4

Comparison of Calculated Average Wind Power Density Between 1983 OLADE Atlas and 1995 NREL Analysis for Several Locations in Mexico

		Wind Power Density in Watts/meter²	
Region in Mexico	**Data Site**	**OLADE Atlas (1983)**	**NREL Data (1995)**
Yucatan Peninsula	Merida	22	165
	Campeche	23	120
	Chetumal	28	205
Northern Gulf Plain	Tampico	8	205
	Ciudad Victoria	32	170
	Matamoros	32	165
Central Highlands	Durango	8	140
	San Luis Potosi	35	155
	Zacatecas	94	270
Northwest	Chihuahua	27	120
	Hermosillo	3	80
	La Paz	10	85

Data at 10 meters height.

winds. In contrast, the new NREL wind data comes from a large number of stations, including open airport locations, and has incorporated the terrain effect in the final numbers. The message here is clear. It is important to have reliable data on the annual wind speed distribution over at least a few years before siting a wind farm with a high degree of confidence.

4.7 Wind Speed Prediction

The available wind energy depends on the wind speed, which is a random variable. For the wind-farm operator, this poses difficulty in the system scheduling and energy dispatching, as the schedule of the wind-power availability is not known in advance. However, if the wind speed can be reliably forecasted up to several hours in advance, the generating schedule can efficiently accommodate the wind generation. Alexiadis et al[2] have proposed a new technique for forecasting wind speed and power output up to several hours in advance. The technique is based on cross-correlation at neighboring sites and artificial neutral networks. The proposed technique can significantly improve forecasting accuracy compared to the persistence forecasting model. The new proposed method is calibrated at different sites over a one-year period.

4.8 Wind Resource Maps

The wind resource of a vast region or a country is mapped in terms of the wind speed, the wind power density in watts per square meter of the rotor swept area, or the wind energy potential in kWh/m^2 per year. Often the wind resource is mapped in all three forms. The data is usually represented by the contour curves, as it is the most useful and easily understood mapping technique. Along the contour line, the plotted parameter remains constant. For example, the isovent map plots the contour lines connecting the sites having the same annual wind speed. The equipotential map plots the contour lines connecting the sites having the same annual wind energy capture potential in kWh/m^2. The wind resource maps of many countries have been prepared in such contour form. Some of them are presented in the following sections:

4.8.1 The U.S.A.

The U.S. wind resource is large enough to produce more than 4.4 trillion kWh of electricity each year. This is more than the entire nation will use in the year 2000. Figure 4-17 is the wind speed map of the U.S.A., whereas Figure 4-18 is the wind power density map.[3-4] Although most of the country's installed wind capacity is presently in California, the Department of Energy estimates that almost 90 percent of the usable wind resource in the U.S.A.

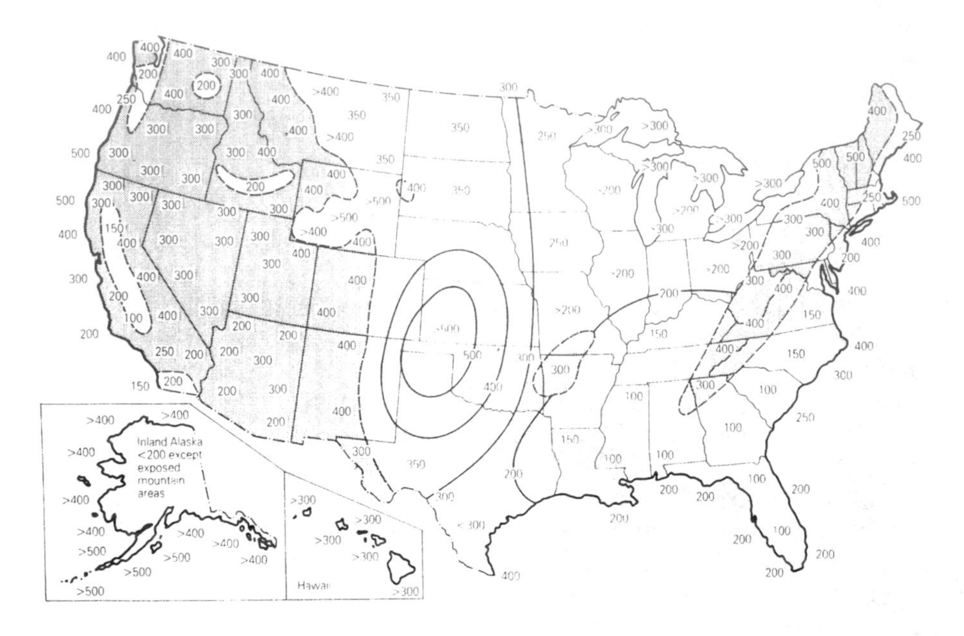

FIGURE 4-17
Annual average wind-power density in watts/m^2 in the U.S.A. at a 50-meter tower height. (Source: DOE/NREL.)

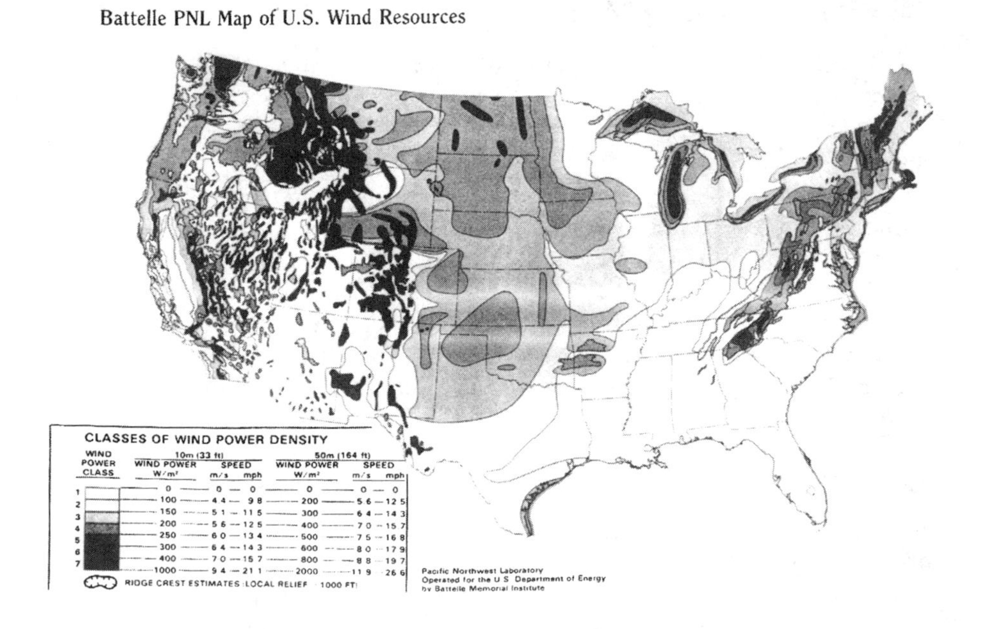

FIGURE 4-18
The U.S. wind resource map. (Source: DOE/Battelle Pacific Northwest Laboratory.)

lies in the wind-belt spanning eleven Great Plains states. These states stretch from Montana, North Dakota and Minnesota in the north, to Texas in the south. The wind resource of this region has remained virtually untapped until recently. However, significant wind-power generating capacity is being added in the wind-belt states as of the late 1990s.

4.8.2 Minnesota

Having a large wind-power potential in the southwestern part of the state, Minnesota started the Wind Resource Assessment Program[1] in 1985. The wind monitoring stations were installed at 41 sites with high wind potential. Detailed information on hourly average wind speed, standard deviation, direction, and temperature at many sites in Minnesota were recorded. The Northern States Power Company initially recorded this data on cassette tape recorders powered from batteries charged by photovoltaic modules. Later, data loggers were installed by utilities throughout the state.

In 1993, a new study of wind and solar capabilities was funded by the U.S. Department of Energy, the Northern States Power, and the Minnesota Department of Public Services. Figure 4-19 is a 1995 compilation of the Minnesota wind power potentials at various locations.[5] As of the late 1990s, many utilities in that region have become active in developing the wind resource (Figure 4-20).

4.8.3 The United Kingdom

The isovent map of the United Kingdom is shown in Figure 4-21 in terms of the annual average wind speed in knots at open sites excluding hilltops.[6] Scotland and the western shore of Ireland are the high wind zones. The central portion of England has relatively low wind speeds.

The wind energy map in MWh/m^2 per year is shown in Figure 4-22.[6] The northwestern offshore sites show high energy potential of 5 MWh/m^2 per year.

4.8.4 Europe

The wind speed and the energy maps of Europe are combined into one that is published by the Rios National Laboratory in Denmark. The map (Figure 4-23) is prepared for a wind tower height of 50 meters above ground with five different terrains. In the map, the darker areas have higher wind energy density.

4.8.5 Mexico

The wind resource map for Mexico has been recently prepared under the Mexico-U.S. Renewable Energy Cooperation Program.[1] The data was collected

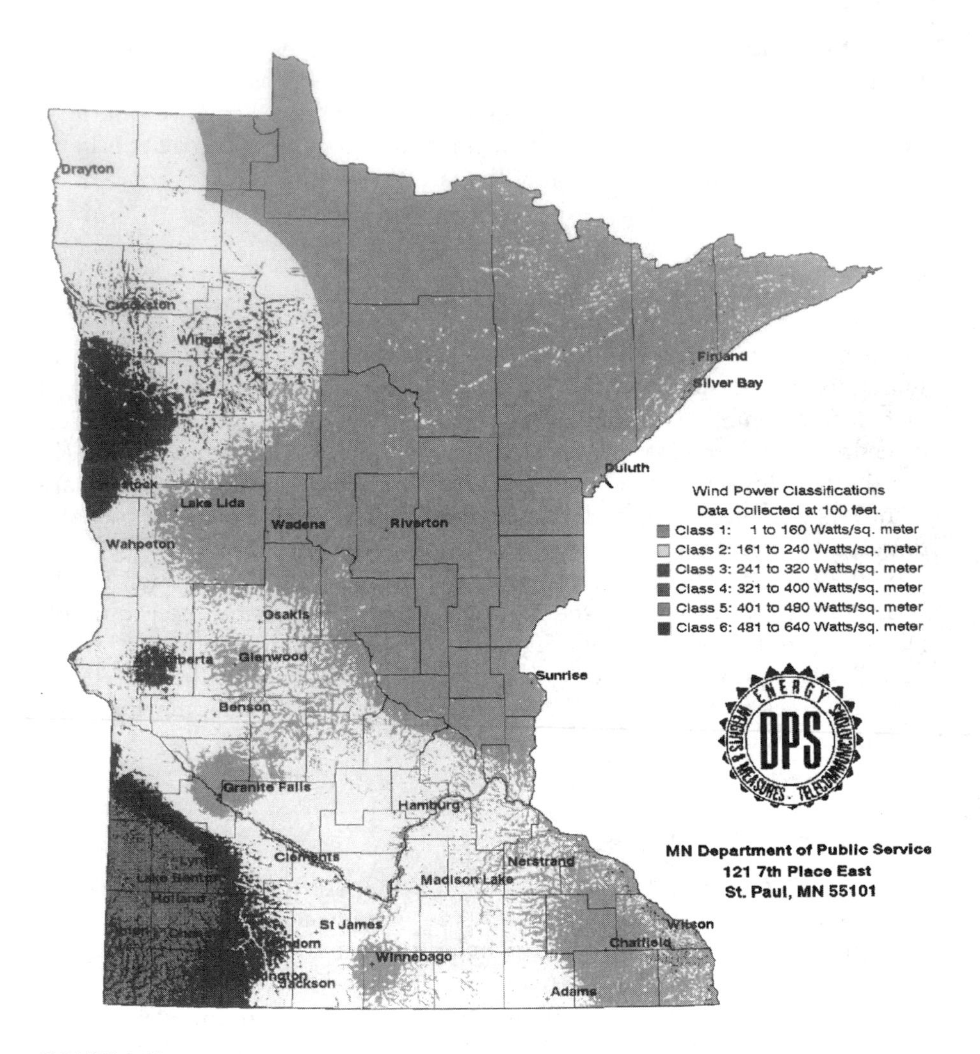

FIGURE 4-19
Minnesota wind resource map. (Source: Minnesota wind resources assessment program, Department of Public Service Report, 1994.)

at numerous city and airport sites spread throughout the country as shown in Figure 4-24. The maps, based on the extensive data collected at those locations, are separately reported for the utility-scale power generation and for the stand-alone remote power generation. This distinction is useful because the wind speed needed to make the wind farm economical for rural application is less than that required for the utility scale application.

Figure 4-25 is the utility-scale wind resource map for Mexico. The highest wind potential for grid-connected power plants is in the Zacatecas and the southern Isthmus of Tehuantepec areas. The rural wind resource map in

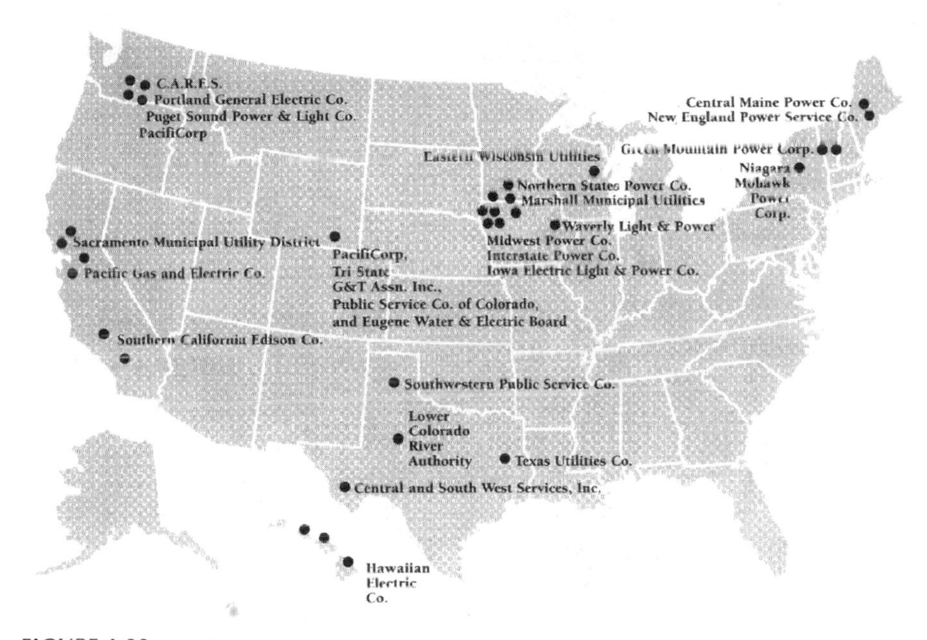

FIGURE 4-20
U.S. utilities gaining experience in developing wind power generation projects. (Source: Wind energy program overview, DOE Report No. 10095-071, March 1995.)

Figure 4-26 shows that areas suitable for stand-alone remote wind farms are widespread, and include about one-half of Mexico.

4.8.6 India

The wind speed measurement stations in India have been in operation since early 1980 at numerous locations. Sites with high wind speeds are shown in Figure 4-27, and their annual mean speed listed in Table 4-5. The data has been collected at 10-m or 25-m tower heights as indicated.[7] It is seen that most of these sites have the annual mean wind speed exceeding 18 km/h, which generally makes the wind power plant economically viable. The states with high wind potentials are Tamilnadu, Gujarat (Figure 4-28) and Andhra Pradesh.

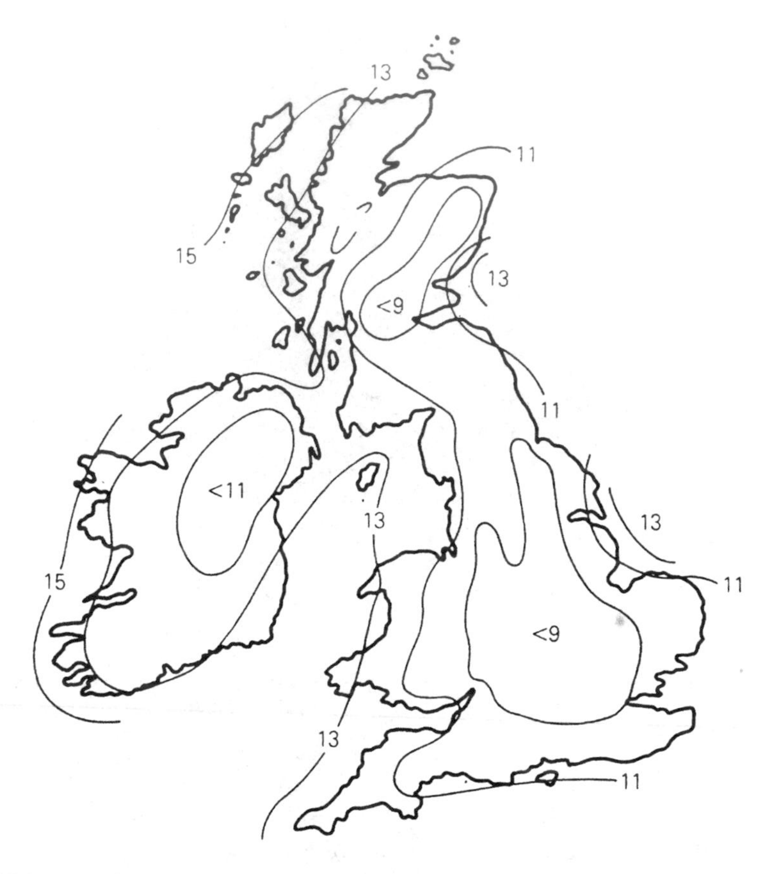

FIGURE 4-21
Isovent map of the United Kingdom in knots at open sites, excluding hilltops. (Source: Freris, L. L., Wind Energy Conversion Systems, Prentice Hall, London, 1990.)

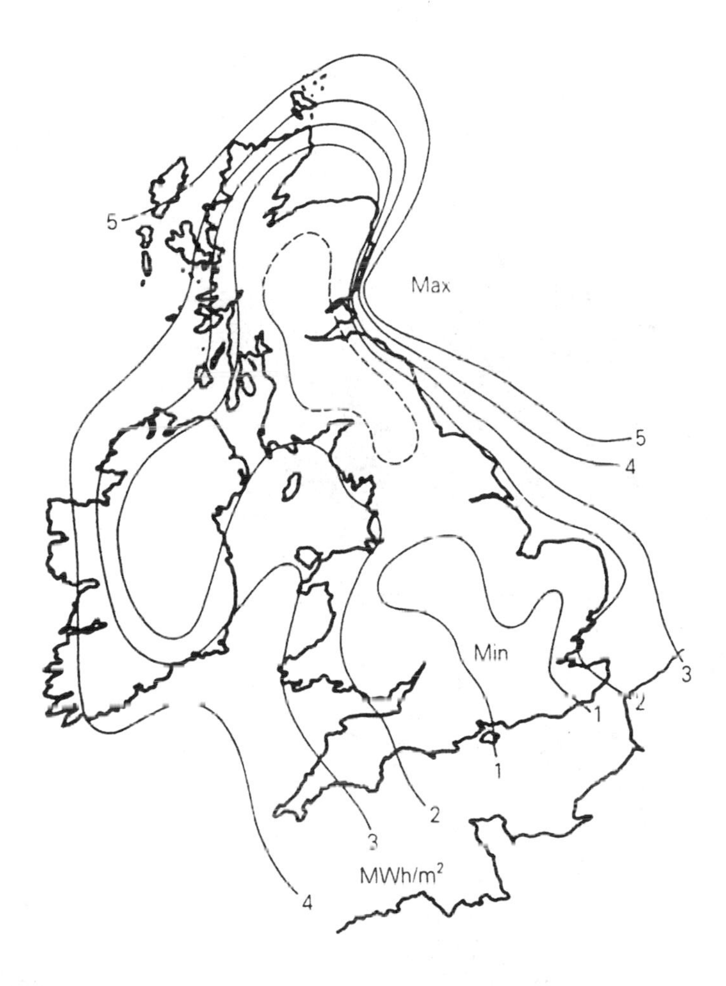

FIGURE 4-22
Wind energy map of the United Kingdom in MWh/m2 per year at open sites, excluding hilltops. (Source: Freris, L. L., Wind Energy Conversion Systems, Prentice Hall, London, 1990.)

Risø Map of European Wind Resources

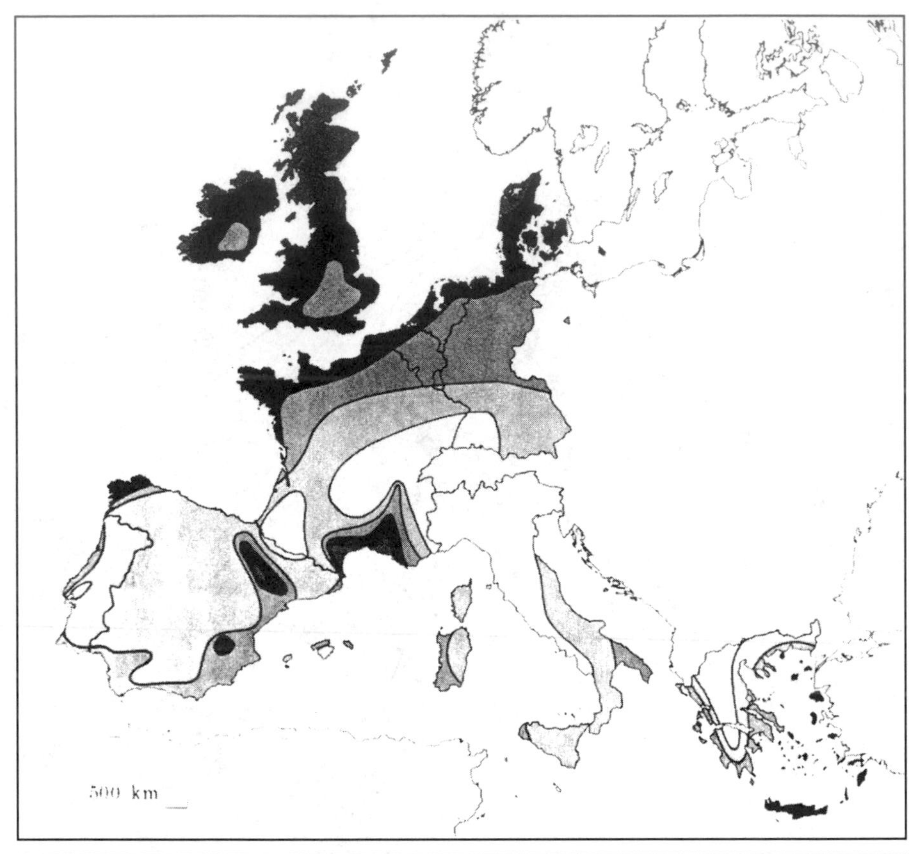

Wind resources[1] at 50 metres above ground level for five different topographic conditions										
	Sheltered terrain[2]		Open plain[3]		At a sea coast[4]		Open sea[5]		Hills and ridges[6]	
	m s⁻¹	Wm⁻²	m s⁻¹	Wm⁻²	m s⁻¹	Wm⁻²	m s⁻¹	Wm⁻²	m s⁻¹	Wm⁻²
	> 6.0	> 250	> 7.5	> 500	> 8.5	> 700	> 9.0	> 800	> 11.5	> 1800
	5.0-6.0	150-250	6.5-7.5	300-500	7.0-8.5	400-700	8.0-9.0	600-800	10.0-11.5	1200-1800
	4.5-5.0	100-150	5.5-6.5	200-300	6.0-7.0	250-400	7.0-8.0	400-600	8.5-10.0	700-1200
	3.5-4.5	50-100	4.5-5.5	100-200	5.0-6.0	150-250	5.5-7.0	200-400	7.0-8.5	400-700
	< 3.5	< 50	< 4.5	< 100	< 5.0	< 150	< 5.5	< 200	< 7.0	< 400

FIGURE 4-23
European wind resource map. (Source: Risø Laboratory, Denmark.)

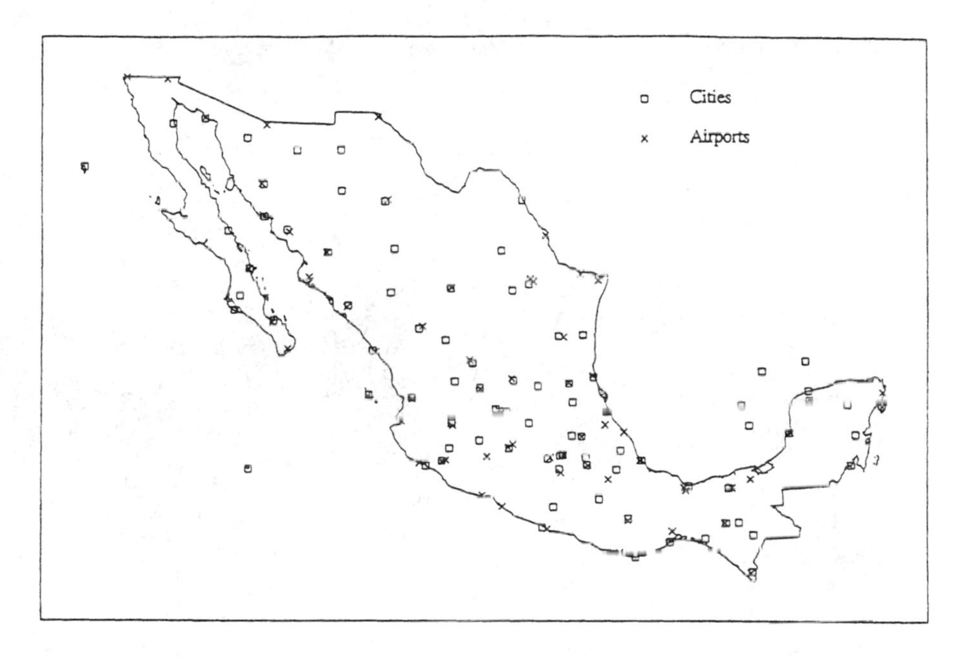

FIGURE 4-24
Locations in Mexico with surface wind data in DATSAV2 data set. (Source: Schwartz, M. N. and Elliott, D. L., Mexico wind resource assessment project, DOE/NREL Report No. DE95009202, March 1995.)

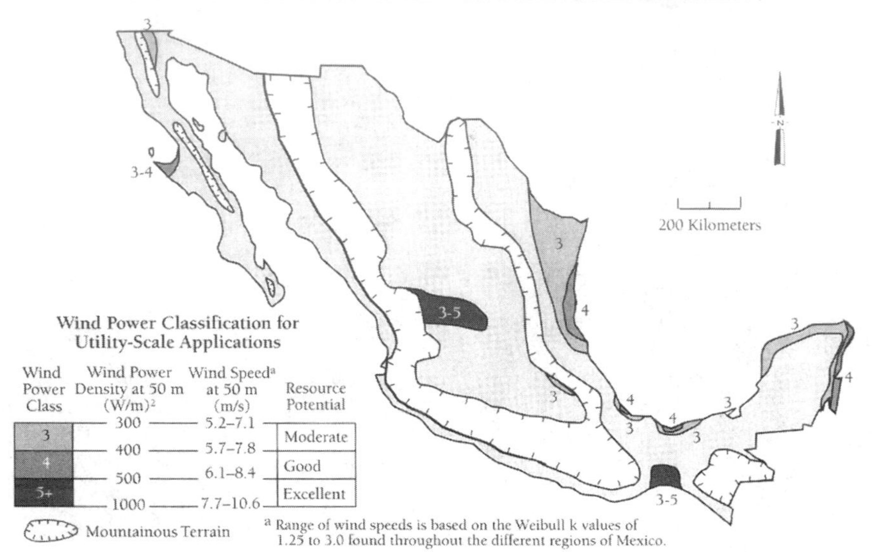

FIGURE 4-25
Annual average wind resource map of Mexico for utility scale applications. (Source: Schwartz, M. N. and Elliott, D. L., Mexico wind resource assessment project, DOE/NREL Report No. DE95009202, March 1995.)

MEXICO—PRELIMINARY WIND RESOURCE MAP FOR RURAL POWER APPLICATIONS

200 Kilometers

Wind Power Classification for Rural Power Applications

Wind Power Class	Wind Power Density at 30 m (W/m)²	Wind Speed[a] at 30 m (m/s)	Resource Potential
	0	0	
1	75	3.5–4.5	Poor
2	150	4.1–5.6	Useful
3	250	4.9–6.7	Good
4	500	6.1–8.4	Excellent

Mountainous Terrain

[a] Range of wind speeds is based on the Weibull k values of 1.25 to 3.0 found throughout the different regions of Mexico.

FIGURE 4-26
Annual average wind resource map of Mexico for rural power applications. (Source: Schwartz, M. N. and Elliott, D. L., Mexico wind resource assessment project, DOE/NREL Report No. DE95009202, March 1995.)

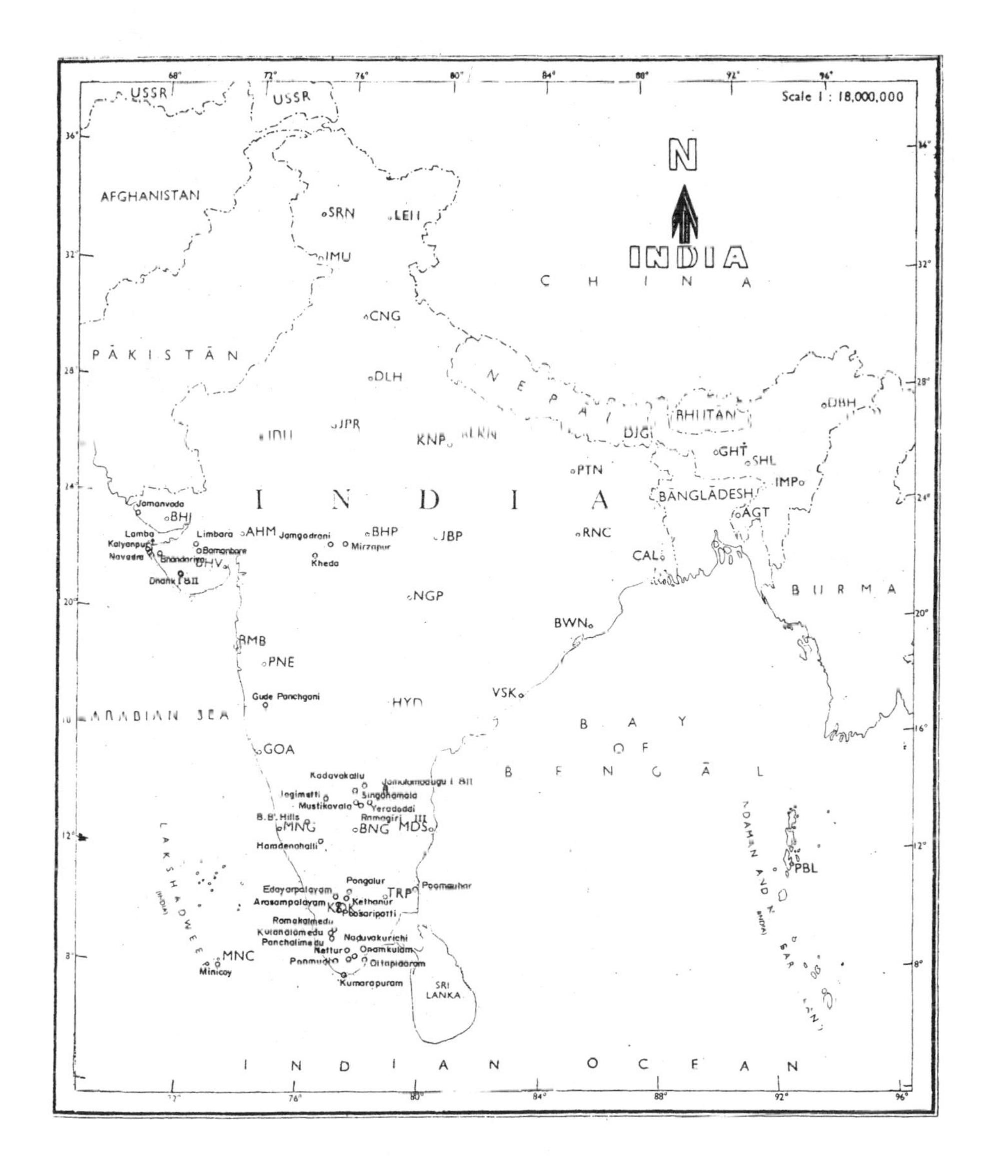

FIGURE 4-27
Wind monitoring sites in India. (Source: Tata Energy Research Institute, New Delhi, India.)

TABLE 4-5

Mean Annual Wind Speed at Selected Sites in India

Site	km/h	Site	km/h
Sites with 25–30 km/h wind speed		*Sites with 15–20 km/h wind speed*	
Ramakalmedu	30.3	Onamkulam*	19.9
Jogmatti	30.0	Nettur*	19.9
B.& H. Hills	27.1	Gude Panchgani	119.8
Dhank-2	25.1	Bhandariya	19.5
		Ramagiri-3	19.4
Sites with 20–25 km/h wind speed		Poosaripatti*	19.3
		Pongalur	19.1
Dhank-1	24.4	Limbara	19.1
Singanamala	23.8	Jamalamadugu-2*	18.7
Edayarpalayam	22.4	Ottapidaram	18.5
Kadavakallu*	22.1	Ponmudi	18.3
Kalyanpur	22.1	Hardenahalli	18.2
Kumarapuram*	22.0	Kheda	18.2
Kethanur	21.1	Jamgodrani	18.0
Navadra	20.8	Jamalamadunga-1*	17.5
Arasampalayam	20.5	Poompuhar	17.2
Bamambore-2	20.3	Minicoy	17.0
Mustikovala	20.2	Kolahalamendu	16.9
Pancjalimedu	20.2	Naduvakkurichi	16.8
Lamda	20.0	Yeradoddi*	15.8
		Mirzapur	15.4

* Data at 25-m height, all others at 10-m height

FIGURE 4-28
Lamda wind farm in Gujarat, India. (Source: Vestas Wind Systems, Denmark. With permission.)

References

1. Schwartz, M. N. and Elliott, D. L. 1995. "Mexico Wind Resource Assessment Project," *DOE/NREL Report No. DE95009202*, National Renewable Energy Laboratory, Golden, Colorado, March 1995.
2. Alexiadis, M. C., Dokopoulos, P. S., and Sahsamanogdou, H. S. 1998. "Wind Speed and Power Forecasting Based on Spatial Correlation Models," *IEEE Paper No. PE-437-EC-0-04-1998*.
3. Elliott, D. L., Holladay, C. G., Barchet, W. R., Foote, H. P., and Sandusky, W. F. 1991. "Wind Energy Resources Atlas of the United States," *DOE/Pacific Northwest Laboratory Report No. DE-86004442*, April 1991.
4. Elliott, D. L. 1997. "Synthesis of National Wind Assessment," *DOE/Pacific Northwest Laboratory, NTIS Report No. BNWL-2220 Wind-S*, 1997.
5. Rory, A. 1994. "Minnesota Wind Resources Assessment Program," *Minnesota Department of Public Service Report*, 1994.
6. Freris, L. L. 1990. "Wind Energy Conversion Systems," London, Prentice Hall, 1990.
7. Gupta, A. K. 1997. "Power Generation from Renewables in India," *Ministry of Non-Conventional Energy Sources*, New Delhi, India, 1997.

5

Wind Power System

The wind power system is fully covered in this and the following two chapters. This chapter covers the overall system level performance, design considerations and trades. The electrical generator is covered in the next chapter and the speed control in Chapter 7.

5.1 System Components

The wind power system is comprised of one or more units, operating electrically in parallel, having the following components.

- the tower.
- the wind turbine with two or three blades.
- the yaw mechanism such as the tail vane.
- the mechanical gear.
- the electrical generator.
- the speed sensors and control.

The modern system often has the following additional components:

- the power electronics.
- the control electronics, usually incorporating a computer.
- the battery for improving the load availability in stand-alone mode.
- the transmission link connecting to the area grid.

Because of the large moment of inertia of the rotor, the design challenges include the starting, the speed control during the power producing operation, and stopping the turbine when required. The eddy current or other type of brake is used to halt the turbine when needed for emergency or for routine maintenance. In the multiple tower wind farm, each turbine must

FIGURE 5-1
Control center at Baix Ebre wind farm in Catalonia, Spain. (Source: Institut Catalia d'Energia, Barcelona, Spain. With permission.)

have its own control system for operational and safety functions from a remote location (Figure 5-1).

5.1.1 Tower

The wind tower supports the turbine and the nacelle containing the mechanical gear, the electrical generator, the yaw mechanism, and the stall control. The nacelle component details and the layout are shown in Figure 5-2. Figure 5-3 shows a large nacelle during installation. The height of tower in the past has been in the 20 to 50-meter range. For medium and large size turbines, the tower is slightly taller than the rotor diameter, as seen in the dimension drawing of a 600 kW wind turbine (Figure 5-4). Small turbines are generally mounted on the tower a few rotor diameters high. Otherwise, they would suffer due to the poor wind speed found near the ground surface (Figure 5-5).

Both steel and concrete towers are available and are being used. The construction can be tubular or lattice.

The main issue in the tower design is the structural dynamics. The tower vibration and the resulting fatigue cycles under wind speed fluctuation are avoided by design. This requires careful avoidance of all resonance frequencies of the tower, the rotor and the nacelle from the wind fluctuation frequencies.

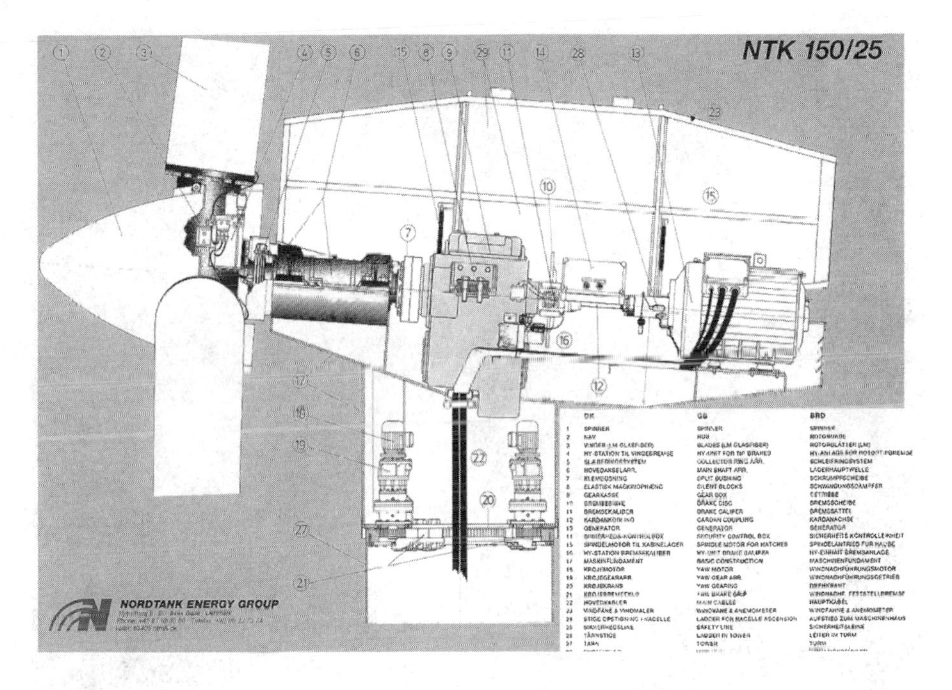

FIGURE 5-2
Nacelle details of a 150 kW/25 meter diameter turbine. (Source: Nordtank Energy Group/NEG Micon, Denmark. With permission.)

Sufficient margin must be maintained between the two sets of frequencies in all vibrating modes.

The resonance frequencies of the structure are determined by complete modal analyses, leading to the eigenvectors and eigenvalues of complex matrix equations representing the motion of the structural elements. The wind fluctuation frequencies that are found form the measurements at the site under consideration. Experience on a similar nearby site can bridge a gap on the required information.

Big cranes are generally required to install wind towers. Gradually increasing tower height, however, is bringing a new dimension in the installation (Figure 5-6). Large rotors add to the transportation problem as well. Tillable towers to nacelle and rotors moving upwards along with the tower are among some of the newer developments in the wind tower installation. The offshore installation comes with its own challenge that must be met.

5.1.2 Turbine Blades

The turbine blades are made of high-density wood or glass fiber and epoxy composites. Modern wind turbines have two or three blades. The steady

FIGURE 5-3
A large nacelle under installation. (Source: Nordtank Energy Group / NEG Micon, Denmark. With permission.)

mechanical stress due to centrifugal forces and fatigue under continuous vibrations make the blade design the weakest mechanical link in the system. Extensive design effort is needed to avoid premature fatigue failure of the blade.

The mechanical stress in the blade under gusty wind is kept under the allowable limit. This is achieved by controlling the rotor speed below the set limit. This not only protects the blades, but also protects the electrical generator from overloading and overheating. One method that has been used from the early designs and continues to be used today is the stall control. At stall, the wind flow ceases to be smooth around the blade contour, but separates before reaching the trailing edge. This always happens at high pitch angle. The blades experience high drag, thus lowering the rotor power output. The high pitch angle also produces high lift. The resulting load on the blade can cause a high level of vibration and fatigue, possibly leading to the mechanical failure. Regardless of the fixed or variable speed, the design engineer must deal with the stall forces. Researchers are moving from

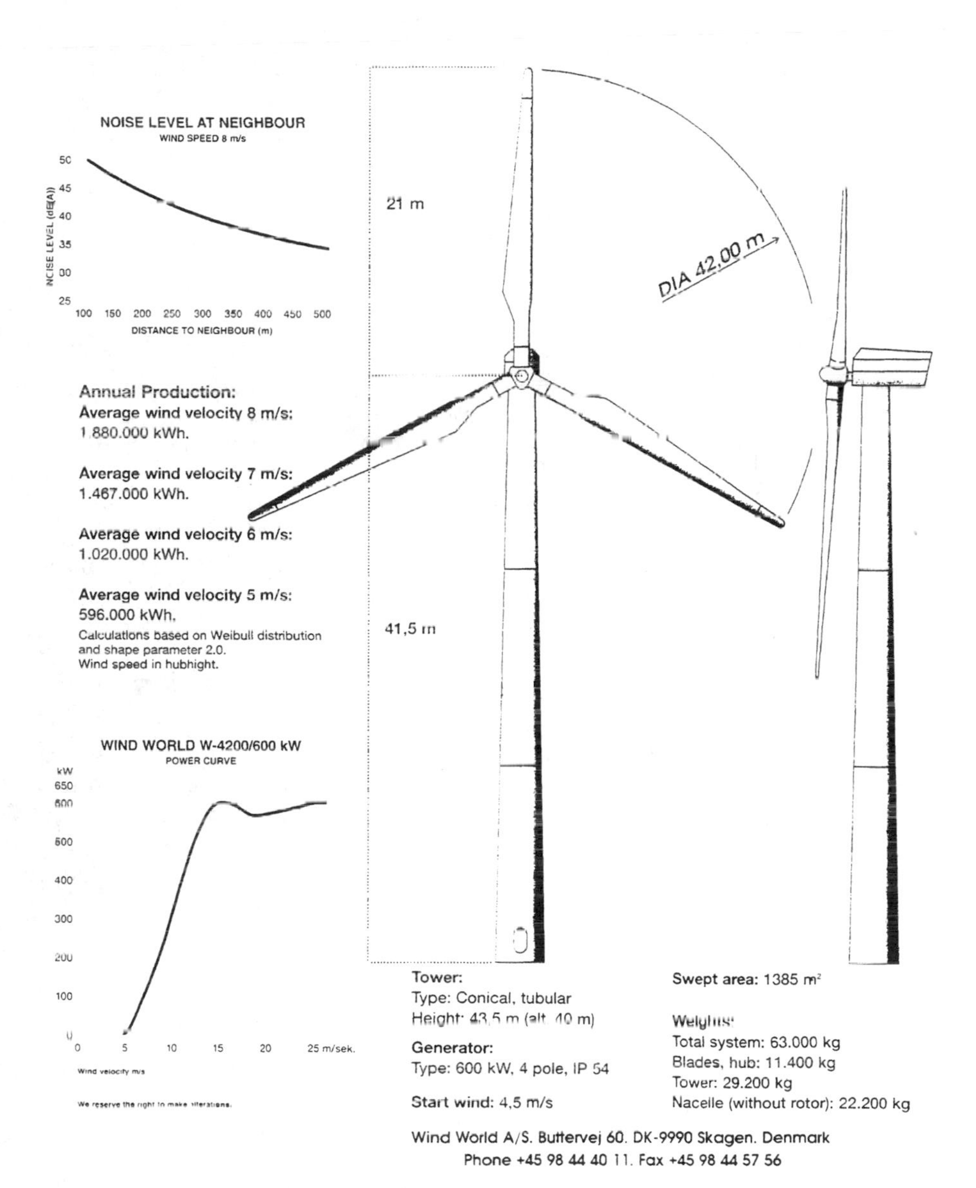

FIGURE 5-4
A 600 kW wind turbine and tower dimensions with specifications. (Source: Wind World Corporation, Denmark. With permission.)

the 2D to 3D stress analyses to better understand and design for such forces. As a result, the blade design is continually changing, particularly at the blade root where the loading is maximum due to the cantilever effect.

The aerodynamic design of the blade is important, as it determines the energy capture potential. The large and small machine blades have significantly different design philosophies. The small machine sitting on the tower

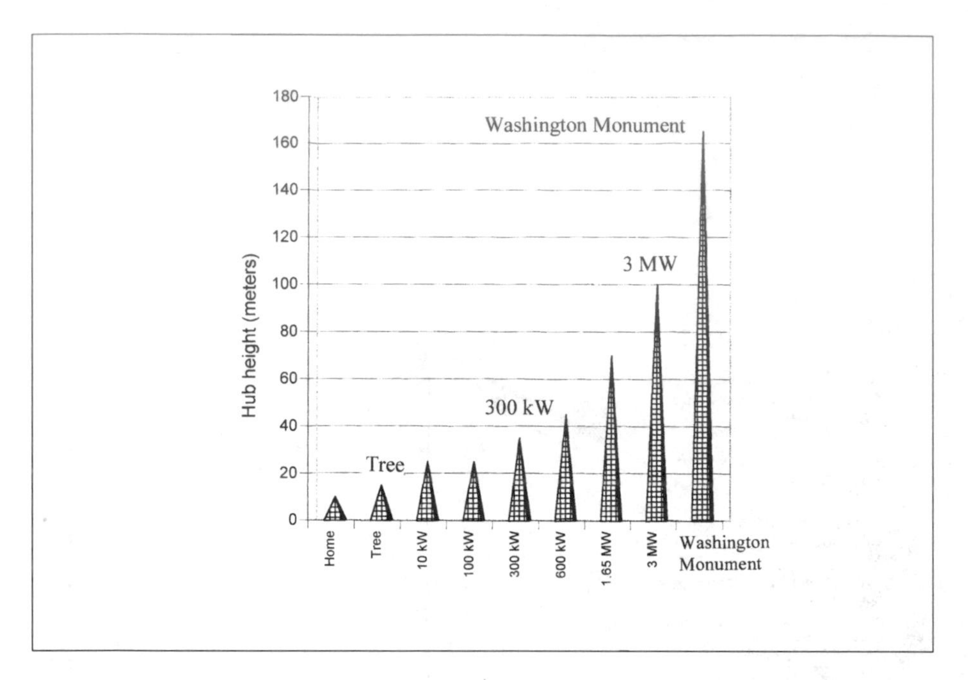

FIGURE 5-5
Tower heights of various capacity wind turbines.

relatively taller than the blade diameter, and generally unattended, requires low maintenance design. On the other hand, the large machine tends to optimize the aerodynamic performance for the maximum possible energy capture. In either case, the blade cost is generally kept below 10 percent of the total installed cost.

5.1.3 Yaw Control

The yaw control continuously orients the rotor in the direction of the wind. It can be as simple as the tail vane, or more complex on modern towers. Theoretical considerations dictate free yaw as much as possible. However, rotating blades with large moments of inertia produce high gyroscopic torque during yaw, often resulting in loud noise. Too rapid yaw may generate noise exceeding the local ordinance limit. Hence, a controlled yaw is often required and is used.

5.1.4 Speed Control

The wind turbine technology has changed significantly in the last 25 years.[1] Large wind turbines being installed today tend to be of variable speed

FIGURE 5-6
WEG MS-2 wind turbine installation at Myers Hill. (Source: Wind Energy Group, a Taylor Woodrow subsidiary and ETSU/DTI, U.K.)

design, incorporating the pitch control and the power electronics. Small machines on the other hand must have simple, low cost power and speed control. The speed control methods fall into the following categories:

- no speed control whatsoever. In this method, the turbine, the electrical generator, and the entire system is designed to withstand the extreme speed under gusty wind.
- yaw and tilt control, in which the rotor axis is shifted out of the wind direction when the wind speed exceeds the design limit.
- pitch control, which changes the pitch of the blade with the changing wind speed to regulate the rotor speed.
- stall control. In this method of speed control, when the wind speed exceeds the safe limit on the system, the blades are shifted into a position such that they stall. The turbine has to be restarted after the gust has gone.

Figure 5-7 depicts the distribution of the control methods used in small wind turbine designs. Large machines generally use the power electronic speed control, which is covered in Chapters 7 and 11.

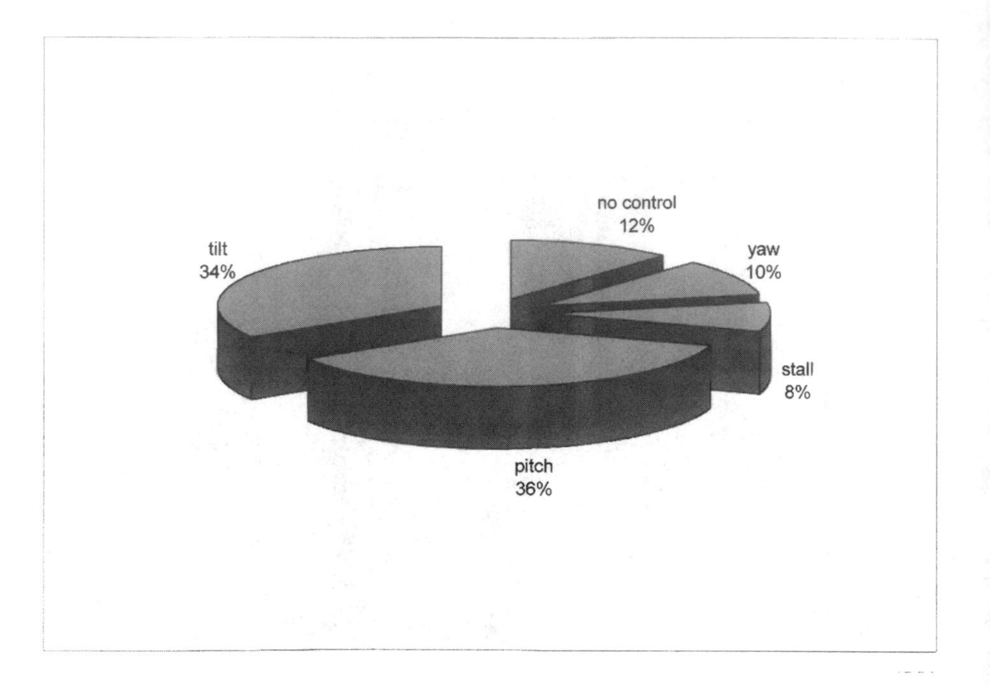

FIGURE 5-7
Speed control methods used in small to medium size turbines.

5.2 Turbine Rating

The wind turbines are manufactured in sizes ranging from a few kW for stand-alone remote applications to a couple of MW each for utility scale power generation. The grid-connected turbine as large as 2 MW capacity was installed in 1979 on Howard Knob Mountain in the United States, and 3 MW capacity was installed in 1988 at Berger Hill in the United Kingdom.

The method of assessing the nominal rating of the wind turbine has no globally acceptable standard. The difficulty arises because the power output of the turbine depends on the square of the rotor diameter and the cube of the wind speed. The rotor of a given diameter will, therefore, generate different power at different wind speed. The turbine that can generate 300 kW at 7 m/s would produce 450 kW at 8 m/s wind. What rating should then be assigned to this turbine? Should we also specify the "rated speed"? Early wind turbine designers created a rating system that specified the power output at some arbitrary wind speed. This method did not work well as everyone could not agree on one speed to specify the power rating. The "rated" wind speeds varied from 10 to 15 m/s under this practice. Manufacturers are lax on providing the higher side of the wind speed, claiming greater output from the same design.

To avoid such rating confusion, some European manufacturers refer to only the rotor diameter. But the confusion continues as to the maximum power the machine can generate under the highest wind speed the turbine can continuously operate. Many manufacturers have, therefore, adopted the combined rating designations — the wind turbine diameter following the generator peak electrical rating. For example, the 300/30 wind system means 300 kW electrical generator and 30-meter diameter turbine.

The specific rated capacity (SRC) is often used as a comparative index of the wind turbine designs. It is defined as follows:

$$SRC = \frac{Generator\ electrical\ capacity}{Rotor\ swept\ area} \tag{5-1}$$

For the 300/30 wind turbine, the specific rated capacity is 300/π 15^2 = 0.42 kW/m^2. The specific rated capacity increases with the diameter, giving a favorable economy of scale to large machine. It ranges from approximately 0.2 kW/m^2 for 10-meter diameter rotor to 0.5 kW/m^2 for 40-meter diameter rotor. Some aggressively rated turbines have SRC of 0.7 kW/m^2, and some reaching as high as 1.0 kW/m^2. The operating stresses in rotor blades of the high SRC are high, generally resulting in shorter fatigue life. All stress concentration regions are carefully identified and eliminated in high SCR designs. Modern design tools, such as the finite element stress analysis and the modal vibration analysis can be of great help in the rotor design.

The turbine rating is important as it indicates to the system designer how to size the induction generator, the plant's transformer, connecting cables to the substation, and the transmission link interfacing the grid. The power system must be sized on the peak capacity of the generator, and the generator is rated in a different manner than the wind turbine. The turbine power depends on the cube of the wind speed. The system design engineer is, therefore, required to match the turbine and the generator performance characteristics. This means selecting the rated speed of the turbine to match with the generator. Since the gearbox and the generator are manufactured only in discrete sizes, selecting the turbine rated speed can be complex. The selection process goes through several iterations, trading the cost with benefit of the available speeds. Selecting a low rated speed would result in wasting much energy at high winds. On the other hand, if the rated speed is high, the rotor efficiency will suffer most of the times.

Figure 5-8 is an example of the summary data sheet of the 550/41 kW/m wind turbine manufactured by Nordtank Energy Group of Denmark. Such data is used in the preliminary design of the overall system. The specific rated capacity of this machine is 0.415. It has the cut-in wind speed of 5 m/s, the cutout speed of 25 m/s and it reaches the peak power at 15 m/s.

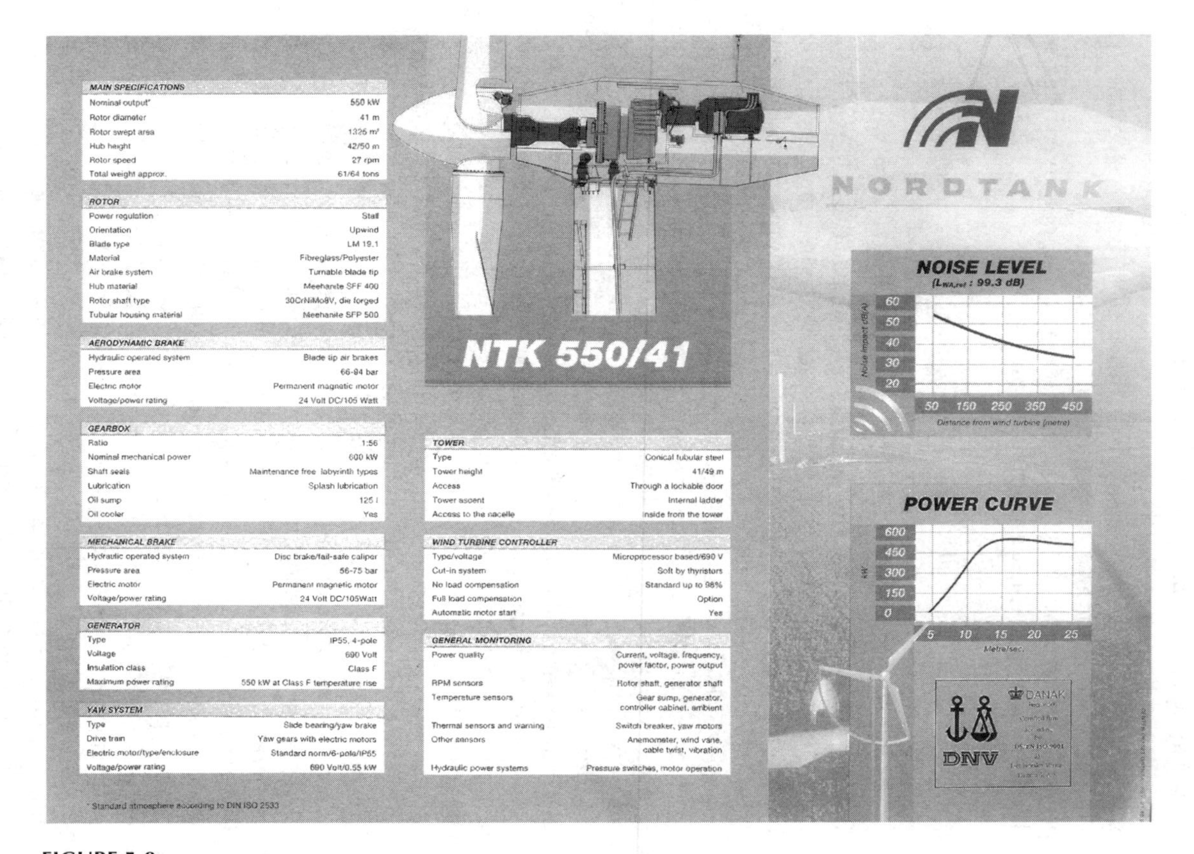

MAIN SPECIFICATIONS	
Nominal output*	550 kW
Rotor diameter	41 m
Rotor swept area	1326 m²
Hub height	42/50 m
Rotor speed	27 rpm
Total weight approx.	61/64 tons

ROTOR	
Power regulation	Stall
Orientation	Upwind
Blade type	LM 19.1
Material	Fibreglass/Polyester
Air brake system	Turnable blade tip
Hub material	Meehanite SFF 400
Rotor shaft type	30CrNiMo8V, die forged
Tubular housing material	Meehanite SFP 500

AERODYNAMIC BRAKE	
Hydraulic operated system	Blade tip air brakes
Pressure area	66-84 bar
Electric motor	Permanent magnetic motor
Voltage/power rating	24 Volt DC/105 Watt

GEARBOX	
Ratio	1:56
Nominal mechanical power	600 kW
Shaft seals	Maintenance free labyrinth types
Lubrication	Splash lubrication
Oil sump	125 l
Oil cooler	Yes

MECHANICAL BRAKE	
Hydraulic operated system	Disc brake/fail-safe caliper
Pressure area	56-75 bar
Electric motor	Permanent magnetic motor
Voltage/power rating	24 Volt DC/105Watt

GENERATOR	
Type	IP55, 4-pole
Voltage	690 Volt
Insulation class	Class F
Maximum power rating	550 kW at Class F temperature rise

YAW SYSTEM	
Type	Slide bearing/yaw brake
Drive train	Yaw gears with electric motors
Electric motor/type/enclosure	Standard norm/6-pole/IP55
Voltage/power rating	690 Volt/0.55 kW

* Standard atmosphere according to DIN ISO 2533

TOWER	
Type	Conical tubular steel
Tower height	41/49 m
Access	Through a lockable door
Tower ascent	Internal ladder
Access to the nacelle	Inside from the tower

WIND TURBINE CONTROLLER	
Type/voltage	Microprocessor based/690 V
Cut-in system	Soft by thyristors
No load compensation	Standard up to 98%
Full load compensation	Option
Automatic motor start	Yes

GENERAL MONITORING	
Power quality	Current, voltage, frequency, power factor, power output
RPM sensors	Rotor shaft, generator shaft
Temperature sensors	Gear sump, generator, controller cabinet, ambient
Thermal sensors and warning	Switch breaker, yaw motors
Other sensors	Anemometer, wind vane, cable twist, vibration
Hydraulic power systems	Pressure switches, motor operation

FIGURE 5-8
Technical data sheet of a 550 kW/41-meter diameter wind turbine, with power level and noise level. (Source: Nordtank Energy Group, Denmark. With permission.)

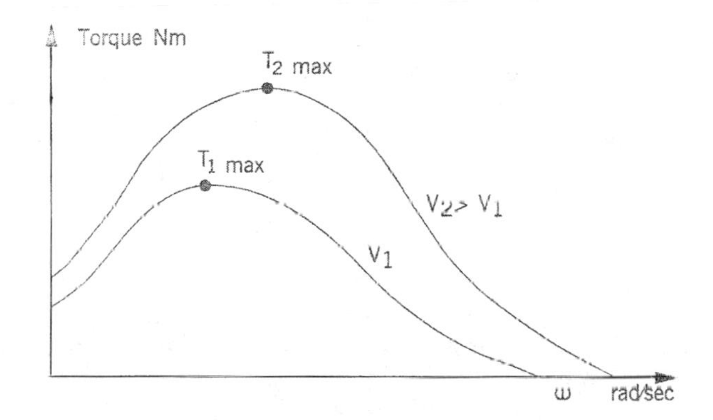

FIGURE 5-9
Wind turbine torque versus rotor speed characteristic at two wind speeds V_1 and V_2.

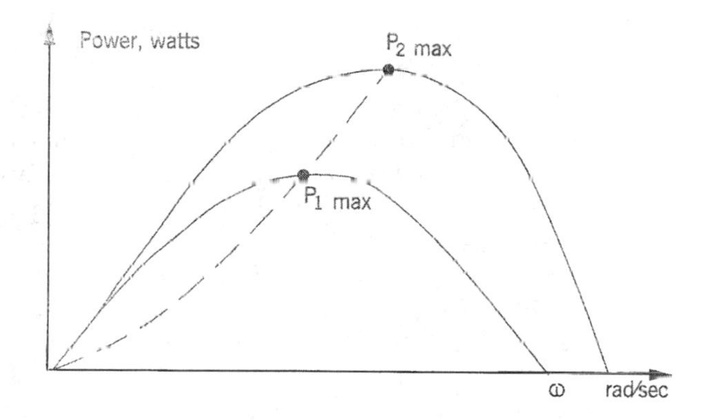

FIGURE 5-10
Wind turbine power versus rotor speed characteristic at two wind speeds V_1 and V_2.

5.3 Electrical Load Matching

The typical turbine torque versus rotor speed is plotted in Figure 5-9. It shows a small torque at zero speed, rising to a maximum value before falling to nearly zero when the rotor just floats with the wind. Two such curves are plotted for different wind speeds V_1 and V_2, with V_2 being higher than V_1. The corresponding powers versus rotor speed at the two wind speeds are plotted in Figure 5-10. As the mechanical power converted into the electrical power is given by the product of the torque T and the angular speed ω, the power is zero at zero speed, and again at high speed with zero torque. The maximum power is generated at rotor speed somewhere in between, as marked by P_{1max} and P_{2max} for speeds V_1 and V_2, respectively. The speed at the maximum power is not the same speed at which the torque is maximum.

The operating strategy of a well-designed wind power system is to match the load on the electrical generator so that the rotor operates continuously at speeds as close as possible to the P_{max} points. Since the P_{max} point changes with the wind speed, the rotor speed must therefore be adjusted in accordance with the wind speed to make the rotor work continuously at P_{max}.

5.4 Variable-Speed Operation

At a given site, the wind speed can vary from zero to high gust. As discussed in Chapter 4, the Rayleigh statistical distribution is found to be the best approximation to represent the wind speed at most sites. It varies over a wide range. Earlier in Chapter 4, we defined the tip-speed ratio as follows:

$$TSR = \frac{\textit{Linear speed of the blade outer most tip}}{\textit{Free upstream wind velocity}} = \frac{\omega \cdot R}{V} \tag{5-2}$$

where R and ω are the rotor radius and the angular speed, respectively.

For a given wind speed, the rotor efficiency C_p varies with TSR as shown in Figure 5-11. The maximum value of C_p occurs approximately at the same wind speed that gives peak power in the power distribution curve of Figure 5-10. To capture the high power at high wind, the rotor must also turn at high speed, keeping the TSR constant at the optimum level.

Three system performance attributes are related to the TSR:

- The centrifugal mechanical stress in the blade material is proportional to the TSR. The machine working at a higher TSR is stressed more. Therefore, if designed for the same power in the same wind speed, the machine operating at a higher TSR would have slimmer rotor blades.
- The ability of a wind turbine to start under load is inversely proportional to the design TSR. As this ratio increases, the starting torque produced by the blade decreases.
- As seen above, the TSR is also related to the operating point for extracting the maximum power. The maximum rotor efficiency C_p is achieved at a particular TSR, which is specific to the aerodynamic design of a given turbine. The TSR needed for the maximum power extraction ranges from nearly one for multiple-blade, slow-speed machines to nearly six for modern high-speed, two-blade machines.

The wind power system design must optimize the annual energy capture at a given site. The only operating mode for extracting the maximum energy is to vary the turbine speed with varying wind speed such that at all times the TSR is continuously equal to that required for the maximum power

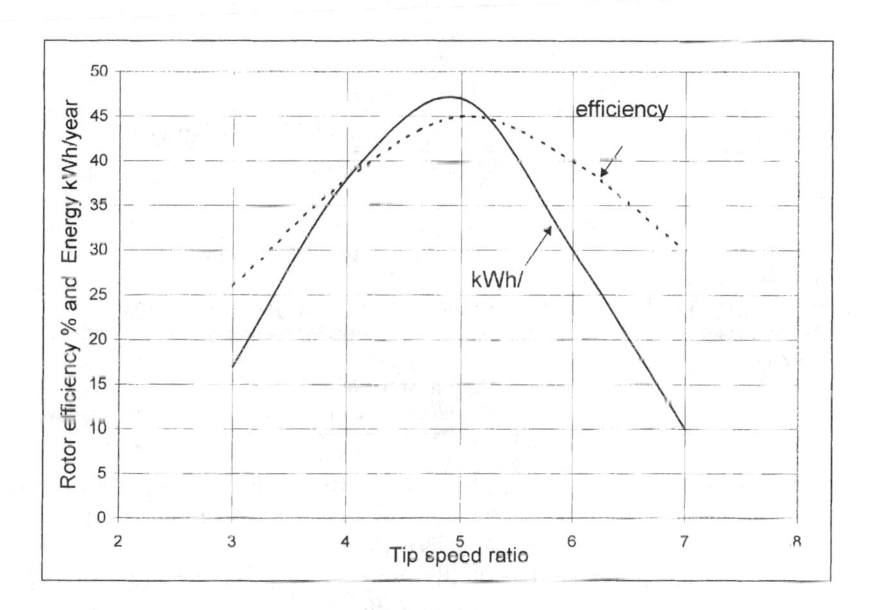

FIGURE 5-11
Rotor efficiency and annual energy production versus rotor tip-speed ratio.

coefficient Cp. The theory and field experience indicates that the variable-speed operation yields 20 to 30 percent more energy than with the fixed-speed operation. Nevertheless, cost of the variable-speed control is added. In the system design, this trade-off between the energy increase and the cost increase has to be optimized. In the past, the added costs of designing the variable-pitch rotor, or the speed control with power electronics, outweighed the benefit of the increased energy capture. However, falling prices of the power electronics for speed control, and availability of high-strength fiber composites for constructing high-speed rotors, have made it economical to capture more energy when the speed is high. In the fixed-speed operation, on the other hand, the rotor is shut off during high-wind speeds, losing significant energy. The pros and cons of the variable and the fixed-speed operations are listed in Table 5-1.

5.5 System Design Features

The following additional design trade-offs are available to the system engineer:

5.5.1 Number of Blades

This is the first determination the design engineer must make. Wind machines have been built with the number of blades ranging from 2 to 40

TABLE 5-1

Advantages of Fixed and Variable Speed Systems

Fixed-Speed System	Variable-Speed System
Simple and inexpensive electrical system	Higher rotor efficiency, hence, higher energy capture per year
Fewer parts, hence higher reliability	Low transient torque
Lower probability of excitation of mechanical resonance of the structure	Fewer gear steps, hence inexpensive gear box
No frequency conversion, hence, no current harmonics present in the electrical system	Mechanical damping system not needed, the electrical system could provide damping if required
Lower capital cost	No synchronization problems
	Stiff electrical controls can reduce system voltage sags

or more. The high number of blades was used in old low, tip-speed ratio rotors for water pumping, the application which needs high starting torque. The modern high, tip-speeds ratio rotors for generating electrical power have two or three blades, many of them with just two. The major factors involved in deciding the number of blades are as follows:

- the effect on power coefficient.
- the design tip-speeds ratio.
- the cost.
- the nacelle weight.
- the structural dynamics.
- the means of limiting yaw rate to reduce gyroscopic fatigue.

Compared to the two-blade design, the three-blade machine has smoother power output and balanced gyroscopic force. There is no need to teeter the rotor, allowing the use of simple rigid hub. Three blades are more common in Europe, where large machines up to 1 MW are being developed using the three-blade configuration. The American practice, however, has been in the two blade designs. Adding the third blade increases the power coefficient only by about 5 percent, thus giving a diminished rate of return for the 50 percent more weight and cost. The two-blade rotor is also simpler to erect, since it can be assembled on the ground and lifted to the shaft without complicated maneuvers during the lift. The number of blades is often viewed as the blade solidity. Higher solidity ratio gives higher starting torque and operates at low speed. For electrical power generation, the turbine must run at high speed since the electrical generator weighs less and operates more efficiently at high speed. That is why all large-scale wind turbines have low solidity ratio, with just two or three blades.

5.5.2 Rotor Upwind or Downwind

Operating the rotor upwind of the tower produces higher power as it eliminates the tower shadow on the blades. This also results in lower noise, lower blade fatigue, and smoother power output. The downwind blades, on the other hand, allow the use of free yaw system. It also allows the blades to deflect away from the tower when loaded. Both types are used at present with no clear trend.

5.5.3 Horizontal Axis Versus Vertical Axis

Most wind turbines built at present have a horizontal axis. The vertical axis Darrieus machine has several advantages. First of all, it is omnidirectional and requires no yaw mechanism to continuously orient itself toward the wind direction. Secondly, its vertical drive shaft simplifies the installation of the gearbox and the electrical generator on the ground, making the structure much simpler. On the negative side, it normally requires guy wires attached to the top for support. This could limit its applications, particularly for the offshore sites. Overall, the vertical axis machine has not been widely used because its output power cannot be easily controlled in high winds simply by changing the blade pitch. With modern low-cost, variable-speed power electronics emerging in the wind power industry, the Darrieus configuration may revive, particularly for large capacity applications.

5.5.4 Spacing of the Towers

When installing a cluster of machines in a wind farm, certain spacing between the wind towers must be maintained to optimize the power cropping. The spacing depends on the terrain, the wind direction, the speed, and the turbine size. The optimum spacing is found in rows 8 to 12-rotor diameters apart in the wind direction, and 1.5 to 3-rotor diameters apart in the crosswind direction (Figure 5-12). A wind farm consisting of 20 towers rated at 500 kW each need 1 to 2 square kilometers of land area. Of this, only a couple of percent would actually occupy the tower and the access roads. The remaining land could continue its original use (Figure 5-13). The average number of machines in wind farms varies greatly, ranging from several to hundreds depending on the required power capacity.

When the land area is limited or is at a premium price, one optimization study that must be conducted in an early stage of the wind farm design is to determine the number of turbines, their size, and the spacing for extracting the maximum energy from the farm. The trades in such a study are as follows:

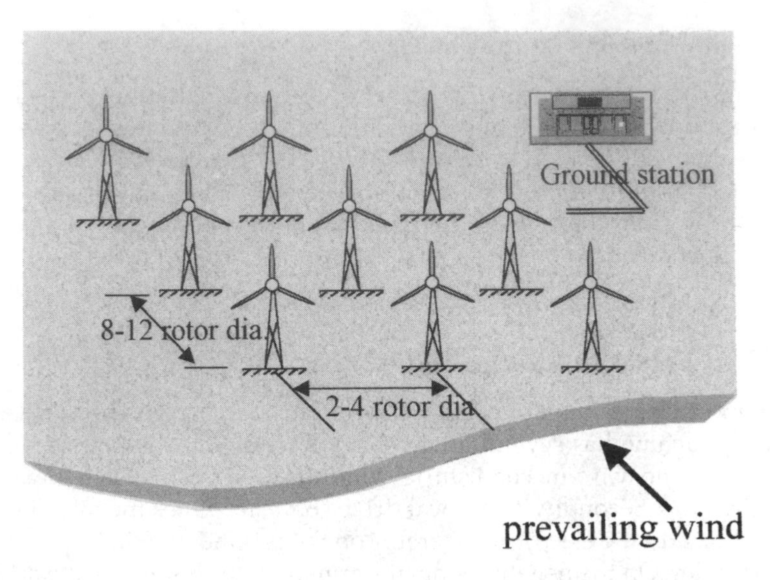

FIGURE 5-12
Optimum tower spacing in wind farms in flat terrain.

- larger turbines cost less per MW capacity and occupy less land area.
- fewer large machines can reduce the MWh energy crop per year, as downtime of one machine would have larger impact on the energy output.
- the wind power fluctuations and electrical transients on fewer large machines would cost more in electrical filtering of the power and voltage fluctuations, or would degrade the quality of power, inviting penalty from the grid.

The optimization method presented by Roy[2] takes into account the above trades. Additionally, it includes the effect of tower height that goes with the turbine diameter, the available standard ratings, cost at the time of procurement, and the wind speed. The wake interaction and tower shadow are ignored for simplicity.

Such optimization leads to a site specific number and size of the wind turbines that will minimize the energy cost.

5.6 Maximum Power Operation

As seen earlier, operating the wind turbine at a constant tip-speed ratio corresponding to the maximum power point at all times can generate 20 to 30 percent more electricity per year. However, this requires a control scheme

FIGURE 5-13
Original land use continues in a wind farm in Germany. (Source: Vestas Wind Systems, Denmark With permission.)

to operate with variable speed. Two possible schemes used with the variable-speed operation are as follows:

5.6.1 Constant Tip-Speed Ratio Scheme

This scheme is based on the fact that the maximum energy is extracted when the optimum tip-speed ratio is maintained constantly at all wind speeds. The optimum TSR is a characteristic of the given wind turbine. This optimum value is stored as the reference TSR in the control computer. The wind speed is continuously measured and compared with the blade tip speed. The error signal is then fed to the control system, which changes the turbine speed to minimize the error (Figure 5-14). At this time the rotor must be operating at the reference TSR generating the maximum power. This scheme has a disadvantage of requiring the local wind speed measurements, which could

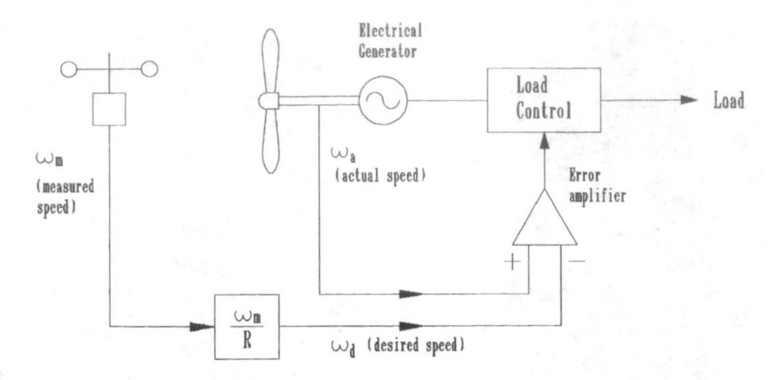

FIGURE 5-14
Maximum power operation using rotor tip-speed control scheme.

have significant error particularly in a large wind farm with shadow effects. Being sensitive to the changes in the blade surface, the optimum TSR gradually changes. The computer reference TSR must be changed accordingly many times over the life. This is expensive. Besides, it is difficult to determine the new optimum tip-speed ratio with changes that are not fully understood, nor easily measured.

5.6.2 Peak Power Tracking Scheme

The power versus speed curve has a single well-defined peak. If we operate at the peak point, a small increase or decrease in the turbine speed would result in no change in the power output, as the peak point locally lies in a flat neighborhood. Therefore, a necessary condition for the speed to be at the maximum power point is as follows:

$$\frac{dP}{d\omega} = 0 \tag{5-3}$$

This principle is used in the control scheme (Figure 5-15). The speed is increased or decreased in small increments, the power is continuously measured, and $\Delta P / \Delta\omega$ is continuously evaluated. If this ratio is positive, meaning we get more power by increasing the speed, the speed is further increased. On the other hand, if the ratio is negative, the power generation will reduce if we change the speed any further. The speed is maintained at the level where $\Delta P / \Delta\omega$ is close to zero. This method is insensitive to the errors in local wind speed measurement, and also to the wind turbine design. It is, therefore, the preferred method. In a multiple machine wind farm, each turbine must be controlled by its own control loop with operational and safety functions incorporated.

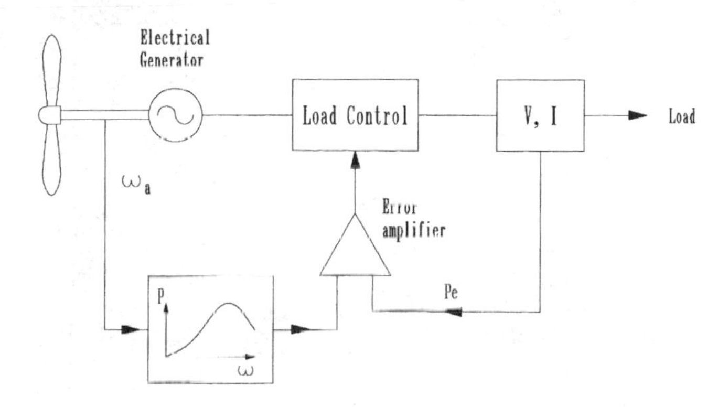

FIGURE 5-15
Maximum power operation using power control scheme.

5.7 System Control Requirements

5.7.1 Speed Control

The rotor speed must be controlled for three reasons:

- to capture more energy, as seen above.
- to protect the rotor, the generator and the power electronic equipment from overloading at high wind.
- when the generator is disconnected accidentally or for a scheduled event, losing the electrical load. Under this condition, the rotor speed may run away, destroying it mechanically, if it is not controlled.

The speed control requirement of the rotor has five separate regions (Figure 5-16):

1. The cut-in speed at which the turbine starts producing power. Below this speed, it is not efficient to turn on the turbine.
2. The constant maximum C_p region where the rotor speed varies with the wind-speed variation to operate at the constant TSR corresponding to the maximum C_p value.
3. During high winds, the rotor speed is limited to an upper constant limit based on the design limit of the system components. In the constant speed region, the C_p is lower than the maximum C_p, and the power increases at a lower rate than that in the first region.
4. At still higher wind speeds, such as during a gust, the machine is operated at constant power to protect the generator and the power

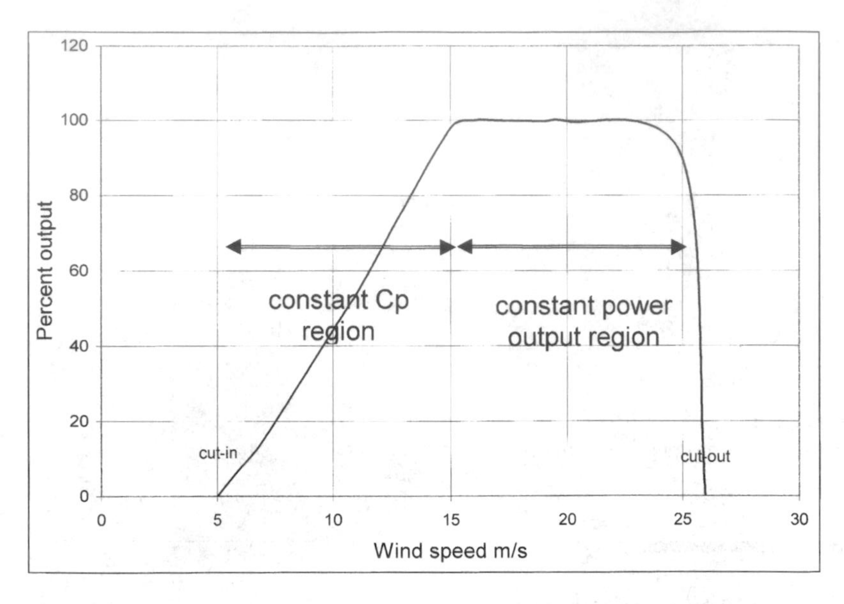

FIGURE 5-16
Five regions of the turbine speed control.

electronics from overloading. This can be achieved by lowering the rotor speed. If the speed is decreased by increasing electrical load, the generator will be overloaded, defeating the purpose. To avoid the generator overloading, some sort of brake, eddy current, or other type, must be installed on the rotor.

5. The cutout speed. Beyond certain wind speed, the rotor is shut off producing power in order to protect the blades, the electrical generator, and other components of the systems.

5.7.2 Rate Control

The large rotor inertia of the blades must be taken into account in controlling the speed. The acceleration and deceleration must be controlled to limit the dynamic mechanical stress on the rotor blades and the hub, and the electrical load on the generator and the power electronics. The instantaneous difference between the mechanical power produced by the blades and the electrical power delivered by the generator will change the rotor speed as follows:

$$J\frac{d\omega}{dt} = \frac{P_m - P_e}{\omega} \tag{5-4}$$

where J = polar moment of inertia of the rotor
ω = angular speed of the rotor
P_m = mechanical power produced by the blades
P_e = electrical power delivered by the generator.

Integrating Equation 5-4, we obtain:

$$\frac{1}{2} J \cdot \left(w_2^2 - \omega_1^2\right) = \int_{t_1}^{t_2} \left(P_m - P_e\right) \cdot dt \tag{5-5}$$

Let us examine this aspect for an example rotor with the moment of inertia J = 7500 kg.m². Changing this rotor speed from 100 to 95 revolutions per minute in five seconds requires ΔP of 800 kW. The resulting torque of 80 Nm would produce the torsional stress on the rotor structure and the hub components. If the same speed change is made in one second, the required power would be 4,000 kW, and the torque 400 Nm. Such high torque can overstress and damage the rotor parts or shorten the life significantly. For this reason, the acceleration and deceleration must be controlled within the design limits with adequate margins.

The strategy for controlling the speed of the wind turbine varies with the type of the electrical machine used, i.e., the induction machine, the synchronous machine or the DC machine.

5.8 Environmental Aspects

5.8.1 Audible Noise

The wind turbine is generally quiet. It poses no objectionable noise disturbance in the surrounding area. The wind turbine manufacturers generally supply the machine noise level data in dB versus the distance from the tower. A typical 600 kW machine noise level is shown in Figure 5-8. This machine produces 55 dBA noise at a 50-meter distance from the turbine and 40 dBA at a 250-meter distance. Table 5-2 compares the turbine noise level with other generally known noise levels. The table indicates that the turbine at a 50-meter distance produces noise no higher than the average factory. This

TABLE 5-2

Noise Level of Some Commonly Known Sources Compared with Wind Turbine

Source	Noise level
Elevated train	100 dB
Noisy factory	90 dB
Average street	70 dB
Average factory	60 dB
Average office	50 dB
Quiet conversation	30 dB

noise, however, is a steady noise. The turbine makes loud noise while yawing under the changing wind direction. The local noise ordinance must be complied with. There have been cases of noise complaints reported by the nearby communities.

5.8.2 Electromagnetic Interference (EMI)

Any stationary or moving structure in the proximity of a radio or TV tower interferes with the signals. The wind turbine towers, being large structures, can cause objectionable electromagnetic interference on the performance of the nearby transmitters or receivers. Additionally, rotor blades of an operating wind turbine may reflect impinging signals so that the electromagnetic signals in the neighborhood may experience interference at the blade passage frequency. The exact nature and magnitude of such EMI depend on a number of parameters. The primary parameters are the location of the wind turbine tower relative to the radio or TV tower, physical and electrical properties of the rotor blades, the signal frequency modulation scheme, and the high-frequency electromagnetic wave propagation characteristics in the local atmosphere.[3]

In other aspects, the visual impact of the wind farm along with the effect on the wind life can be of concern to some. The breeding and feeding patterns of the birds may be disturbed. They may even be injured and killed if they collide with the blades. Under these concerns, obtaining permission from the local planning authorities can take considerable time and effort.

References

1. Roy, S. 1997. "Optimal planning of wind energy conversion systems over an energy scenario," *IEEE Transaction on Energy Conversion,* September 1997.
2. Gijs, van Kuik. 1998. "Wind turbine technology — 25 years' progress" *Wind Directions,* April 1998.
3. Spera, D. A. 1994. "Wind Turbine Technology," *American Society of Mechanical Engineers,* New York, 1994.

6

Electrical Generator

6.1 Electromechanical Energy Conversion

The conversion of the mechanical power of the wind turbine into the electrical power can be accomplished by any one of the following types of the electrical machines:

- the direct current (DC) machine.
- the synchronous machine.
- the induction machine.

These machines work on the principles of the electromagnetic actions and reactions. The resulting electromechanical energy conversion is reversible. The same machine can be used as the motor for converting the electrical power into mechanical power, or as the generator converting the mechanical power into the electrical power.

Figure 6-1 depicts common features of the electrical machines. Typically, there is an outer stationary member (stator) and an inner rotating member (rotor). The rotor is mounted on bearings fixed to the stator. Both the stator and the rotor carry cylindrical iron cores, which are separated by an air gap. The cores are made of magnetic iron of high permeability, and have conductors embedded in slots distributed on the core surface. Alternatively, the conductors are wrapped in the coil form around salient magnetic poles. Figure 6-2 is the cross-sectional view of the rotating electrical machine with the stator with salient poles and the rotor with distributed conductors. The magnetic flux, created by the excitation current in one of the two members, passes from one core to the other in the combined magnetic circuit always forming a closed loop. The electromechanical energy conversion is accomplished by interaction of the magnetic flux produced by one member with the electric current in the other member. The latter may be externally supplied or electromagnetically induced. The induced current is proportional to the rate of change in the flux linkage due to rotation.

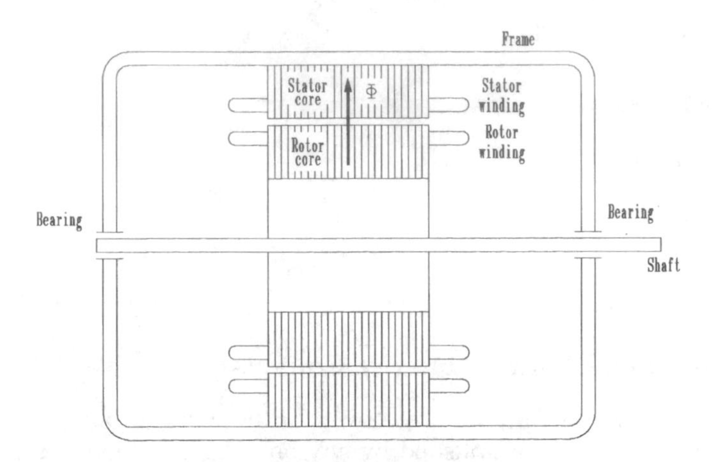

FIGURE 6-1
Common constructional features of the rotating electrical machines.

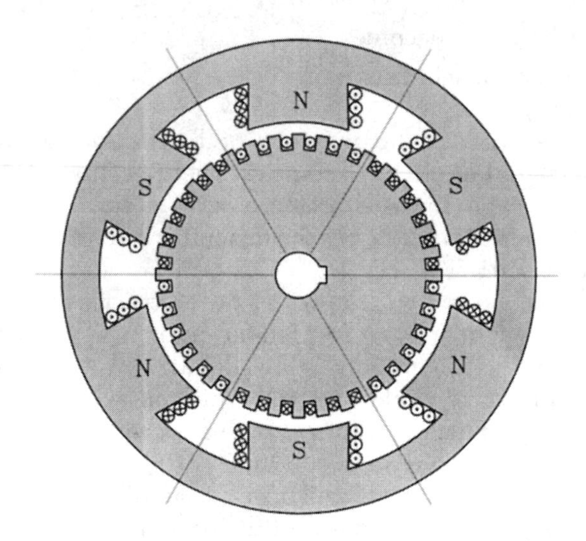

FIGURE 6-2
Cross section of the electrical machine stator and rotor.

The various types of machines differ fundamentally in the distribution of the conductors forming the windings, and in whether the elements have continuous slotted cores or salient poles. The electrical operation of any given machine depends on the nature of the voltages applied to its windings. The narrow annular air gap between the stator and the rotor is the critical region of the machine operation, and the theory of performance is mainly concerned with the conditions in or near the air gap.

6.1.1 DC Machine

All machines are internally alternating current (AC) machines because of the conductor rotation in the magnetic flux of alternate north and south polarity. The DC machine must convert the AC into DC, and does so by using the mechanical commutator. The commutator performs this function by sliding carbon brushes on a series of copper segments. The positive output terminal is, thus, continuously switched to the conductor generating the positive polarity voltage, as is the negative polarity terminal. The sliding contacts inherently result in low reliability and high maintenance cost. Despite this disadvantage, the DC machine had been used extensively until early 1980s because of its extremely easy speed control. It has been used in a limited number of wind power installations of small capacity, particularly where electricity can be locally used in the DC form. However, the conventional DC machine with mechanical commutator has fallen out of favor.

The conventional DC machine is either self-excited by shunt or series coils carrying DC current to produce a magnetic field. The DC machine of the present day is often designed with permanent magnets to eliminate the field current requirement, hence, the commutator. It is designed in the "inside-out" configuration. The rotor carries the permanent magnet poles and the stator carries the wound armature which produces AC current. The AC is then rectified using the solid state rectifiers. Such machines do not need the commutator and the brushes, hence, the reliability is greatly improved. The permanent magnet DC machine is used with small wind turbines, however, due to limitation of the permanent magnet capacity and strength. The brushless DC machine is expected to be limited to ratings below one hundred kW.

6.1.2 Synchronous Machine

Most of the electrical power consumed in the world is generated by the synchronous generator. For this reason, the synchronous machine is an established machine. The machine works at a constant speed related to the fixed frequency. Therefore, it is not well suited for variable-speed operation in the wind plants. Moreover, the synchronous machine requires DC current to excite the rotor field, which needs sliding carbon brushes on slip rings on the rotor shaft. This poses a limitation on its use. The need of the DC field current and the brushes can be eliminated by using the reluctance rotor, where the synchronous operation is achieved by the reluctance torque. The reliability is greatly improved while reducing the cost. The machine rating, however, is limited to tens of kW. The reluctance synchronous generator is being investigated at present for small wind generators.[1]

The synchronous machine is ideally suited in constant-speed systems, such as in the solar thermal power plants. The machine is, therefore, covered in some detail in Chapter 9.

Unlike the induction machine covered later in this chapter, the synchronous machine, when used in the grid-connected system, has some advantage. It does not require the reactive power from the grid. This results in a better quality of power at the grid interface. This advantage is more pronounced when the wind farm is connected to a small capacity grid using long low voltage lines. For this reason, early California plants used synchronous generators. Today's wind plants generally connect to larger grids using shorter lines, and almost universally use the induction generator.

6.1.3 Induction Machine

Most of the electrical power in the industry is consumed by the induction machine driving the mechanical load. For this reason, the induction machine represents a well established technology. The primary advantage of the induction machine is the rugged brushless construction and no need for separate DC field power. The disadvantages of both the DC machine and the synchronous machine are eliminated in the induction machine, resulting in low capital cost, low maintenance, and better transient performance. For these reasons, the induction generator is extensively used in small and large wind farms and small hydroelectric power plants. The machine is available in numerous power ratings up to several megawatts capacity, and even larger.

The induction machine needs AC excitation current. The machine is either self-excited or externally excited. Since the excitation current is mainly reactive, a stand-alone system is self-excited by shunt capacitors. The induction generator connected to the grid draws the excitation power from the network. The synchronous generators connected to the network must be capable of supplying this reactive power.

For economy and reliability, many wind power systems use induction machines as the electrical generator. The remaining part of this chapter is, therefore, devoted to the construction and the theory of operation of the induction generator.

6.2 Induction Generator

6.2.1 Construction

In the electromagnetic structure of the induction generator, the stator is made of numerous coils with three groups (phases), and is supplied with three-phase current. The three coils are physically spread around the stator periphery and carry currents which are out of time-phase. This combination produces a rotating magnetic field, which is a key feature of the working of

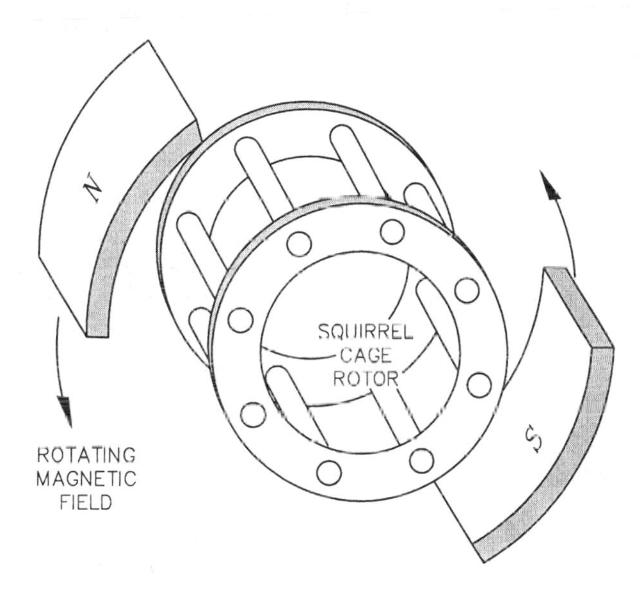

FIGURE 6-3
Squirrel cage rotor of the induction machine under rotating magnetic field

the induction machine. The angular speed of the rotating magnetic field is called the synchronous speed. It is denoted by N_s and is given by the following:

$$N_S = 60 \cdot \frac{f}{p} \text{ revolutions per minute (rpm)} \tag{6-1}$$

where f = frequency of the stator excitation
p = number of magnetic pole pairs.

The stator coils are embedded in slots of high-permeability magnetic core to produce the required magnetic field intensity with low exciting current.

The rotor, however, has a completely different structure. It is made of solid conducting bars embedded in the slots of a magnetic core. The bars are connected together at both ends by the conducting end rings (Figure 6-3). Because of its resemblance, the rotor is called the squirrel cage rotor, or the cage rotor in short.

6.2.2 Working Principle

The stator magnetic field is rotating at the synchronous speed determined by Equation 6-1. This field is conceptually represented by the rotating magnets in Figure 6-3. The relative speed between the rotating field and the rotor induces the voltage in each rotor turn linking the stator flux ϕ. The magnitude of the induced voltage is given by Faraday's law of electromagnetic induction, namely:

$$e = -\frac{d\phi}{dt} \tag{6-2}$$

where ϕ = the magnetic flux linking the rotor turn.

This voltage in turn sets up the circulating current in the rotor. The electromagnetic interaction of the rotor current and the stator flux produces the torque. The amplitude of this torque is given by the following:

$$T = K \cdot \Phi \cdot I_2 \cdot \cos\phi_2 \tag{6-3}$$

where K = constant of proportionality
Φ = amplitude of the stator flux wave
I_2 = amplitude of induced current in the rotor bars
ϕ_2 = phase angle by which the rotor current lags the rotor voltage.

The rotor will accelerate under this torque. If the rotor was on frictionless bearings with no mechanical load attached, it is completely free to rotate with zero resistance. Under this condition, the rotor will attain the same speed as the stator field, namely, the synchronous speed. At this speed, the current induced in the rotor speed is zero, no torque is produced and none is required. The rotor finds equilibrium at this speed and will continue to run at the synchronous speed.

If the rotor is now attached to a mechanical load such as a fan, it will slow down. The stator flux, which always rotates at the constant, synchronous speed, will have relative speed with respect to the rotor. As a result, the electromagnetically induced voltage, current, and torque are produced in the rotor. The torque produced must equal that needed to drive the load at that speed. The machine works as the motor in this condition.

If we attach the rotor to a wind turbine and drive it faster than the synchronous speed, the induced current and the torque in the rotor reverse the direction. The machine now works as the generator, converting the mechanical power of the turbine into electrical power delivered to the load connected to the stator terminals. If the machine was connected to a grid, it would feed power into the grid.

Thus, the induction machine can work as the electrical generator only at speeds higher than the synchronous speed. The generator operation, for that reason, is often called the super-synchronous speed operation of the induction machine. .

As described above, the induction machine needs no electrical connection between the stator and the rotor. Its operation is entirely based on the electromagnetic induction, hence, the name. The absence of rubbing electrical contacts and simplicity of its construction make the induction generator very robust, reliable, and a low-cost machine. For this reason, it is widely used in numerous applications in the industry.

The working principle of the induction machine can be seen as the transformer. The high voltage coil on the stator is excited and the low voltage coil on the rotor is shorted on itself. The power from one to the other can flow in either direction. The theory of operation of the transformer, therefore, holds true when modified to account for the relative motion between the stator and the rotor. This motion is expressed in terms of the slip of the rotor relative to the synchronously rotating magnetic field.

6.2.3 Rotor Speed and Slip

The slip of the rotor is defined as the ratio of the speed of the rotating magnetic field sweeping past the rotor and the synchronous speed of the stator magnetic field. That is,

$$s = \frac{N_s - N_r}{N_s} \tag{6-4}$$

where s = slip of the rotor
N_0 = synchronous speed = $60 \cdot f/p$
N_r = rotor speed.

The slip is generally considered positive in the motoring operation. In the generator mode, the slip would therefore be negative. In both the motoring mode and the generating mode, higher rotor slips induce higher current in the rotor and higher electromechanical power conversion. In both modes, the value of the slip is generally a few to several percent. Higher slips result in greater electrical loss, which must be effectively dissipated from the rotor to keep the operating temperature below the allowable limit.

The heat is removed from the machine by the fan blades attached to one end-ring of the rotor. The fan is enclosed in a shroud at the end. The forced air travels axially along the machine exterior, which has fins to increase the dissipation area. Figure 6-4 is an exterior view of a 150 kW induction machine showing the end shroud and the cooling fins running axially. Figure 6-5 is a cutaway view of the machine interior of a 2 MW induction machine.

The induction generator feeding the 60 or 50 Hz grid must run at speed higher than 3,600 rpm in a two-pole design, 1,800 rpm in a six-pole design, and 1,200 rpm in a six-pole design. The wind turbine speed, on the other hand, varies from a few hundred rpm in the kW range machines to a few tens of rpm in the MW rage machines. The wind turbine therefore must interface the generator via a mechanical gear. Since this somewhat degrades the efficiency and reliability, many small stand-alone plants operate with custom designed generators operating at lower speed without the mechanical gear.

FIGURE 6-4
A 150 kW induction machine. (Source: General Electric Company, Fort Wayne, IN.)

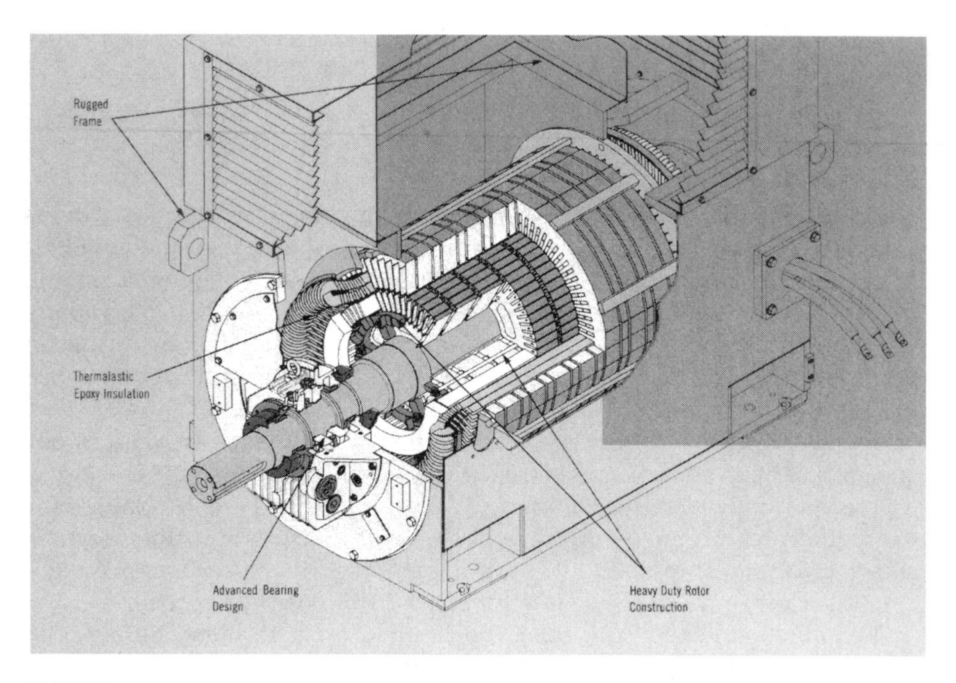

FIGURE 6-5
Two MW induction machine. (Source: Teco Westinghouse Motor Company, Round Rock, Texas, With permission.)

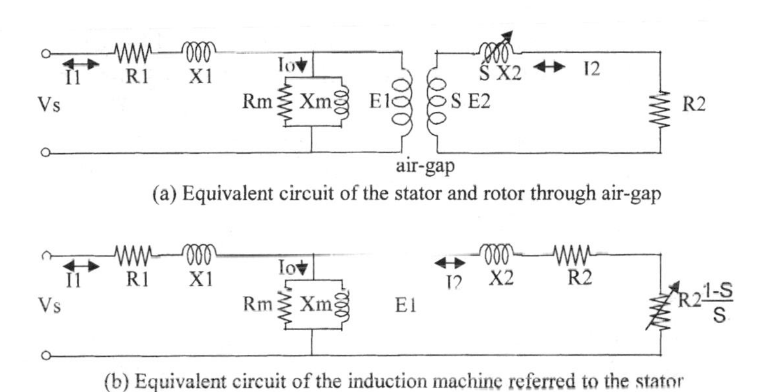

FIGURE 6-6
Equivalent electrical circuit of induction machine for performance calculations.

Under the steady state operation at slip "s", the induction generator has the following operating speeds:

- stator flux wave speed N_s
- rotor mechanical speed $N_r = (1 - s) \cdot N_s$
- stator flux speed with respect to rotor $s \cdot N_s$
- rotor flux speed with respect to stator $N_r + s \cdot N_s = N_s$ (6-5)

6.2.4 Equivalent Circuit for Performance Calculations

The theory of operation of the induction machine is represented by the equivalent circuit shown in Figure 6-6. It is similar to that of the transformer. The left-hand side of the circuit represents the stator and the right hand side, the rotor. The stator and the rotor currents are represented by I_1 and I_2, respectively. The vertical circuit branch at the junction carries the magnetizing (or excitation) current I_o, which sets the magnetic flux required for the electromagnetic operation of the machine. The total stator current is then the sum of the rotor current and the excitation current. The air-gap separation is not shown, nor is the difference in the number of turns in the stator and rotor windings. This essentially means that the rotor is assumed to have the same number of turns as the stator and has an ideal 100 percent magnetic coupling. We calculate the performance parameters taking the stator winding as the reference. The actual rotor voltage and current would be related with the calculated values through the turn ratio between the two windings. Thus, the calculations are customarily performed in terms of the stator, as we shall do in this chapter. This matches the practice, as the performance measurements are always done on the stator side. The rotor is inaccessible for any routine measurements.

Most of the flux links both the stator and the rotor. The flux which does not link both is called the leakage flux. The leakage flux is represented by the leakage reactance. One-half of the total leakage reactance is attributed to each side, namely the stator leakage reactance X_1 and the rotor leakage reactance X_2 in Figure 6-6(b). The stator and the rotor conductor resistance are represented by R_1 and R_2, respectively. The magnetizing parameters X_m and R_m represent the permeability and losses (hysteresis and eddy current) in the magnetic circuit of the machine.

The slip dependent rotor resistance $R_2 \cdot (1\text{-}s)/s$ represents the electromechanical power conversion. The power conversion per phase of the three-phase machine is given by $I_2^2 R_2 \cdot (1 - s)/s$. The three-phase power conversion is then as follows:

$$P_{em} = 3 \cdot I_{2\cdot}^2 R_2 \cdot (1 - s)/s \text{ watts.} \tag{6-6}$$

The machine capacity rating is the power developed under rated conditions, that is as follows:

$$\textit{Machine Rating} = \frac{P_{em\ rated}}{1000}\ kW \ \ or \ \ \frac{P_{em\ rated}}{746}\ horsepower. \tag{6-7}$$

The electromechanical power conversion given by Equation 6-6 is physically appreciated as follows. If the machine is not loaded and has zero friction, it runs at the synchronous speed, the slip is zero and the value of $R_2(1-s)/s$ becomes infinite. The rotor current is then zero, and so is P_{em}, as it should be. When the rotor is standing still, the slip is unity and the value of $R_2 \cdot (1-s)/s$ is zero. The rotor current is not zero, but the P_{em} is zero, as the mechanical power delivered by the standstill rotor is zero.

At any slip other than zero or unity, neither the rotor current nor the speed is zero, resulting in a non-zero value of P_{em}.

The mechanical torque is given by the power divided by the angular speed, that is as follows:

$$T_{em} = P_{em}/\omega \tag{6-8}$$

where T_{em} = electromechanical torque developed in the rotor in newton-meters

ω = angular speed of the rotor = $2\pi . N_s \cdot (1-s)/60$ in mechanical radians/sec.

Combining the above equations, we obtain the torque at any slip s, as follows:

$$T_{em} = (180/2\pi N_s) . I_2^2 R_2/s \text{ Newton-meters} \tag{6-9}$$

The value of 1_2 in equation (6-9) is determined by the equivalent circuit parameters, and is slip dependent. The torque developed by the induction machine rotor is, therefore, highly slip-dependent, as is discussed later in this chapter.

We take a note here that the performance of the induction machine is completely determined by the equivalent circuit parameters. The circuit parameters are supplied by the machine manufacturer, but can be determined by two basic tests on the machine. The full-speed test under no load and the zero-speed test with blocked rotor determine the complete equivalent circuit of the machine.[2,3]

The equivalent circuit parameters are generally expressed in perunit of their respective rated values per phase. The rated impedance per phase is defined as the following:

$$Z_{rated} = \frac{Rated\ voltage\ per\ phase}{Rated\ current\ per\ phase}\ \text{ohms.} \tag{6-10}$$

For example, the perunit (pu) stator resistance is expressed as the following:

$$R_{1\,perunit} = \frac{R_1\ in\ ohms}{Z_{rated}\ in\ ohms} \tag{6-11}$$

and similar expressions for all other circuit parameters. When expressed as such, X_1 and X_2 are equal, each a few to several percent. The R_1 and R_2 are approximately equal, each a few percent of the rated impedance. The magnetizing parameters X_m and R_m are usually large, in several hundred percent of Z_{rated}, hence, drawing negligible current compared to the rated current. For this reason, the magnetizing branch of the circuit is often ignored in making approximations of the machine performance calculations.

All of the above performance equations hold true for both the induction motor and the induction generator by taking the proper sign of the slip. In the generator mode, the value of the slip is negative in the performance equations wherever it appears. We must also remember that the real power output is negative, that is the shaft receives power instead of delivering it. The reactive power drawn from the stator terminals remains lagging with respect to the line voltage, hence, we say that the induction generator delivers leading reactive power. Both of these mean that the magnetizing volt-amperes are supplied by an external source.

6.2.5 Efficiency and Cooling

The values of R_1 and R_2 in the equivalent circuit represent electrical losses in the stator and the rotor, respectively. As will be seen later, for a well-designed machine, the magnetic core loss must equal the conductor loss. Therefore, with R_1 and R_2 expressed in perunit of the base impedance, the induction machine efficiency is approximately equal to the following:

$$\eta = 1 - 2 \cdot (R_1 + R_2) \tag{6-12}$$

For example, a machine with R_1 and R_2 each 2 percent, we write $R_1 = R_2 = 0.02$ perunit. The efficiency is then simply 1–2(0.02+0.02) = 0.92 perunit, or 92 percent. Eight percent of the input power in this machine is lost in the conversion process.

The losses generated in the machine are removed by providing adequate cooling. Small machines are generally air-cooled. Large generators located inside the nacelle can be difficult to cool by air. Water cooling, being much more effective than air-cooling, can be advantageous in three ways:

- for the same machine rating, the water cooling reduces the generator weight on the nacelle, thus benefiting the structural design of the tower.
- it absorbs and thus reduces the noise and vibrations.
- it eliminates the nacelle opening by mounting the weather-air heat exchanger outside, making the nacelle more weather-proof.
- overall, it reduces the maintenance requirement, a significant benefit in large machines usually sitting on tall towers.

6.2.6 Self-Excitation Capacitance

As the generator, the induction machine has one drawback of requiring reactive power for excitation. The exciting power can be provided by an external capacitor connected to the generator terminals (Figure 6-7). No separate AC supply is needed in this case. In the grid-connected generator, the reactive power is supplied from the synchronous generators working at the other end of the grid. Where the grid capacity of supplying the reactive power is limited, local capacitors can be used to partly supply the needed reactive power.

The induction generator will self-excite using the external capacitor only if the rotor has an adequate remnant magnetic field. In the self-excited mode, the generator output frequency and voltage are affected by the speed, the

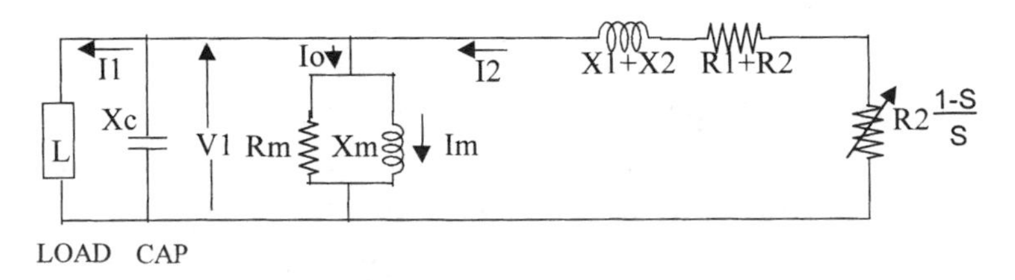

FIGURE 6-7
Self-excited induction generator with external capacitor.

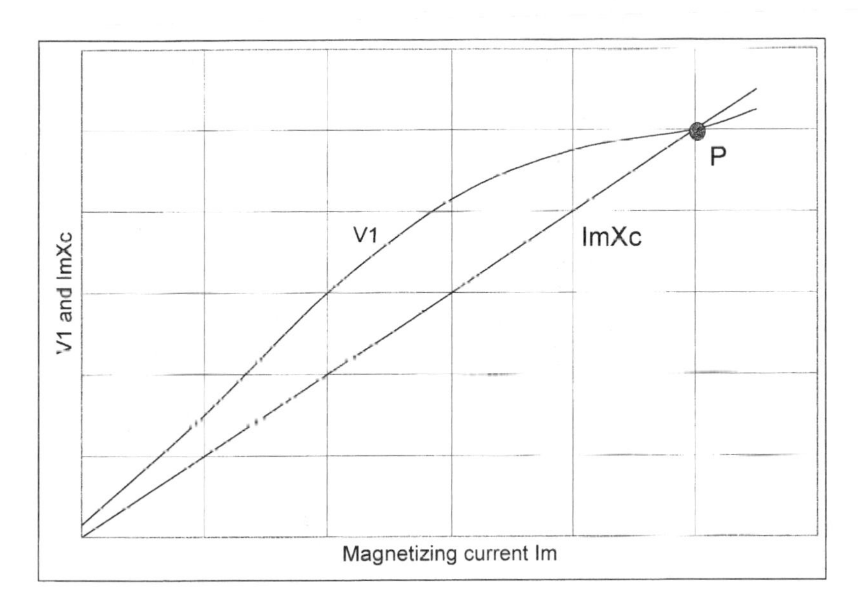

FIGURE 6-8
Determination of stable operation of self-excited induction generator.

load, and the capacitance value in farads. The operating voltage and frequency are determined below in terms of the approximate equivalent circuit of Figure 6-7.

On no load, the capacitor current $I_c = V_1/X_c$ must be equal to the magnetizing current $I_m = V_1/X_m$. The voltage V_1 is a function of I_m, linearly rising until the saturation point of the magnetic core is reached (Figure 6-8). The stable operation requires the line $I_m X_c$ to intersect the V_1 versus I_m curve. The operating point is fixed where V_1/X_c equal V_1/X_m, that is when $1/X_c = 1/X_m$, where $X_c = 1/\omega C$. This settles the operating frequency in hertz. With the capacitor value C, the output frequency of the self-excited generator is therefore:

$$f = \frac{1}{2\pi C X_m} \tag{6-13}$$

Under load conditions, the generated power $V_1 I_2 \cos\phi_2$ provides for the power in the load resistance R and the loss in R_m. The reactive currents must sum to zero, i.e.:

$$\frac{V_1}{X} + \frac{V_1}{X_m} + I_2 \cdot \sin\phi_2 = \frac{V_1}{X_c} \tag{6-14}$$

Equation 6-14 determines the output voltage of the machine under load.

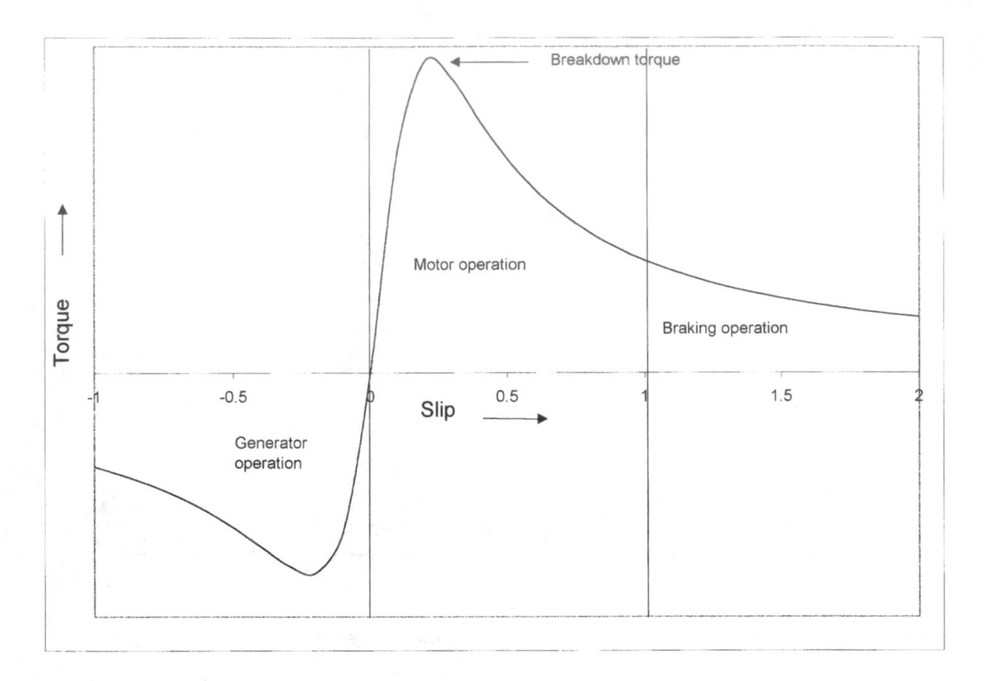

FIGURE 6-9
Torque versus speed characteristic of the induction machine in three operating modes.

The Equations 6-13 and 6-14 determine the induction generator output frequency and voltage with a given value of the capacitance. Inversely, they can be used to determine the required value of the capacitance for the desired frequency and the voltage.

6.2.7 Torque-Speed Characteristic

If we vary the slip over a wide range in the equivalent circuit, we get the torque-speed characteristic as shown in Figure 6-9. In the region of negative slip, the machine works as the generator powering the electrical load connected to its terminals. In the region of positive slip, it works as the motor turning the mechanical load connected to its shaft. In addition to the motoring and the generating regions, the induction machine has yet a third operating mode, and that is the braking mode. If the machine is operated at slips >1 by turning it backward, it absorbs power without putting anything out. That is, it works as a brake. The power in this case is converted into I^2R loss in the rotor conductors, which must be dissipated as heat. The eddy current brake works on this principle. As such, in case of emergencies, the grid-connected induction generator can be used as brake by reversing the three-phase voltage sequence at the stator terminals. This reverses the direction

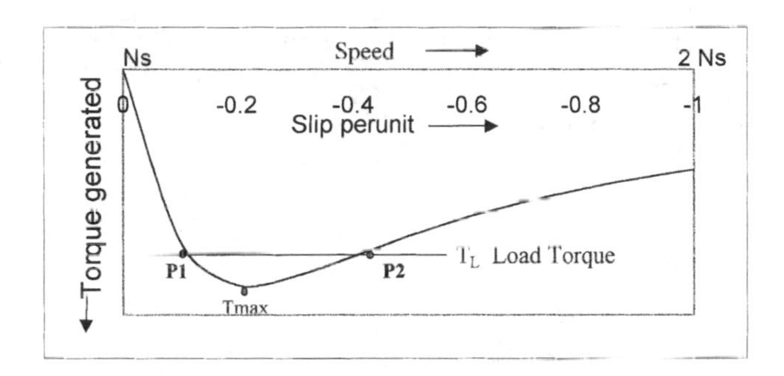

FIGURE 6-10
Torque versus slip characteristic of induction generator under load.

of rotation of the magnetic flux wave with respect to the rotor. The torsional stress on the turbine blades and the hub, however, may limit the braking torque.

The torque-slip characteristic in the generating mode is separately shown in Figure 6-10. If the generator is loaded at constant load torque T_L, it has two possible points of operation, P_1 and P_2. Only one of these two pints, P1, is stable. Any perturbation in speed around point P_1 will produce stabilizing torque to bring it back to P1. The figure also shows the limit to which the generator can be loaded. The maximum torque it can support is called the breakdown torque, which is shown as T_{max}. If the generator is loaded under a constant torque above T_{max}, it will become unstable and stall, draw excessive current, and destroy itself thermally if not properly protected.

6.2.8 Transients

The induction generator may experience the following three types of transient currents:

Starting Transient: In the grid-connected system, the induction generator is started as the motor in starting the turbine from rest to the super-synchronous speed. Then only it is switched to the generating mode, feeding power to the grid. If full voltage is applied during starting, the motor draws high starting current at zero speed when the slip is one and the rotor resistance is the least. The starting inrush current can be five to seven times the rated current, causing overheating problems, particularly in large machines. Moreover, as seen in Figure 6-11, the torque available to accelerate the rotor may be low, taking a long time to start. This also adds into the heating problem. For this reason, the large induction machine is often started with a soft-start circuit, such as the voltage reducing autotransformer or the star-delta starter. The modern method of starting is to apply reduced voltage of variable frequency maintaining a constant volts/hertz ratio. This method starts the machine with the least mechanical and thermal stresses.

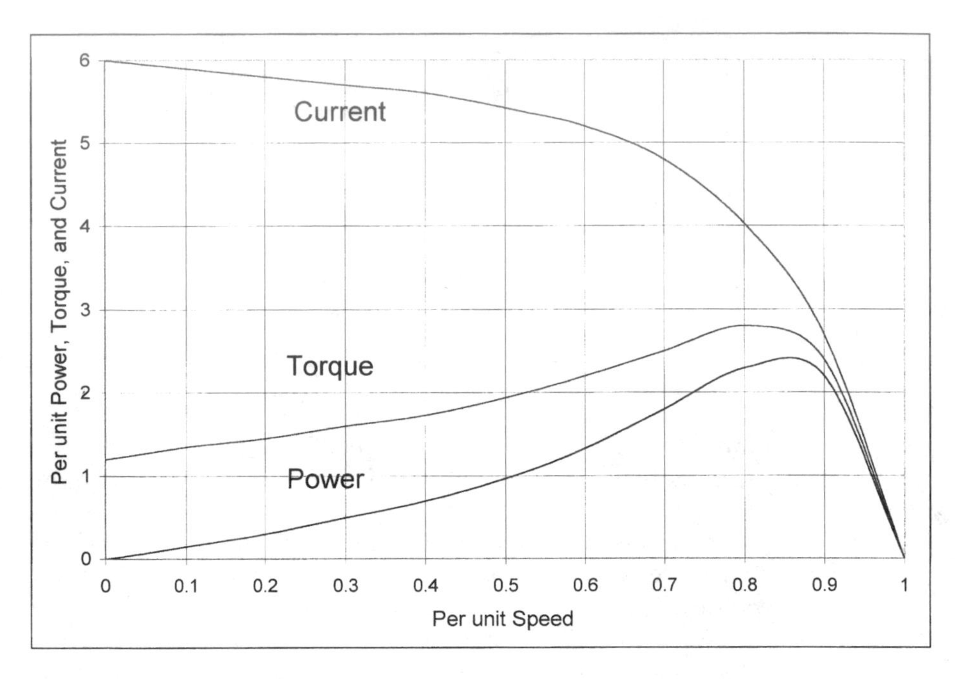

FIGURE 6-11
Induction machine starting and accelerating characteristic.

Reswitching Transient: A severe transient current can flow in the system if the induction generator operating in a steady state suddenly gets disconnected due to a system fault or any other reason, and then reconnected by an automatic reswitching. The magnitude of the current depends on the instant of the voltage wave when the generator gets reconnected to the grid. The physical appreciation of this transient comes from the constant flux linkage theorem. A coil having no resistance keeps its flux linkage constant. Since the winding resistance is small compared to the inductance in most electrical machines, the theorem of the constant flux linkage holds, at least in the beginning of the fault. If the reswitching was done when the stator and rotor voltages were in phase opposition, large transient currents are established to maintain the flux linkage, which then decays slowly to small values after tens of milliseconds. Meanwhile, the transient electromechanical torque may be large enough to give the machine and the tower a severe jolt. The actual amplitude and sign of the first peak of the transient torque are closely dependent on the rotor speed and duration of the interruption. In the worst case, the first peak may reach 15 times the rated full-load torque. Frequent faults of this nature can cause shaft breakage due to fatigue stresses, particularity at the coupling with the wind turbine.

Short Circuit: When a short circuit fault occurs at or near the generator terminals, the machine significantly contributes to the system fault current, particularly if it is running on light load. The short circuit current is always more severe for a single-phase fault than a three-phase fault. The most important quantity is the first peak current as it determines the rating of the protective circuit breaker needed to protect the generator against such faults. The short circuit current has a slowly decaying DC component, and an AC component. The latter is larger than the direct on-line starting inrush current, and may reach 10–15 times the full load rated current.

The transient current and torque, in any case, are calculated using the generalized equivalent circuit of the machine in terms of the d-axis and q-axis transient and subtransient reactance and time constants.[4-6] The q-axis terms do not enter in the induction generator transient analysis, as the d-axis and the q-axis terms are identical due to the perfect circular symmetry in the electromagnetic structure.

References

1. Rahim, Y. H. A. 1997. "Controlled Power Transfer from Wind Driven Reluctance Generators," *IEEE Winter Power Meeting, Paper No. PE-230-EC-1-09,* New York, 1997.
2. Say, M. G. 1983. "Alternating Current Machines," New York, John Wiley & Sons, 1983.
3. Alger, P. L. 1965. "The Nature of Induction Machines." New York, Gordon and Breach. 1965.
4. Adkins, B. 1964. "The General Theory of Electrical Machines," London, Chapman and Hall. 1964.
5. Kron, G. 1967. "Equivalent Circuits of Electrical Machines," New York, Dover Publications, 1967.
6. Yamayee, Z. A. and Bala, J. L. 1994. "Electromechanical Devices and Power Systems," New York, John Wiley & Sons, 1994.

7

Generator Drives

The turbine speed is generally much lower than the desired speed for the electrical generator. For this reason, the turbine speed in most wind systems is stepped up using a drive system. The system can be fixed-speed or variable-speed as described in this chapter.

The wind-power equation as derived in Chapter 3 is as follows:

$$P = \frac{1}{2}\rho AV^3 \cdot C_p \tag{7-1}$$

where C_p = rotor power coefficient.

As seen earlier, the value of C_p varies with the ratio of the rotor tip-speed to the wind speed, termed as the tip-speed-ratio TSR. Figure 7-1 depicts a typical relationship between the power coefficient and the tip-to-speed ratio. As the wind speed changes, the TSR and the power coefficient will vary. The C_p characteristic has single maximum at a specific value of the TSR. Therefore, when operating the rotor at constant speed, the power coefficient will be maximum at only one wind speed.

For achieving the highest annual energy yield, the value of the rotor power coefficient must be maintained at the maximum level all the time, regardless of the wind speed. The theoretical maximum value of C_p is 0.59, but the practical limit is 0.5. Attaining C_p above 0.4 is considered good. Whatever value is attainable with a given wind turbine, it must be maintained constant at that value. Therefore, the rotor speed must change in response to the changing wind speed. To achieve this, the speed control must be incorporated in the system design to run the rotor at high speed in high wind and at low speed in low wind. This is illustrated in Figure 7-2. For given wind speeds V_1, V_2, or V_3, the rotor power curves versus the turbine speed are plotted in solid lines. In order to extract the maximum possible energy over the year, the turbine must be operated at the peak power point at all wind speeds. In the figure, this happens at points P_1, P_2, and P_3 for the wind speed V_1, V_2, and V_3, respectively. The common factor among the peak power production points P_1, P_2, and P_3 is the constant high value of TSR, close to 0.5.

Operating the machine at the constant tip-speed ratio corresponding to the peak power point means high rotor speed in gusty winds. The centrifugal

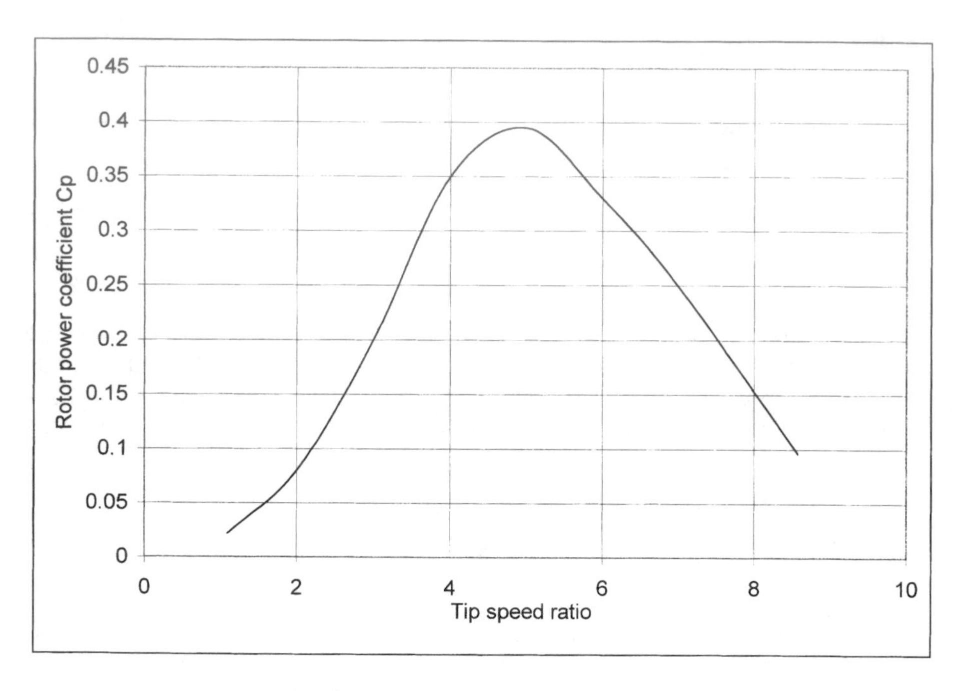

FIGURE 7-1
Rotor power coefficient versus tip-speed ratio has a single maximum.

forces produced in the rotor blades under such speeds can mechanically destroy the rotor. Moreover, the generator producing power above its rated capacity may electrically destroy the generator. For these reasons, the turbine speed and the generator power output must be controlled.

7.1 Speed Control Regions

The speed and the power controls in the wind power systems have three distinct regions:

- the optimum constant Cp region.
- the speed-limited region.
- the power-limited region.

These regions are shown in Figure 7-3. Typically the turbine starts operating (cut in) when the wind speed exceeds 4-5 m/s, and is shut off at speeds exceeding 25 to 30 m/s. In between, it operates in one of the above regions. At a typical site, the wind-turbine may operate about 70 to 80 percent of the time. Other times, it is off due to wind speed too low or too high.

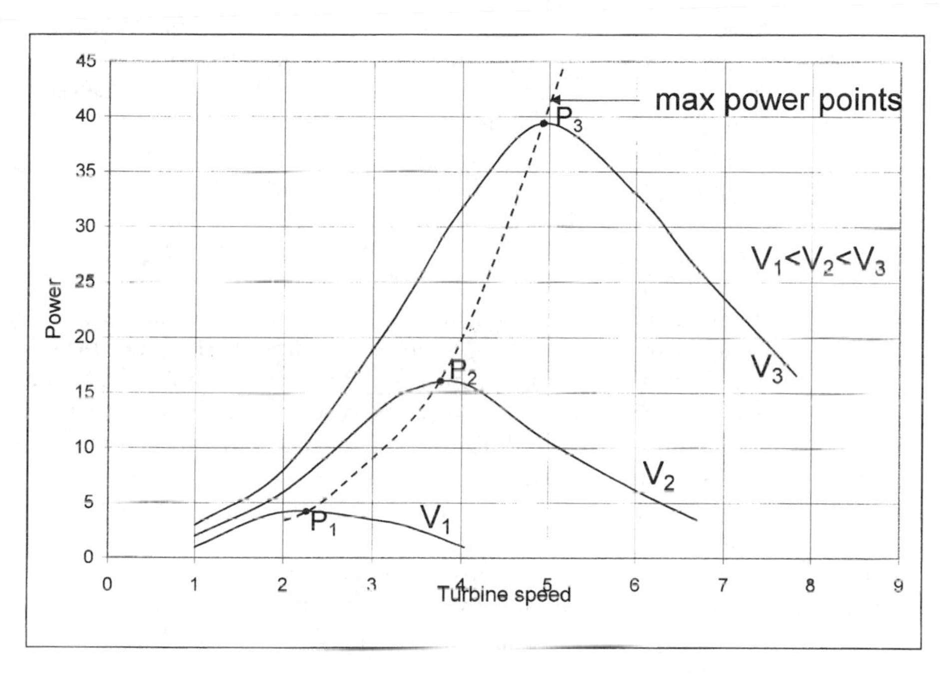

FIGURE 7-2
Turbine power versus rotor-speed characteristics at different wind speeds. The peak power point moves to the right at higher wind speed.

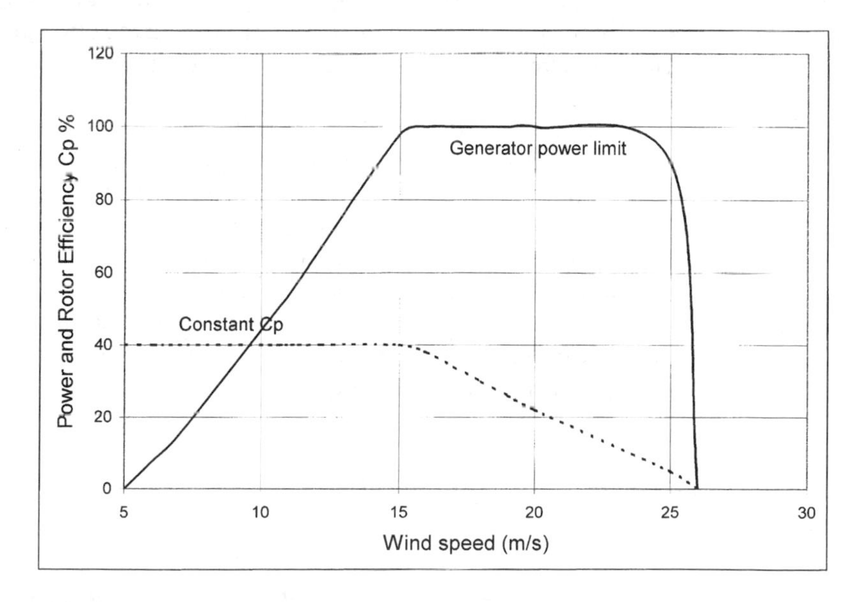

FIGURE 7-3
Three distinct rotor-speed control regions.

The maximum C_p region is the normal mode of operation, where the speed controller operates the system at the optimum constant C_p value stored in the system computer. Two alternative schemes of controlling the speed in this region were described in Section 5.6.

In the constant C_p region, the control system increases the rotor speed in response to the increasing wind speed only up to a certain limit (Figure 7-4). When this limit is reached, the control shifts into the speed-limiting region. The power coefficient C_p is no longer at the optimum value, and the rotor power efficiency suffers.

If the wind speed continues to rise, the systems will approach the power limitation of the electrical generator. When this occurs, the turbine speed is reduced, and the power coefficient C_p moves farther away from the optimum value. The generator output power remains constant at the design limit. When the speed limit and power limit cannot be maintained under extreme gust of wind, the machine is cut out of the power producing operation.

Two traditional methods of controlling the turbine speed and generator power output are as follows:

(1) The pitch control in which the turbine speed is controlled by controlling the blade pitch by mechanical and hydraulic means. The power fluctuates above and below the rated value as the blade pitch mechanism adjusts with changing wind speed. This takes some time because of the large inertia of the rotor. Figure 7-4 depicts variations in the wind speed, the pitch angle of the blades, the generator speed and the power output with respect to time in a fluctuating wind. The curves represent actual measurements on Vestas 1.65 MW wind turbine with OptiSlip® (registered tradename of Vestas Wind Systems, Denmark). The generator power output is held constant even with 10 percent fluctuation in the generator speed, thus, minimizing the undesired fluctuations on grid. The elasticity of the system also reduces the stress on the turbine and the foundation.

(2) The stall control in which the turbine uses the aerodynamic stall to regulate speed in high winds. The power generation peaks somewhat higher than the rated limit, then declines until the cut-out wind speed is reached. Beyond that point, the turbine stalls and the power production drops to zero (Figure 7-5).

In both methods of speed regulation, the power output of most machines in practice is not as smooth. The theoretical considerations give only approximations of the power produced at any given instant. For example, the turbine can produce different power at the same speed depending on whether the speed is increasing or decreasing.

7.2 Generator Drives

Selecting the operating speed of the generator and controlling it with changing wind speed must be determined early in the system design. This is

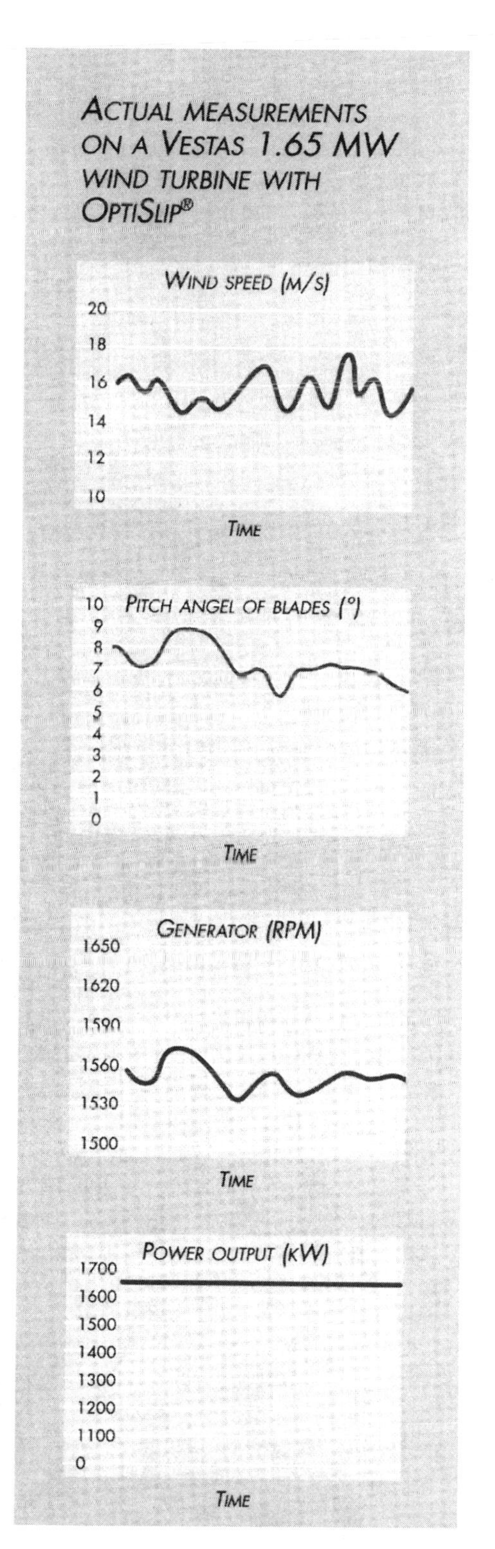

FIGURE 7-4
Wind speed, pitch angle, generator speed, and power output under fluctuating wind speed in 1650 kW turbine. (Source: Vestas Wind Systems, Denmark. With permission.)

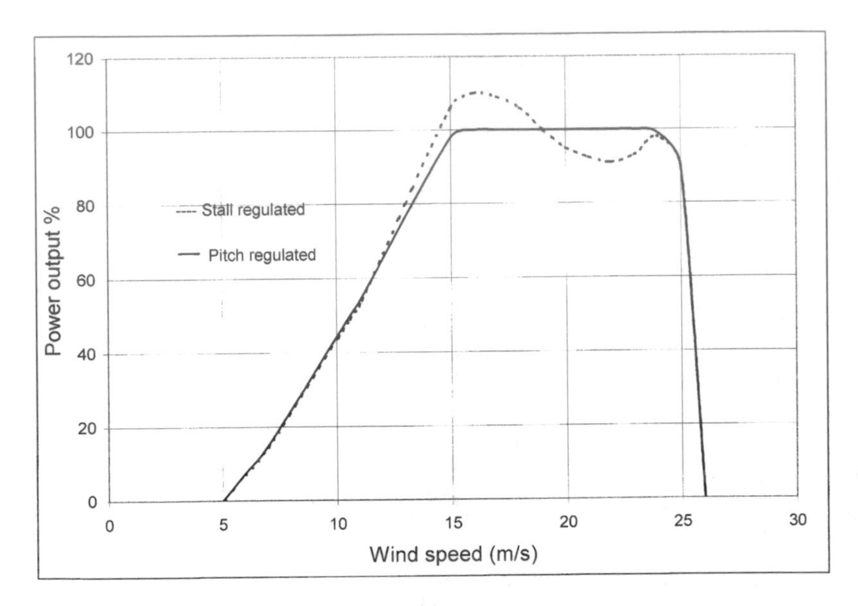

FIGURE 7-5
Generator output power variation with wind speed in the blade pitch-regulated and stall-regulated turbines.

important, as it determines all major components and their ratings. The alternative strategies and the corresponding speed control methods fall in the following categories.

7.2.1 One Fixed-Speed Drive

The fixed-speed operation of the generator naturally fits well with the induction generator, which is inherently a fixed-speed machine. However, the turbine speed is generally low, whereas the electrical generator works more efficiently at high speed. The speed match between the two is accomplished by the mechanical gear. The gearbox reduces the speed and increases the torque, thus improving the rotor power coefficient C_p. Under varying wind speed, the increase and decrease in electromagnetically converted torque and power are accompanied by the corresponding increase or decrease in the rotor slip with respect to the stator. The wind generator generally works at 1 to 2 percent slip. The higher value benefits the drive gear, but increases the electrical loss in the rotor, which leads to cooling difficulty.

The annual energy yield for a fixed-speed wind turbine must be analyzed with the given wind speed distribution at the site of interest. Since the speed is held constant, the turbine running above the rated speed is not a design concern. However, it is possible to generate electrical power above the rated capacity of the generator. When this happens, the generator is shut off by

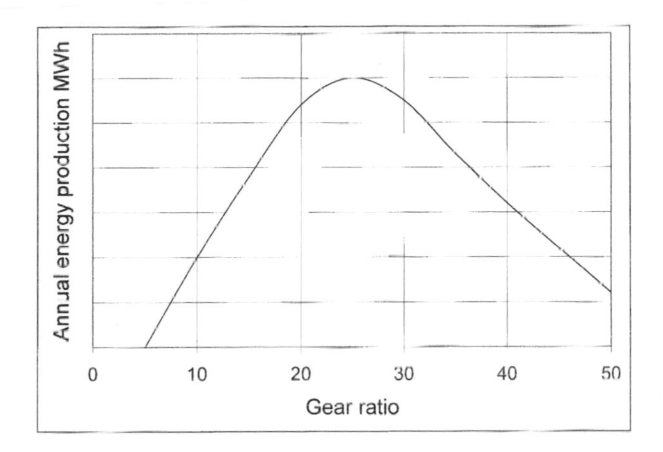

FIGURE 7-6
Annual energy production strongly varies with gear ratio for a given wind speed of 8 m/s.

opening the circuit breaker, thus shedding the load and dropping the system power generation to zero.

The major disadvantage of one fixed-speed operation is that it almost never captures the wind energy at the peak efficiency in terms of the rotor power coefficient C_p. The wind energy is wasted when the wind speed is higher or lower than the certain value selected as the optimum.

With the generator operating at constant speed, the annual energy production depends on the wind speed and the gear ratio. Figure 7-6 depicts the annual energy versus gear ratio relation typical of such systems. It is seen that the annual energy yield is highly dependent on the selected gear ratio. For the given wind-speed distribution in the figure, the energy production is maximum at the gear ratio of 20. When choosing the gear ratio, it is therefore important to consider the average wind speed at the specific site. The optimum gear ratio for the operation of the wind turbine varies from site to site.

Because of the low energy yield over the year, the fixed-speed drives are limited to small machines.

7.2.2 Two Fixed-Speeds Drive

The two-speed machine increases the energy capture, reduces the electrical loss in the rotor and reduces the gear noise. The speed is changed by changing the gear ratio. The two operating speeds are selected to optimize the annual energy production with the wind speed distribution expected at the site. The annual power production varies with the gear ratio and the wind speed as seen in Figure 7-7. It is obvious from the figure that the peak power point wind speeds V_1 and V_2 with two-gear ratios must be on the opposite

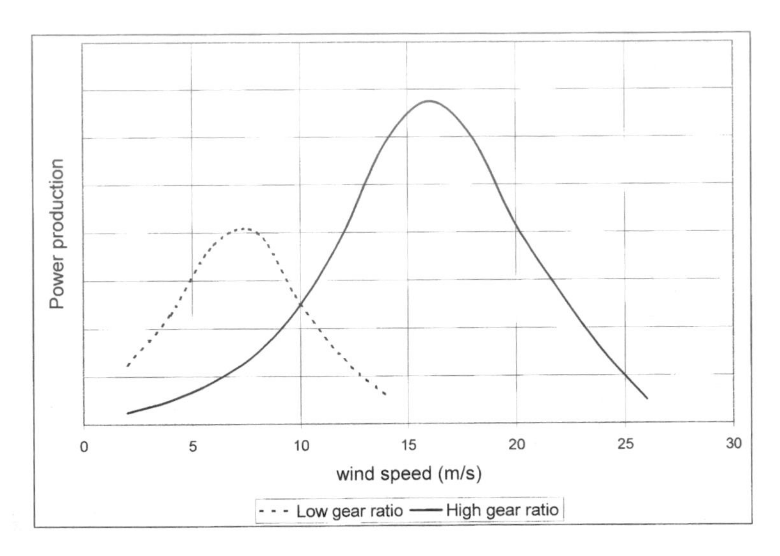

FIGURE 7-7
Power production probability distribution with wind speed with low and high gear ratios.

side of the expected annual average wind speed. For the specific example of Figure 7-7, the system is operated on the low-gear ratio for wind speed below 10 m/s, and on the high-gear ratio for wind speed above 10 m/s. The gear ratio is changed at 10 m/s in this example.

In some early American designs, two speeds were achieved by using two separate generators and switching between the generators by a belt drive.

An economic and efficient method is to design the induction generator to operate at two speeds. The cage motor with two separate stator windings of different pole numbers can run at two or more integrally related speeds. The pole-changing motor, on the other hand, has a single winding, the connection of which is changed to give different numbers of poles. Separate windings matching with the system requirement may be preferred where the speed change must be made without losing control of the machine. Separate windings are, however, difficult to accommodate.

In the pole-changing method with one winding, the stator is wound with coils that can be connected either in P or 2P number of poles. No changes are needed, nor possible, in the squirrel cage rotor. The stator connection which produces a higher pole number for low-speed operation is changed to one-half as many poles for high-speed operation. This maintains the tip-speed ratio near the optimum to produce high rotor power coefficient C_p. The machine, however, operates with only one-speed ratio of 2:1.

Figure 7-8 shows one phase of the pole changing stator winding. For the higher pole number, the coils are in series. For the lower number, they are in series-parallel. The resulting magnetic flux pattern corresponds to eight and four poles, respectively. It is common to use double layer winding, with 120° electrical span for the higher pole number. An important design

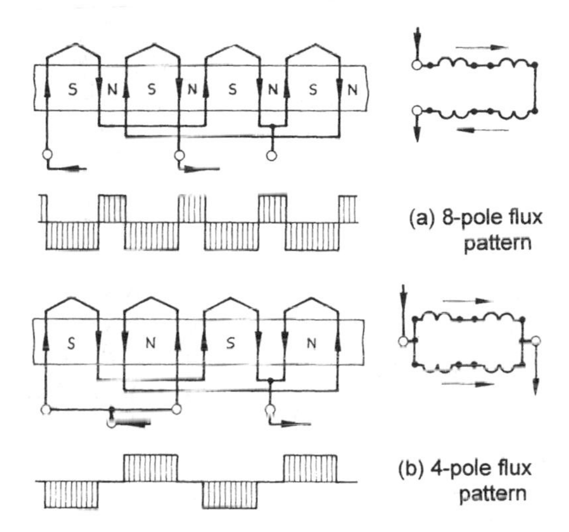

FIGURE 7-8
Pole-changing stator winding for speed ratio 2:1.

consideration in such windings is to limit the space harmonics, which may decrease the efficiency as the generator, and may also produce a tendency to crawl when using the machine as the motor during the start up operation.

The coil pitch of the stator winding is fixed once wound, but its electrical span depends on the number of poles. A coil pitch one-eighth of the circumference provides full-pitch coils for an eight-pole connection, two-thirds for a six-pole, and one-half for a four-pole connection. Too narrow a coil span must be avoided. For a 2:1 speed-ratio generator, a possible coil span is 1.33 pole-pitch for the larger and 0.67 for the smaller pole number. In each case, the coil span factor would be 0.87. Using the spans near 1 and 0.5, with the span factor of 1.0 and 0.71, one can avoid excessive leakage reactance in the lower-speed operation.

7.2.3 Variable-Speed Using Gear Drive

The variable-speed operation using variable-gear ratio has been considered in the past, but have been found to add more problems than the benefits. Therefore, such drives are not generally used at present.

7.2.4 Variable-Speed Using Power Electronics

The modern variable-speed drive uses power electronics to convert variable voltage, variable frequency output of the generator into the fixed voltage, fixed frequency output. The technology is similar to that used in the aircraft power system. The trend of using such a system is being propelled by the declining cost of the power semiconductors. Conventional silicon controlled rectifiers and inverters can be used, but the modern design in the wind industry appears to prefer pulse-width-modulated thyristors. The speed ratio is not limited in theory, but practical considerations limit the ratio to 3:1, which is wider than that obtainable using the pole-changing method described above and the Scherbius machine described later. The energy yield of the variable-speed system is higher. However, the added cost and the electrical loss in the power electronics partially offset the benefit. The cost and benefit trade is generally positive for large machines.

The power-electronics based variable-speed system introduces some system level issues not found in other systems. It produces high-frequency harmonics (electrical noise) in the network, which degrades the quality of power. Alternatively, for the same quality of power, it requires a higher degree of electrical filtering to meet the grid-quality requirement.

In addition to high annual energy production, the variable-speed power electronic system offers remotely adjustable and controllable quality of power. This has two major benefits not available in other systems:

- opportunity for remote control. This makes it attractive for offshore applications.
- fine-tuning for superior grid connection, making it better suited for meeting the demand of weak grids in developing countries like China and India.

7.2.5 Scherbius Variable-Speed Drive

Compared to the variable-speed system using power electronics, the Scherbius machine offers lower cost and eliminates the power quality disadvantage. It has been used in hoist applications in factories and mines. Extending the analysis of the equivalent circuit described in Chapter 6, the speed of the induction machine can be changed by changing the rotor resistance or by injecting an external voltage of the frequency corresponding to the desired rotor slip. The squirrel cage construction does not allow such injection. Therefore, the wound rotor construction with slip rings is used (Figure 7-9). The rotor circuit is connected to an external variable frequency source via slip rings, and the stator is connected to the grid system. For this reason, the Scherbius machine is also called the doubly-fed induction machine. It is fed from both the stator and the rotor. The speed is controlled by adjusting the frequency of the external source of the rotor current. The range of variable-speed control using the Scherbius machines is generally limited to 2:1.

The concept was used in early turbines. The decreased reliability due to the rubbing electrical contacts at the slip rings had been a concern. However, some manufacturers appear to have resolved this concern, and are implementing the system in turbines with several hundred kW ratings. The need of the variable frequency source for the rotor adds into the cost and complexity. For large systems, however, the added cost may be less than the benefit of greater energy production of the variable-speed operation.

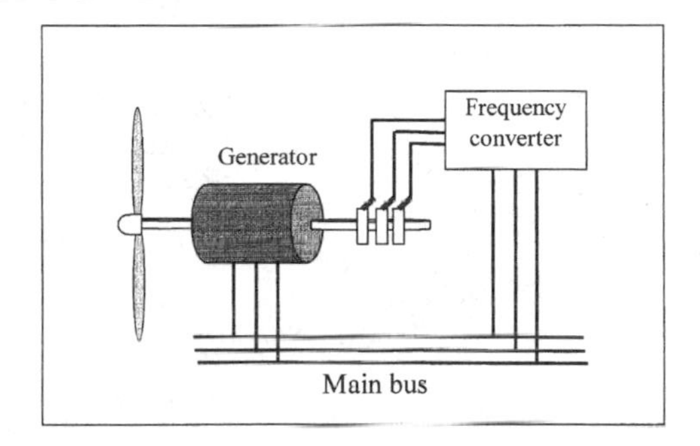

FIGURE 7-9
Scherbius adjustable-speed drive system injects variable frequency voltage in the wound rotor circuit.

7.2.6 Variable-Speed Direct Drive

The generator that operates directly at the turbine speed is extremely attractive. This is possible particularity for small machines where the rotor speed is high. The direct drive eliminates the mechanical gear altogether, and needs no power electronics. This results in multiple benefits:

- lower nacelle weight.
- reduced noise and vibration.
- lower power loss by several percent.
- less frequent servicing requirement at the nacelle.

The last benefit is particularly attractive for offshore installation.

The low rotor-speed requirement for large rotors imposes a design limitation on the electrical machine. That is, the generator must have large numbers of poles. Such machines must have short pole-pitch, resulting in poor magnetic design. To circumvent such limitation, the permanent magnet and wound rotor synchronous machines are being considered for 1.5 MW direct drive generators. Another possible solution is the axial gap induction machine. It can be designed with a large number of poles with less difficulty compared to the conventional radial gap induction machine. The axial gap machine is being considered for direct drive marine propulsion, which is inherently a low-speed system. For small gearless wind drives, the axial-flux permanent magnet generator may find some interest for its simplicity. A 5 kW, 200 revolution-per-minute laboratory prototype of the axial-gap permanent magnet design has been recently tested,[1] however, significant research and development effort is needed before the variable-speed direct drive systems can be commercially made available for large wind power systems.

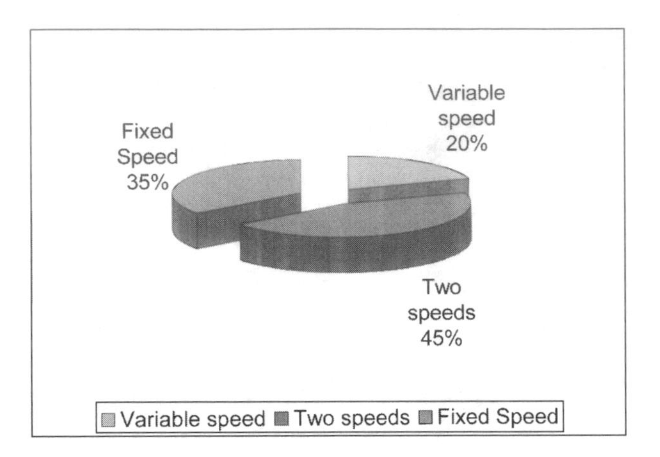

FIGURE 7-10
Design choices in generator drive systems.

7.3 Drive Selection

The variable-speed operation can capture theoretically about one-third more energy per year than the fixed-speed system.[2] The actual improvement reported by the variable-speed systems operators in the field is lower, around 20 to 35 percent. However, the improvement of even 15 to 20 percent in the annual energy yield by variable-speed operation can make the systems commercially viable in low wind region. This can open a whole new market for the wind-power installations, and this is happening at present in many countries. Therefore the newer installations are more likely to use the variable-speed systems.

As of 1997, the distribution of the system design is 35 percent one fixed-speed, 45 percent two fixed-speed and 20 percent variable-speed power electronics systems (Figure 7-10). The market share of the variable-speed systems, however, is increasing every year.[3]

7.4 Cut-Out Speed Selection

In any case, it is important that the machine is operated below its speed and power limits. Exceeding either one above the design limit can damage and even destroy the machine.

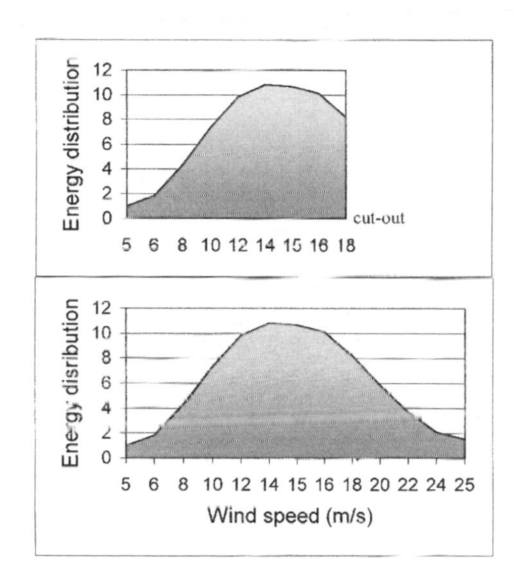

FIGURE 7-11
Probability distribution of annual energy production at various cut out speeds.

In designing the variable-speed system, important decision must be made on the upper limit of the operating speed. For the energy distribution shown in Figure 7 11, if the wind plant is designed to operate at fixed speed V_1, it can capture energy E_1 over the year. On the other hand, if it is designed to operate at a variable speed up to V_2, it can capture energy E_2 over the same period. The latter, however, comes with an added cost of designing the wind turbine and the generator to handle high power. The benefit and cost must be traded-off for the given site to arrive at the optimum upper limit on the rotor speed.

On one side of the trade is additional energy $(E_2 - E_1)$ that can be captured over the year. If the revenue of the generated electricity is valued at p $/kWh, the added benefit per year is $p \cdot (E_2 - E_1)$ dollars. The present worth P of this yearly benefit over the life of n years at the annual cost of capital i is as follows:

$$P = p \cdot (E_2 - E_1) \left[\frac{(1+i)^n - 1}{i(1+i)^n} \right] \tag{7-2}$$

If the variable-speed system requires additional capital cost of C, then the variable-speed system will be financially beneficial if $C < P$.

References

1. Chalmers, B. J. and Spooner, E. 1998. "Axial flux permanent magnet generator for gearless wind energy systems," *IEEE Paper No. PE-P13-EC-02,* July 1998.
2. Zinger, D. S., Muljadi, E. 1997. "Annualized wind energy improvement using variable-speed," *IEEE Transactions on Industry Applications,* Vol. 33-6, p. 1444-47, 1997.
3. Gardner, P. 1997. "Wind turbine generator and drive systems," *Wind Directions, Magazine of the European Wind Energy Association,* London, October, 1997.

8

Solar Photovoltaic Power System

The photovoltaic effect is the electrical potential developed between two dissimilar materials when their common junction is illuminated with radiation of photons. The photovoltaic cell, thus, converts light directly into electricity. The pv effect was discovered in 1839 by French physicist Becquerel. It remained in the laboratory until 1954, when Bell Laboratories produced the first silicon solar cell. It soon found application in the U.S. space programs for its high power capacity per unit weight. Since then it has been an important source of power for satellites. Having developed maturity in the space applications, the pv technology is now spreading into the terrestrial applications ranging from powering remote sites to feeding the utility lines.

8.1 The pv Cell

The physics of the pv cell is very similar to the classical p-n junction diode (Figure 8-1). When light is absorbed by the junction, the energy of the absorbed photons is transferred to the electron system of the material, resulting in the creation of charge carriers that are separated at the junction. The charge carriers may be electron-ion pairs in a liquid electrolyte, or electron-hole pairs in a solid semiconducting material. The charge carriers in the junction region create a potential gradient, get accelerated under the electric field and circulate as the current through an external circuit. The current squared times the resistance of the circuit is the power converted into electricity. The remaining power of the photon elevates the temperature of the cell.

The origin of the photovoltaic potential is the difference in the chemical potential, called the Fermi level, of the electrons in the two isolated materials. When they are joined, the junction approaches a new thermodynamic equilibrium. Such equilibrium can be achieved only when the Fermi level is equal in the two materials. This occurs by the flow of electrons from one material to the other until a voltage difference is established between the two materials which have the potential just equal to the initial difference of the Fermi level. This potential drives the photocurrent.

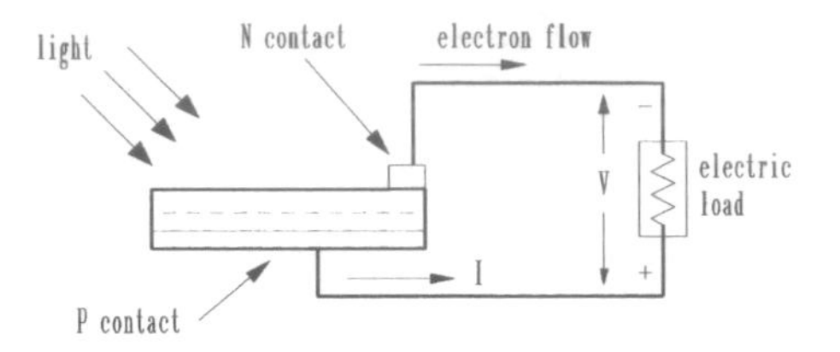

FIGURE 8-1
Photovoltaic effect converts the photon energy into voltage across the p-n junction.

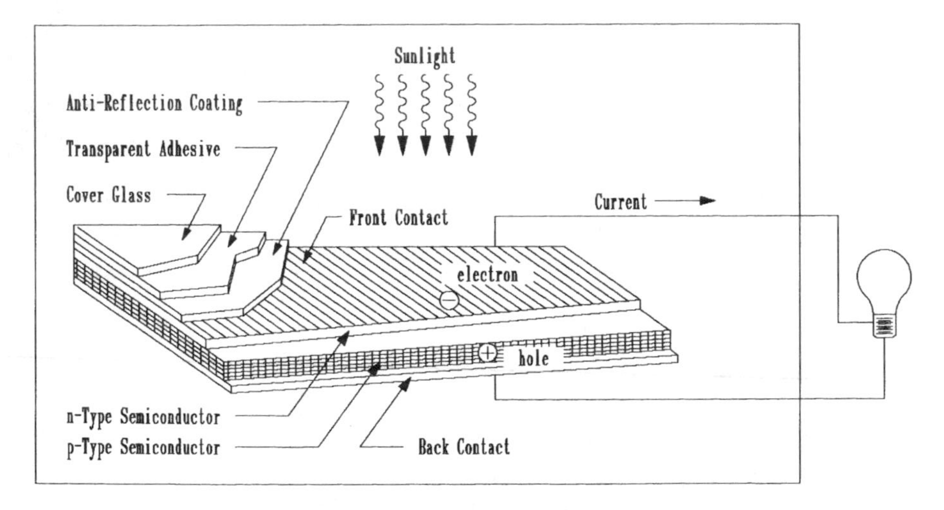

FIGURE 8-2
Basic construction of pv cell with performance enhancing features (current collecting mesh, anti-reflective coating and cover glass protection).

Figure 8-2 shows the basic cell construction.[1] For collecting the photocurrent, the metallic contacts are provided on both sides of the junction to collect electrical current induced by the impinging photons on one side. Conducting foil (solder) contact is provided over the bottom (dark) surface and on one edge of the top (illuminated) surface. Thin conducting mesh on the remaining top surface collects the current and lets the light through. The spacing of the conducting fibers in the mesh is a matter of compromise between maximizing the electrical conductance and minimizing the blockage of the light. In addition to the basic elements, several enhancement features are also included in the construction. For example, the front face of the cell has anti-reflective coating to absorb as much light as possible by minimizing the reflection. The mechanical protection is provided by the coverglass applied with a transparent adhesive.

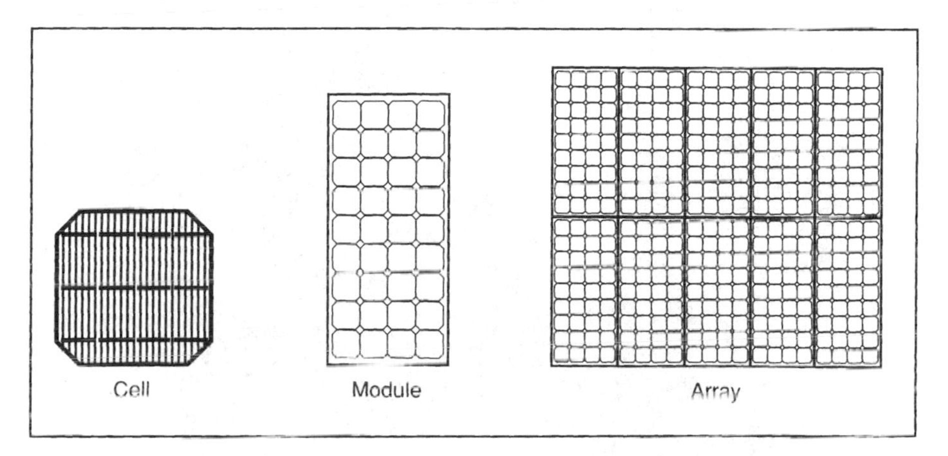

FIGURE 8-3
Several pv cells make a module and several modules make an array.

8.2 Module and Array

The solar cell described above is the basic building block of the pv power system. Typically, it is a few square inches in size and produces about one watt of power. For obtaining high power, numerous such cells are connected in series and parallel circuits on a panel (module) area of several square feet (Figure 8-3). The solar array or panel is defined as a group of several modules electrically connected in series-parallel combinations to generate the required current and voltage. Figure 8-4 shows the actual construction of a module in a frame that can be mounted on a structure.

Mounting of the modules can be in various configurations as seen in Figure 8-5. In the roof mounting, the modules are in the form that can be laid directly on the roof. In the newly developed amorphous silicon technology, the pv sheets are made in shingles that can replace the traditional roof shingles on one-to-one basis, providing a better economy in the material and labor.

8.3 Equivalent Electrical Circuit

The complex physics of the pv cell can be represented by the equivalent electrical circuit shown in Figure 8-6. The circuit parameters are as follows.

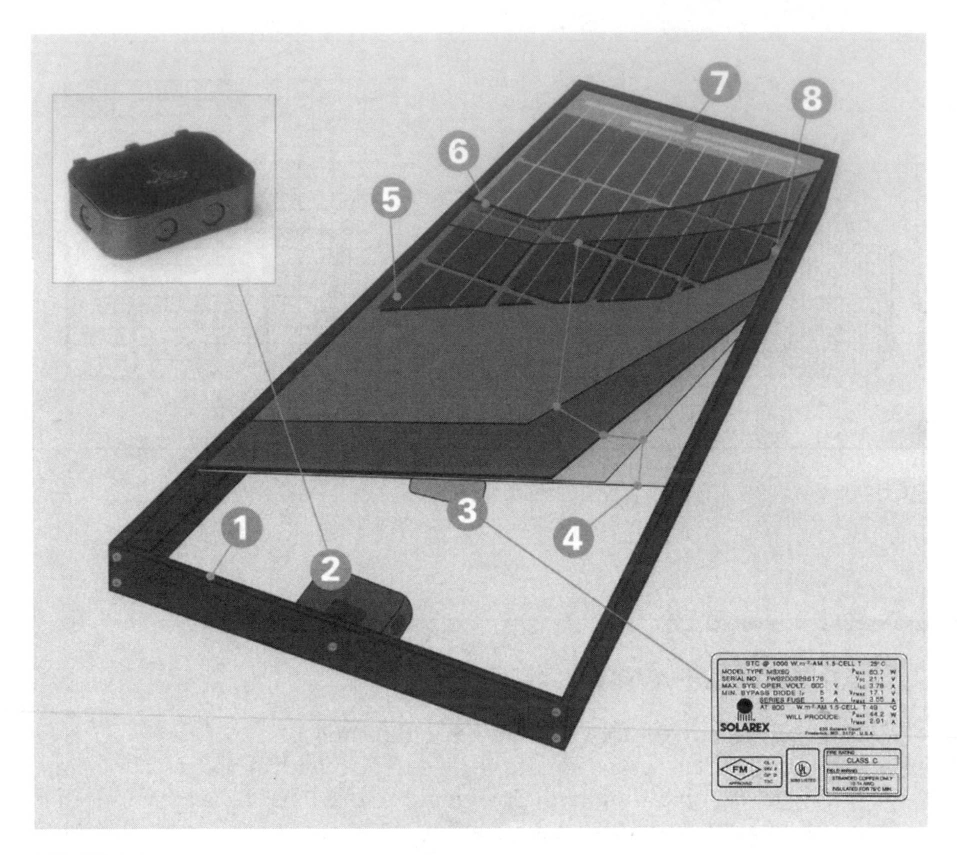

FIGURE 8-4
Construction of pv module: 1) frame, 2) weatherproof junction box, 3) rating plate, 4) weather protection for 30-year life, 5) pv cell, 6) tempered high transmissivity coverglass, 7) outside electrical bus, 8) frame clearance. (Source: Solarex Corporation, Frederick, Maryland, With permission.)

The output-terminal current I is equal to the light-generated current I_L, less the diode-current I_d and the shunt-leakage current I_{sh}. The series resistance R_s represents the internal resistance to the current flow, and depends on the p-n junction depth, the impurities and the contact resistance. The shunt resistance R_{sh} is inversely related with leakage current to the ground. In an ideal pv cell, $R_s = 0$ (no series loss), and $R_{sh} = \infty$ (no leakage to ground). In a typical high quality one square inch silicon cell, $R_s = 0.05$ to 0.10 ohm and $R_{sh} = 200$ to 300 ohms. The pv conversion efficiency is sensitive to small variations in R_s, but is insensitive to variations in R_{sh}. A small increase in R_s can decrease the pv output significantly.

In the equivalent circuit, the current delivered to the external load equals the current I_L generated by the illumination, less the diode current I_d and the ground-shunt current I_{sh}. The open circuit voltage V_{oc} of the cell is obtained when the load current is zero, i.e., when $I = 0$, and is given by the following:

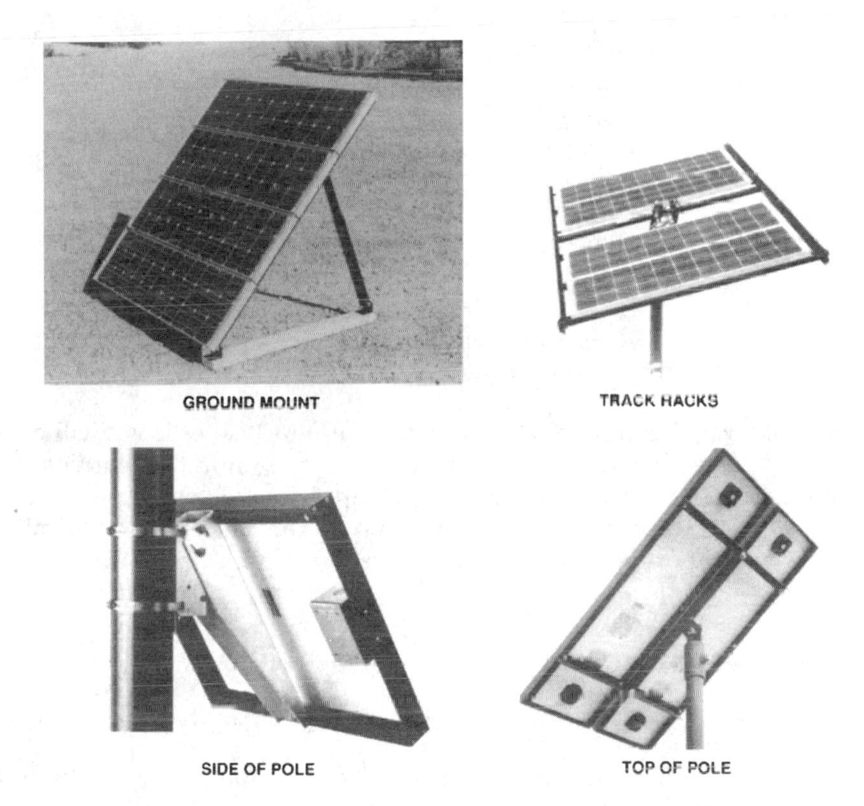

FIGURE 8-5
pv module mounting methods.

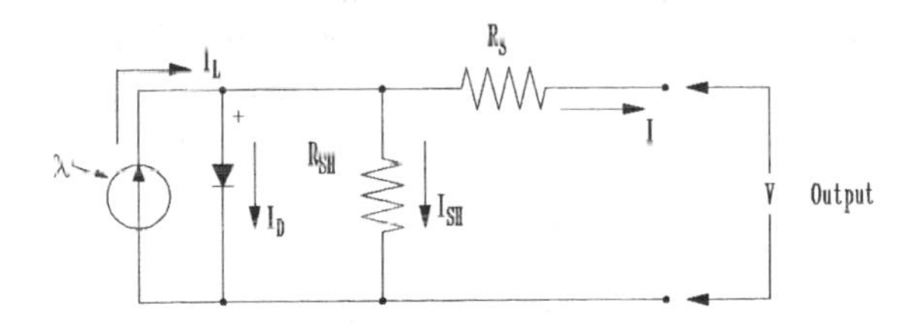

FIGURE 8-6
Equivalent electrical circuit of pv module, showing the diode and ground leakage currents.

$$V_{oc} = V + I\,R_{sh} \tag{8-1}$$

The diode current is given by the classical diode current expression:

$$I_d = I_D \left[\frac{QV_{oc}}{AKT} - 1 \right] \tag{8-2}$$

where I_D = the saturation current of the diode
Q = electron charge = $1.6 \cdot 10^{-19}$ Coulombs
A = curve fitting constant
K = Boltzmann constant = $1.38 \cdot 10^{-23}$ Joule/°K
T = temperature on absolute scale °K

The load current is therefore given by the expression:

$$I = I_L - I_D \left[e^{\frac{QV_{oc}}{AKT}} - 1 \right] - \frac{Voc}{R_{sh}} \tag{8-3}$$

The last term, the ground-leakage current, in practical cells is small compared to I_L and I_D, and can be ignored. The diode-saturation current can, therefore, be determined experimentally by applying voltage V_{oc} in the dark and measuring the current going into the cell. This current is often called the dark current or the reverse diode-saturation current.

8.4 Open Circuit Voltage and Short Circuit Current

The two most important parameters widely used for describing the cell electrical performance is the open-circuit voltage V_{oc} and the short-circuit current I_{sc}. The short-circuit current is measured by shorting the output terminals, and measuring the terminal current under full illumination. Ignoring the small diode and the ground-leakage currents under zero-terminal voltage, the short-circuit current under this condition is the photocurrent I_L.

The maximum photovoltage is produced under the open-circuit voltage. Again, by ignoring the ground-leakage current, Equation 8-3 with I = 0 gives the open-circuit voltage as the following:

$$V_{oc} = \frac{AKT}{Q} Log_n \left(\frac{I_L}{I_D} + 1 \right) \tag{8-4}$$

The constant KT/Q is the absolute temperature expressed in voltage (300°K = 0.026 volt). In practical photocells, the photocurrent is several orders of magnitude greater than the reverse saturation current. Therefore, the open-circuit voltage is many times the KT/Q value. Under condition of constant illumination, I_L/I_D is a sufficiently strong function of the cell temperature, and the solar cell ordinarily shows a negative temperature coefficient of the open-circuit voltage.

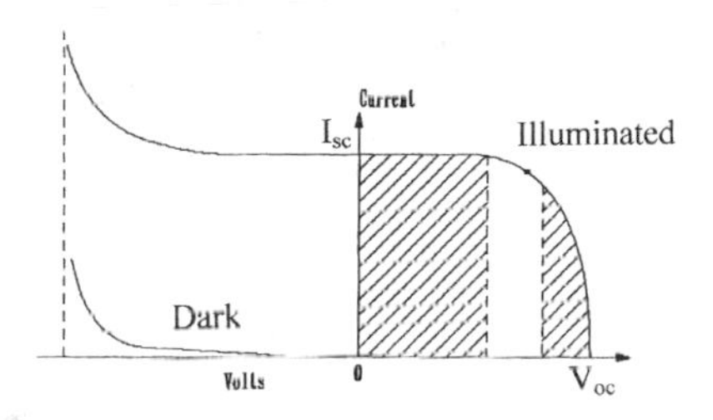

FIGURE 8-7
Current versus voltage (i-v) characteristics of the pv module in sunlight and in dark.

8.5 i-v and p-v Curves

The electrical characteristic of the pv cell is generally represented by the current versus voltage (i-v) curve. Figure 8-7 shows the i-v characteristic of a pv module under two conditions, in sunlight and in dark. In the first quadrant, the top left of the i-v curve at zero voltage is called the short-circuit current. This is the current we would measure with the output terminals shorted (zero voltage). The bottom right of the curve at zero current is called the open-circuit voltage. This is the voltage we would measure with the output terminals open (zero current). In the left shaded region, the cell works like a constant current source, generating voltage to match with the load resistance. In the shaded region on the right, the current drops rapidly with a small rise in voltage. In this region, the cell works like a constant voltage source with an internal resistance. Somewhere in the middle of the two shaded regions, the curve has a knee point.

If the voltage is externally applied in the reverse direction, say during a system fault transient, the current remains flat and the power is absorbed by the cell. However, beyond a certain negative voltage, the junction breaks down as in a diode, and the current rises to a high value. In the dark, the current is zero for voltage up to the breakdown voltage which is the same as in the illuminated condition.

The power output of the panel is the product of the voltage and the current outputs. In Figure 8-8, the power is plotted against the voltage. Notice that

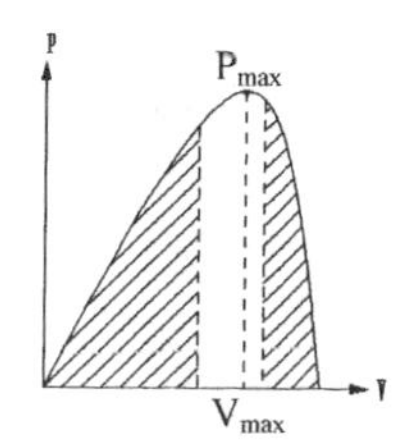

FIGURE 8-8
Power versus voltage (p-v) characteristics of the pv module in sunlight.

the cell produces no power at zero voltage or zero current, and produces the maximum power at voltage corresponding to the knee point of the i-v curve. This is why pv power circuits are designed such that the modules operate closed to the knee point, slightly on the left hand side. The pv modules are modeled approximately as a constant current source in the electrical analysis of the system.

Figure 8-9 is the i-v characteristic of a 22-watts panel under two solar illumination intensities, 1,000 watts/m^2 and 500 watts/m^2. These curves are at AM1.5 (air mass 1.5). The air mass zero (AM0) represents the condition in outer space, where the solar radiation is 1,350 watts/m^2. The AM1 represents the ideal earth condition in pure air on a clear dry noon when the sunlight experiences the least resistance to reach earth. The air we find on a typical day with average humidity and pollution is AM1.5, which is taken as the reference value. The solar power impinging a normal surface on a bright day with AM1.5 is about 1,000 watts/m^2. On a cloudy day, it would be low. The 500 watts/m^2 solar intensity is another reference condition the industry uses to report the i-v curves.

The photoconversion efficiency of the pv cell is defined as the following:

$$\eta = \frac{\textit{electrical power output}}{\textit{solar power impinging the cell}} \tag{8-5}$$

Obviously, the higher the efficiency, the higher the output power we get under a given illumination.

8.6 Array Design

The major factors influencing the electrical design of the solar array are as follows:

- the sun intensity.
- the sun angle.
- the load matching for maximum power.
- the operating temperature.

These factors are discussed below.

8.6.1 Sun Intensity

The magnitude of the photocurrent is maximum under full bright sun (1.0 sun). On a partially sunny day, the photocurrent diminishes in direct proportion to the sun intensity. The i-v characteristic shifts downward at a lower sun

Mechanical Specifications

UPM-880

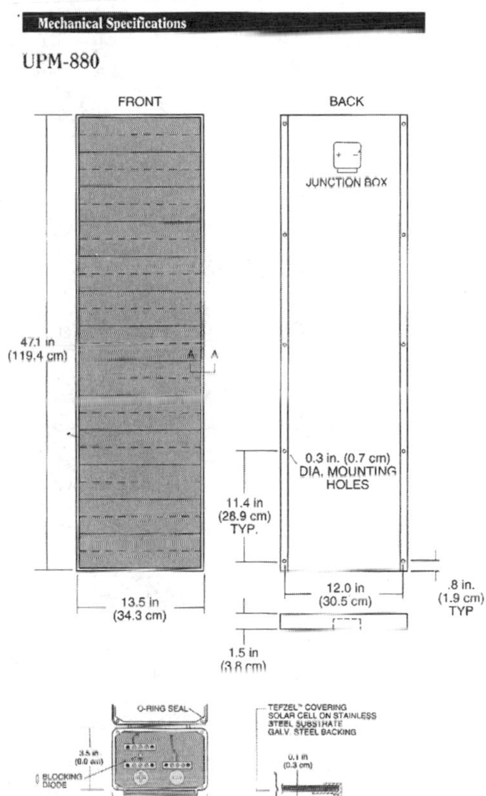

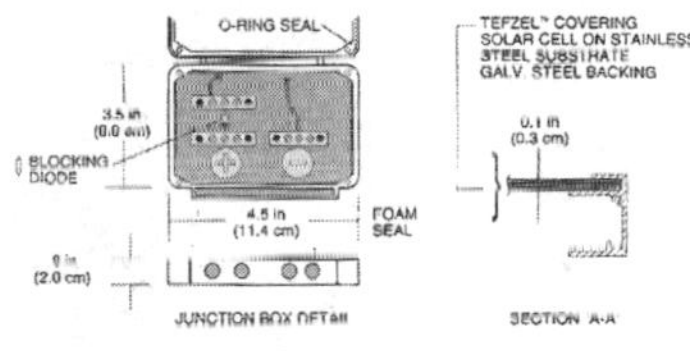

Specifications

Rated Power (Wp)	22 W
Voltage (Vmp)	15.6 V
Current (Imp)	1.4 A
Voltage (Voc)	22 V
Current (Isc)	1.8 A
Size.	
Inches	47.1 x 13.5 x 1.5
Centimeters	119.4 x 34.3 x 3.8
Weight, lbs/kgs	8.3/3.6

Standard Test Conditions: 1000 W/m², Cell Temp. 25°C, Air Mass 1.5.

UPM-880 Product IV Curves

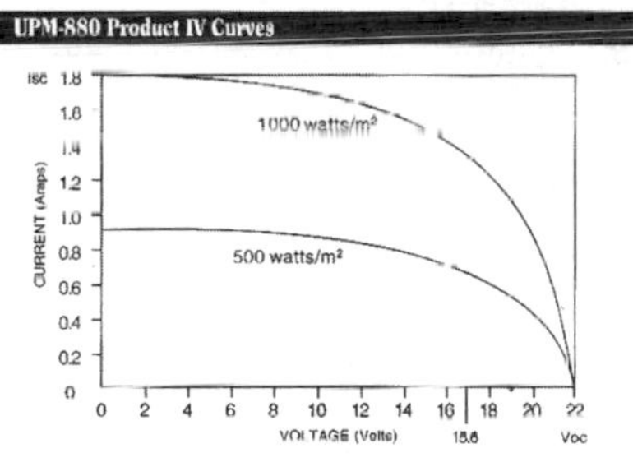

Characteristic performance at rated power conditions of 1000 and 500 W/m², Cell Temp 25°C, Air Mass 1.5. Does not include losses due to blocking diode, wiring or connections. Performance may vary up to 10% from rated power due to low temperature operation, spectral and related effects. Contact United Solar Systems Corp. for detailed performance information.

UNI-SOLAR®
United Solar Systems Corp.
5278 Eastgate Mall
San Diego, CA 92121-2814
619/625-2080
FAX 619/625-2083

FIGURE 8-9
i-v characteristic of 22 watts pv module at full and half sun intensities. (Source: United Solar Systems Corporation, San Diego, California. With permission.)

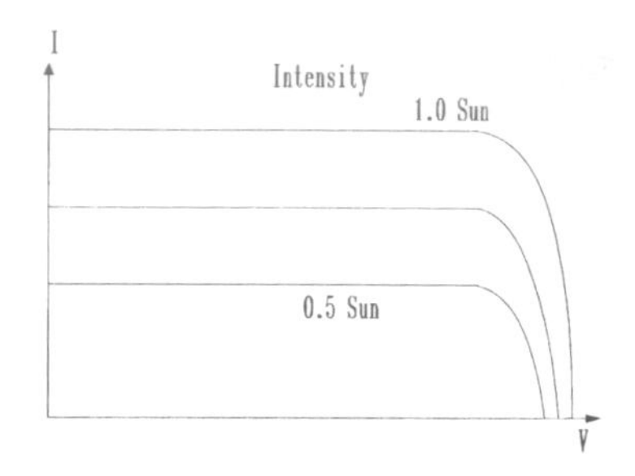

FIGURE 8-10
i-v characteristic of pv module shifts down at lower sun intensity, with small reduction in voltage.

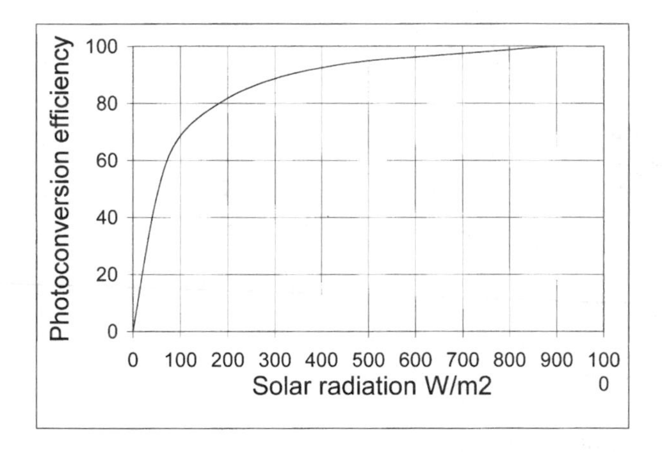

FIGURE 8-11
Photoconversion efficiency versus solar radiation. The efficiency is practically constant over a wide range of radiation.

intensity as shown in Figure 8-10. On a cloudy day, therefore, the short circuit current decreases significantly. The reduction in the open-circuit voltage, however, is small.

The photoconversion efficiency of the cell is insensitive to the solar radiation in the practical working range. For example, Figure 8-11 shows that the efficiency is practically the same at 500 watts/m^2 and 1,000 watts/m^2. This means that the conversion efficiency is the same on a bright sunny day and a cloudy day. We get lower power output on a cloudy day only because of the lower solar energy impinging the cell.

8.6.2 Sun Angle

The cell output current is given by $I = I_0 \cos \theta$, where I_o is the current with normal sun (reference), and θ is the angle of the sunline measured from the normal. This cosine law holds well for sun angles ranging from 0 to about 50°.

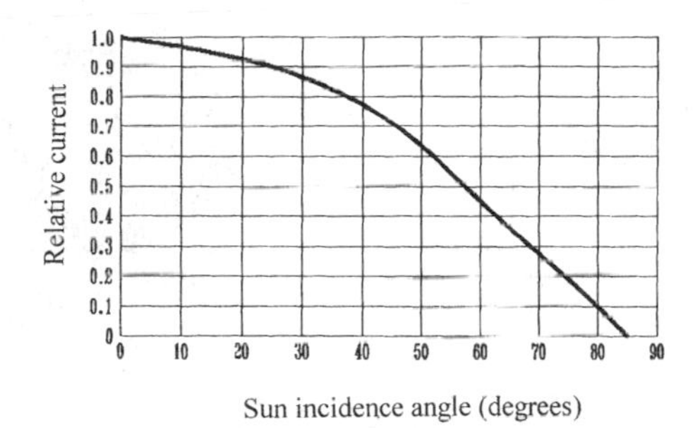

FIGURE 8-12
Kelley cosine curve for pv cell at sun angles from 0 to 90°.

TABLE 8-1
The Kelley Cosine Values of the Photocurrent in Silicon Cells

Sun Angle Degrees	Mathematical Cosine Value	Kelly Cosine Value
30	0.866	0.866
50	0.643	0.635
60	0.500	0.450
80	0.174	0.100
85	0.087	0

Beyond 50°, the electrical output deviates significantly from the cosine law, and the cell generates no power beyond 85°, although the mathematical cosine law predicts 7.5 percent power generation. The actual power-angle curve of the pv cell is called Kelly cosine, and is shown in Figure 8-12 and Table 8-1.

8.6.3 Shadow Effect

The array may consist of many parallel strings of series-connected cells. Two such strings are shown in Figure 8-13. A large array may get partially shadowed due to a structure interfering with the sunline. If a cell in a long-series string gets completely shadowed, it will lose the photovoltage, but still must carry the string current by virtue of its being in series with the other fully operating cells. Without internally generated voltage, it cannot produce power. Instead, it acts as a load, producing local I^2R loss and heat. The remaining cells in the string must work at higher voltage to make up the loss of the shadowed cell voltage. Higher voltage in healthy cells means lower string current as per the i-v characteristic of the string. This is shown in the bottom left of Figure 8-13. The current loss is not proportional to the shadowed area, and may go unnoticed for mild shadow on a small area. However, if more cells are shadowed beyond the critical limit, the i-v curve

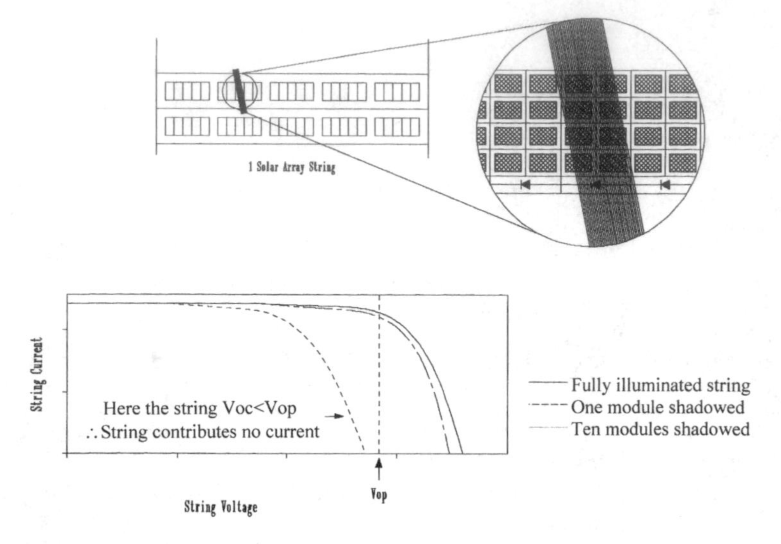

FIGURE 8-13
Shadow effect on one long pv string of an array. The power degradation is small until shadow exceeds the critical limit.

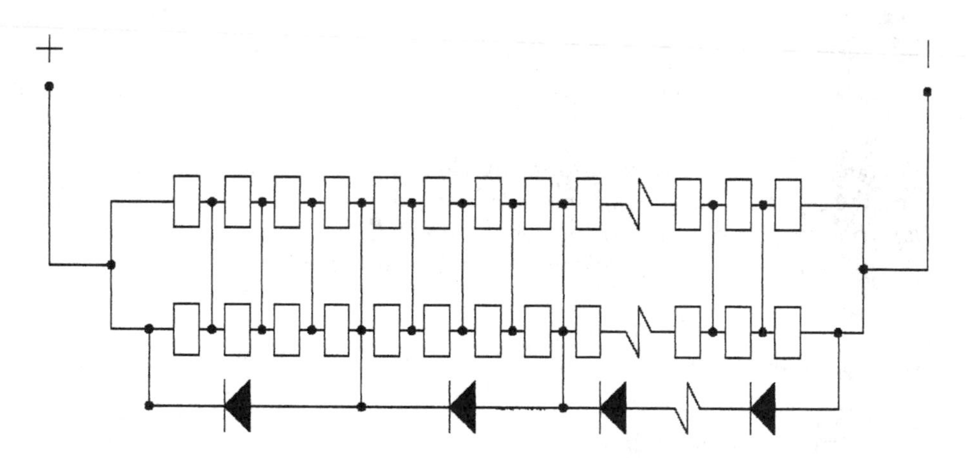

FIGURE 8-14
Bypass diode in pv string minimizes the power loss under heavy shadow.

gets below operating voltage of the string, making the string current fall to zero, losing all power of the string.

The commonly used method to eliminate the loss of string due to shadow effect is to subdivide the circuit length in several segments with bypass diodes (Figure 8-14). The diode across the shadowed segment bypasses only that segment of the string. This causes a proportionate loss of the string voltage and current, without losing the whole string power. Some modern pv modules come with such internally embedded bypass diodes.

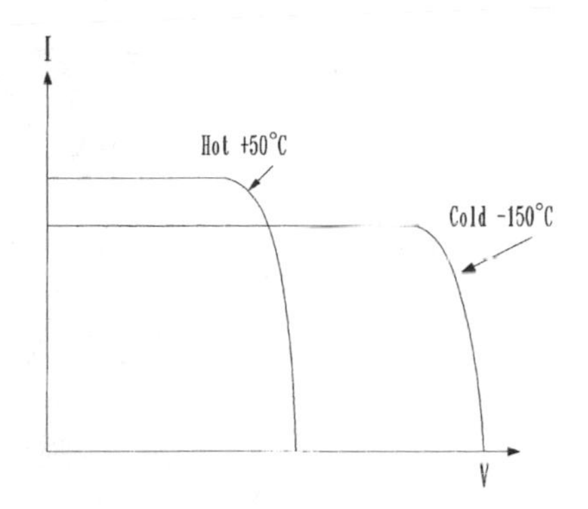

FIGURE 8-15
Effect of temperature on the i-v characteristic. The cell produces less current but greater voltage, with net gain in the power output at cold temperature.

8.6.4 Temperature Effect

With increasing temperature, the short-circuit current of the cell increases, whereas the open-circuit voltage decreases (Figure 8-15). The effect of temperature on the power is quantitatively evaluated by examining the effects on the current and the voltage separately. Say I_o and V_o are the short-circuit current and the open-circuit voltage at the reference temperature T, and α and β are their respective temperature coefficients. If the operating temperature is increased by ΔT, then the new current and voltage are given by the following.

$$I_{sc} = I_o(1+\alpha\cdot\Delta T) \text{ and } V_{oc} = V_o(1-\beta\cdot\Delta T) \tag{8-6}$$

Since the operating current and the voltage change approximately in the same proportion as the short-circuit current and open-circuit voltage, respectively, the new power is as follows:

$$P = V\cdot I = I_o(1+\alpha\cdot\Delta T)\cdot V_o(1-\beta\cdot\Delta T) \tag{8-7}$$

This can be simplified in the following expression by ignoring a small term:

$$P = P_o\cdot\left[1+(\alpha-\beta)\cdot\Delta T\right] \tag{8-8}$$

For typical single crystal silicon cells, α is 500 μu per °C and β is 5 mu per °C. The power is therefore:

$$P = P_o\cdot\left[1+(500\mu u-5mu)\cdot\Delta T\right] \text{or } P_o\cdot\left[1-.0045\Delta T\right] \tag{8-9}$$

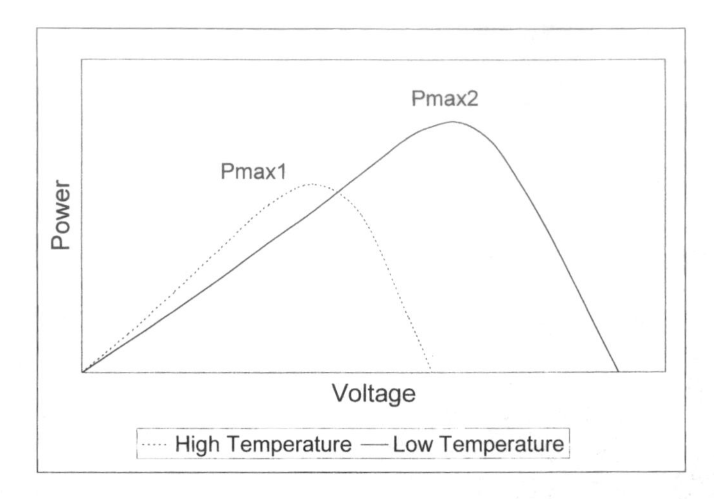

FIGURE 8-16
Effect of temperature on the p-v characteristic. The cell produces more power at cold temperature.

This expression indicates that for every °C rise in the operating temperature above the reference temperature, the silicon cell power output decreases by 0.45 percent. Since the increase in the current is much less than the decrease in the voltage, the net effect is the decrease in power at high operating temperatures.

The effect of varying temperature on the power output is shown in the power versus voltage characteristics at two operating temperatures in Figure 8-16. The figure shows that the maximum power available at lower temperature is higher than that at higher temperature. Thus, cold temperature is actually better for the pv cell, as it generates more power. However, the two P_{max} points are not at the same voltage. In order to extract maximum power at all temperatures, the pv system must be designed such that the module output voltage can increase to V_2 for capturing P_{max2} at lower temperature and can decrease to V_1 for capturing P_{max1} at higher temperature.

8.6.5 Effect of Climate

On a partly cloudy day, the pv module can produce up to 80 percent of their full sun power. Even on an extremely overcast day, it can produce about 30 percent power. Snow does not usually collect on the modules, because they are angled to catch the sun. If snow does collect, it quickly melts. Mechanically, modules are designed to withstand golf ball size hail.

8.6.6 Electrical Load Matching

The operating point of any power system is the intersection of the source line and the load line. If the pv source having the i-v and p-v characteristics

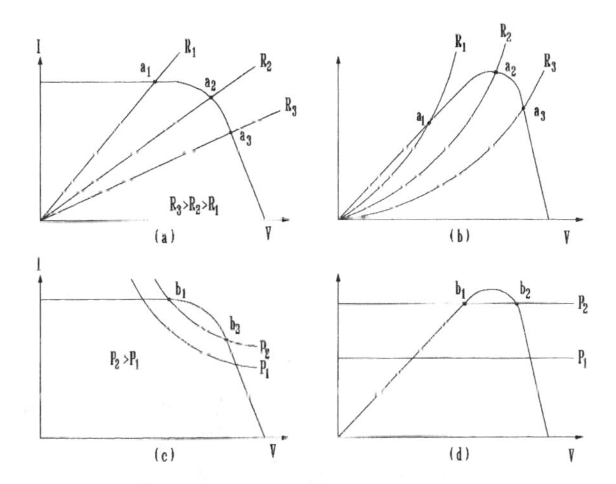

FIGURE 8-17
Operating stability and electrical load matching with resistive load and constant power load.

shown in Figure 8-17 (a) is supplying the power to the resistive load R_1, it will operate at point A_1. If the load resistance increases to R_2 or R_3, the operating point moves to A_2 or A_3, respectively. The maximum power is extracted from the module when the load resistance is R_2 (Figure 8-17b). Such load matching with the source is always necessary for maximum power extraction from the pv module.

The operation with constant power loads is shown in Figure 8-17(c) and (d). The constant power load line has two points of intersection with the source line, denoted by B_1 and B_2. Only point B_2 is stable, as any perturbation from it will generate a restoring power to take the operation back to B_2. Therefore, the system will operate at B_2.

The necessary condition for the electrical operating stability of the solar array is as follows:

$$\left[\frac{dP}{dV}\right]_{load} > \left[\frac{dP}{dV}\right]_{source} \tag{8-10}$$

Some loads such as heaters have constant resistance, with power varying with the voltage squared. On the other hand, some loads such as induction motors behave more like constant power loads, drawing more current at lower voltage. In most large systems with mix loads, the power varies approximately in linear proportion with voltage.

8.6.7 Sun Tracking

More energy is collected by the end of the day if the pv module is installed on a tracker, with an actuator that follows the sun like a sunflower. There are two types of sun trackers:

FIGURE 8-18
Dual-axis suntracker follows the sun like a sunflower around the year. (Source: American Sun Company, Blue Hill, Maine. With permission.)

- one-axis tracker, which follows the sun from east to west during the day.
- two-axis tracker tracks the sun from east to west during the day, and from north to south during the seasons of the year (Figure 8-18). A sun tracking design can increase the energy yield up to 40 percent over the year compared to the fixed-array design. The dual-axis tracking is done by two linear actuator motors, which aim the sun within one degree of accuracy (Figure 8-19). During the day, it tracks the sun east to west. At night it turns east to position itself for the next morning sun. Old trackers did this after the sunset using a small nickel-cadmium battery. The new designs eliminate the battery requirement by doing it in the weak light of the dusk and/or dawn. The Kelley cosine presented in Table 8-1 is useful to assess, accurately, the power available from sun at the evening angles.

FIGURE 8-19
Actuator motor of the suntracker. (Source: American Sun Company, Blue Hill, Maine. With permission.)

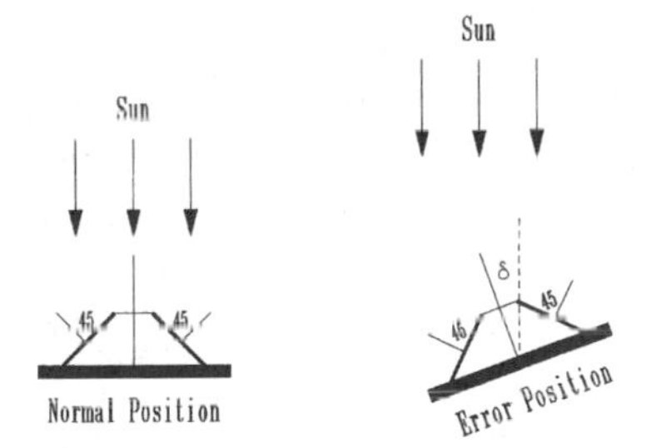

FIGURE 8-20
Sun tracking actuator principle. The two differentially connected sensors at 45° generate signal proportional to the pointing error.

When the sun is obscured by a dark cloud, the tracker may aim at the next brightest object, which is generally the edge of the cloud. When the cloud is gone, the tracker aims at the sun once again. Such sun-hunting is eliminated in newer suntracker design.

One method of designing the suntracker is to use two pv cells mounted on two 45° wedges (Figure 8-20), and connecting them differentially in series through an actuator motor. When the sun is perfectly normal, the current on both cells are equal to $I_o \cdot \cos 45°$. Since they are connected in series opposition, the net current in the motor is zero, and the array stays put. On the other hand, if the array is not normal to the sun, the sun angles on the two cells are different, giving two different currents:

$$I_1 = I_o \cos(45 + \delta), \text{ and } I_2 = I_o \cos(45 - \delta)$$

The motor current is therefore:

$$I_m = I_1 - I_2 = I_o \cos(45 + \delta) - I_o \cos(45 - \delta)$$

Using Taylor series expansion:

$$f(x + h) = f(x) + h \cdot f'(x) + \frac{h^2}{2!} f''(x) + \cdots$$

we can express the two currents as the following:

$$I_1 = I_o \cos 45 - I_o \delta \sin 45, \text{ and}$$

$$I_2 = I_o \cos 45 - I_o \delta \sin 45$$

The motor current is then $I_m = I_1 - I_2 = 2 I_o \delta \sin 45°$ (8-11)

$$= \sqrt{2} I_o \delta \text{ if } \delta \text{ is in radian}$$

Small pole-mounted panels can use one pole-mounted suntracker. Large array, on the other hand, is divided into small modules, each mounted on its own single-axis or dual-axis tracker. This simplifies the structure and eliminates the problems related with large motion.

8.7 Peak Power Point Operation

The suntracker drives the module mechanically to face the sun to collect the maximum solar radiation. However, that, in itself, does not guarantee the maximum power output from the module. As was seen in Figure 8-16, the module must operate electrically at a certain voltage which corresponds to the peak power point under the given operating conditions. First we examine the electrical principle of the peak power operation.

If the array is operating at voltage V and current I on the i-v curve, the power generation is $P = V \cdot I$ watts. If the operation moves away from the above point, such that the current is now $I + \Delta I$, and the voltage is $V + \Delta V$, the new power is as follows:

$$P + \Delta P = (V + \Delta V) \cdot (I + \Delta I) \tag{8-12}$$

Which, after ignoring a small term, simplifies to the following:

$$\Delta P = \Delta V \cdot I + \Delta I \cdot V \tag{8-13}$$

The ΔP should be zero at peak power point, which necessarily lies on a locally flat neighborhood. Therefore, at peak power point, the above expression in the limit becomes as follows:

$$\frac{dV}{dI} = -\frac{V}{I} \tag{8-14}$$

We take note here that dV/dI is the dynamic impedance of the source, and V/I is the static impedance.

There are three electrical methods of extracting the peak power from the module, as described below:

(1) In the first method, a small signal current is periodically injected into the array bus and the dynamic bus impedance $Z_d = dV/dI$ and the static bus impedance $Z_s = V/I$ are measured. The operating voltage is then increased or decreased until $Z_d = -Z_s$. At this point, the maximum power is extracted from the source.

(2) In another electrical method, the operating voltage is increased as long as dP/dV is positive. That is, the voltage is increased as long as we get more power. If dP/dV is sensed negative, the operating voltage is decreased. The voltage is kept put if the dP/dV is near zero within a preset dead band.

(3) The third method makes use of the fact that for most pv cells, the ratio of the voltage at the maximum power point to the open circuit voltage (i.e., V_{mp}/V_{oc}) is approximately constant, say K. For example, for high-quality crystalline silicon cells $K = 0.72$. An unloaded cell is installed on the array and kept in the same environment as the power-producing module, and its open circuit voltage is continuously measured. The operating voltage of the power-producing array is then set at $K \cdot V_{oc}$, which will produce the maximum power.

8.8 pv System Components

The array by itself does not constitute the pv power system. We must also have a structure to mount it, point to the sun, and the components that accept the DC power produced by the array and condition the power in the form that is usable by the load. If the load is AC, the system needs an inverter to convert the DC power into AC, generally at 50 or 60 Hz.

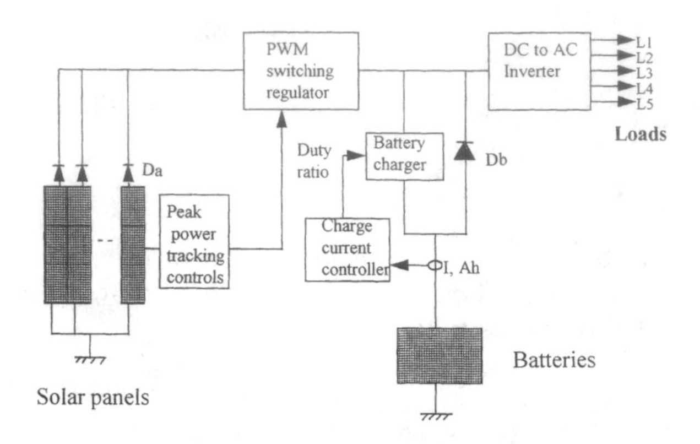

FIGURE 8-21
Peak power tracking photovoltaic power system showing major components.

Figure 8-21 shows the necessary components of a stand-alone pv power system. The peak power tracker senses the voltage and current outputs of the array and continuously adjusts the operating point to extract the maximum power under the given climatic conditions. The output of the array goes to the inverter, which converts the DC into AC. The array output in excess of the load requirement is used to charge the battery. The battery charger is usually a DC-DC buck converter. If excess power is still available after fully charging the battery, it is shunted in dump heaters, which may be space or room heaters in a stand-alone system. When the sun is not available, the battery discharges to the inverter to power the loads. The battery discharge diode Db is to prevent the battery from being charged when the charger is opened after a full charge or for other reasons. The array diode Da is to isolate the array from the battery, thus keeping the array from acting as load on the battery at night. The mode controller collects the system signals, such as the array and the battery currents and voltages, keeps track of the battery state of charge by bookkeeping the charge/discharge ampere-hours, and commands the charger, discharge converter, and dump heaters on or off as needed. The mode controller is the central controller for the entire system.

In the grid-connected system, dump heaters are not required, as all excess power is fed to the grid lines. The battery is also eliminated, except for small critical loads, such as the start up controls and the computers. The DC power is first converted into AC by the inverter, ripples are filtered and then only the filtered power is fed into the grid lines.

For pv applications, the inverter is a critical component, which converts the array DC power into AC for supplying the loads or interfacing with the grid. A new product recently being introduced into the market is the AC-pv modules, which integrates an inverter directly in the module, and is presently available in a few hundred watts capacity. It provides utility grade 60 Hz power directly from the module junction box. This greatly simplifies the pv system design.

References

1. Cook, G., Billman, L. and Adcock R. 1995. "Photovoltaic Fundamental," *DOE/Solar Energy Research Institute Report No. DE91015001*, February 1995.

9

Solar Thermal System

The solar thermal power system collects the thermal energy in solar radiation and uses at high or low temperature. The low temperature applications include water and space heating for commercial and residential buildings.[1] Producing electricity using the steam-turbine-driven electrical generator is a high temperature application discussed in this chapter.

The technology of generating electrical power using the solar thermal energy has been demonstrated at commercial scale. The research and development funding have primarily come from the government, with active participation of some electric utility companies.

Figure 9-1 is a schematic of a large-scale solar thermal power station developed, designed, built, tested, and operated with the U.S. Department of Energy funding. In this plant, the solar energy is collected by thousands of sun-tracking mirrors, called heliostats, that reflect the sun's energy to a single receiver atop a centrally located tower. The enormous amount of energy focused on the receiver is used to generate high temperature to melt a salt. The hot molten salt is stored in a storage tank, and is used, when needed, to generate steam and drive the turbine generator. After generating the steam, the used molten salt at low temperature is returned to the cold salt storage tank. From here it is pumped to the receiver tower to get heated again for the next thermal cycle. The usable energy extracted during such a thermal cycle depends on the working temperatures. The maximum thermodynamic conversion efficiency that can be theoretically achieved with the hot side temperature T_{hot} and the cold side temperature T_{cold} is given by the Carnot cycle efficiency, which is as follows:

$$\eta_{carnot} = \frac{T_{hot} - T_{cold}}{T_{hot}} \tag{9-1}$$

where the temperatures are in absolute scale. The higher the hot side working temperature and lower the cold side exhaust temperature, the higher the plant efficiency of converting the captured solar energy into electricity. The hot side temperature T_{hot}, however, is limited by the properties of the working medium. The cold side temperature T_{cold} is largely determined by the cooling method and the environment available to dissipate the exhaust heat.

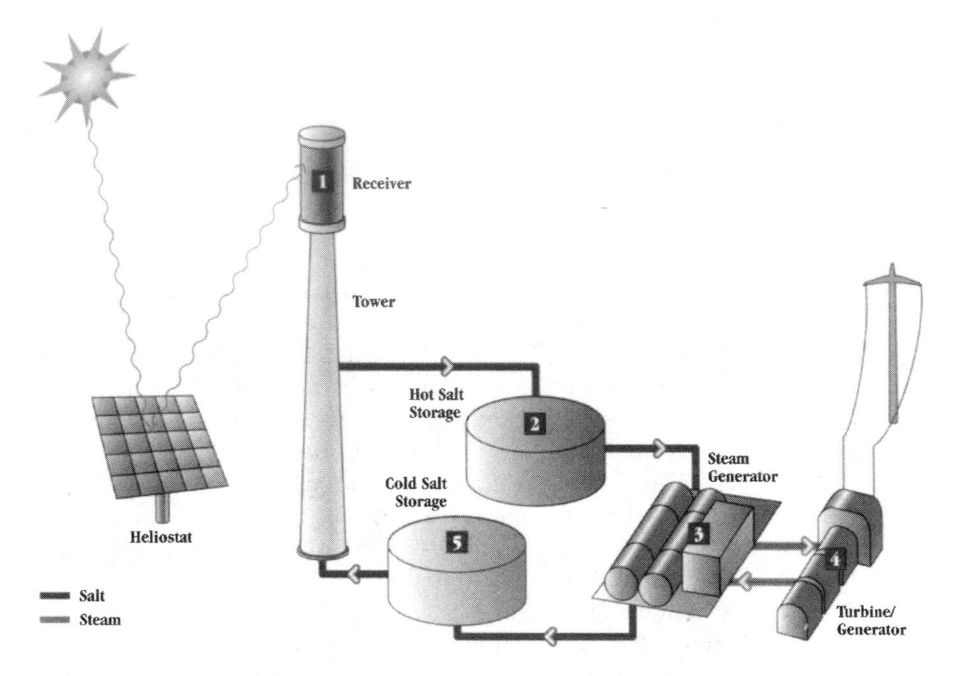

FIGURE 9-1
Solar thermal power plant schematic for generating electricity.

A major benefit of this scheme is that it incorporates the thermal energy storage for duration in hours with no degradation in performance, or longer with some degradation. This feature makes the technology capable of producing high-value electricity for meeting peak demands. Moreover, compared to the solar photovoltaic, the solar thermal system is economical, as it eliminates the costly semiconductor cells.

9.1 Energy Collection

The solar thermal energy is collected by concentrators. Three alternative configurations of the concentrators are shown in Figure 9-2. Their main features and applications are as follows:

9.1.1 Parabolic Trough

The parabolic trough system is by far the most commercially matured of the three technologies. It focuses the sunlight on a glass-encapsulated tube running along the focal line of the collector. The tube carries heat absorbing liquid, usually oil, which in turn, heats water to generate steam. More than 350 MW of parabolic trough capacity is operating in the California Mojave

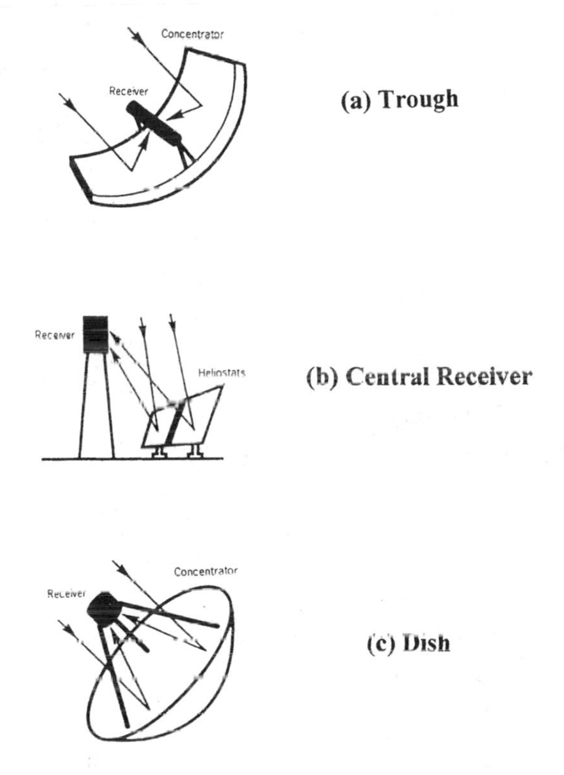

FIGURE 9-2
Alternative thermal energy collection technologies

Desert and is connected to the Southern California Edison's utility grid. This is more than 90 percent of the world's solar thermal capacity at present.

9.1.2 Central Receiver

In the central receiver system, an array of field mirrors focus the sunlight on the central receiver mounted on a tower. To focus the sun on the central receiver at all times, each heliostat is mounted on the dual-axis suntracker to seek position in the sky that is midway between the receiver and the sun. Compared to the parabolic trough, this technology produces higher concentration, and hence, higher temperature working medium, usually a salt. Consequently, it yields higher Carnot efficiency, and is well suited for utility scale power plants in tens or hundreds of megawatt capacity.

9.1.3 Parabolic Dish

The parabolic dish tracks the sun to focus heat, which drives a sterling heat engine-generator unit. This technology has applications in relatively small capacity (tens of kW) due the size of available engines and wind loads on

TABLE 9-1

Comparison of Alternative Solar Thermal Power System Technologies

Technology	Solar Concentration (x Suns)	Operating Temperature on the Hot Side	Thermodynamic Cycle Efficiency
Parabolic Trough Receiver	100	300–500°C	Low
Central Receiver Power Tower	1000	500–1000°C	Moderate
Dish Receiver with Engine	3000	800–1200°C	High

the dish collectors. Because of their small size, it is more modular than other solar thermal power systems, and can be assembled in a few hundred kW to few MW capacities. This technology is particularly attractive for small stand-alone remote applications.

The three alternative solar thermal technologies are compared in Table 9-1.

9.2 Solar II Power Plant

The central receiver technology with power tower is getting new development thrust in the U.S.A. as having a higher potential of generating lower cost electricity at large scale. An experimental 10 MW_e power plant using this technology has been built and commissioned in 1996 by the Department of Energy in partnership with the Solar II Consortium of private investors led by the Southern California Edison, the second largest electrical utility company in the U.S.A. It is connected to the grid, and has enough capacity to power 10,000 homes. The plant is designed to operate commercially for 25 to 30 years. Figure 9-3 is the site photograph of this plant located east of Barstow, California. It uses some components of Solar I plant, which was built and operated at the site using the central receiver power tower technology. The Solar I plant, however, generated steam directly to drive the generator without the thermal storage feature of the Solar II plant.

Solar II central receiver (Figure 9-4) was developed by the Sandia National Laboratory. It raises the salt temperature to 1,050°F. The most important feature of the Solar II design is its innovative energy collection and the storage system. It uses a salt that has excellent heat retention and heat transfer properties. The heated salt can be used immediately to generate steam and electric power. Or, it can be stored for use during cloudy periods or after the sun goes down to meet the evening load demand on the utility grid. Because of this unique energy storage feature, the power generation is decoupled from the energy collection. For electrical utility, this storage capability is crucial in that the energy is collected when available, and is used to generate high-value electricity when it is most needed. The salt selected by

FIGURE 9-3
Solar II plant site view. (Source: U.S. Department of Energy.)

the Sandia laboratory for this plant is sodium and potassium nitrate which works as a single phase liquid, and is colorless and odorless. In addition to having the needed thermal properties up to the operating temperature of 1,050°F, it is inexpensive and safe.

Tables 9-2 and 9-3 give the technical design features of the experimental Solar II power plant. The operating experience to date indicates the overall plant capacity factor of 20 percent, and the overall thermal to electrical conversion efficiency of 16 percent. It is estimated that 23 percent overall efficiency can be achieved in a commercial plant design using this technology.

9.3 Synchronous Generator

The electromechanical energy conversion in the solar thermal power system is accomplished by the synchronous machine, which runs at a constant speed to produce 60 Hz electricity. This power is then directly used to meet the local loads, and/or to feed the utility grid lines.

The electromagnetic features of the synchronous machine are shown in Figure 9-5. The stator is made of conductors placed in slots of magnetic iron

FIGURE 9-4
Experimental 1,050°F thermal receiver tower for Solar II power plant. (Source: DOE/Sandia National Laboratory.)

laminations. The stator conductors are connected in three phase coils. The rotor consists of magnetic poles created by the field coils carrying direct current. The rotor is driven by steam turbine to create a rotating magnetic field. Because of this rotation, the rotor field coils use slip rings and carbon brushes to supply DC power from a stationary source.

The stator conductors are wound in three groups, connected in three-phase configuration. Under the rotating magnetic field of the rotor, the three phase coils generate AC voltages that are 120 electrical degrees out of phase with each other. If the electromagnetic structure of the machine has p pole pairs, and it is required to generate electricity at frequency f, then the rotor must rotate at N revolution per minute given by the following:

$$N = 60 \cdot \frac{f}{p} \tag{9-2}$$

The synchronous machine must operate at this constant speed to generate power at the specified frequency. In a stand-alone solar thermal system, small speed variations could be tolerated within the frequency tolerance band of the AC system. If the generator is connected to the grid, it must be synchronous with the grid frequency, and must operate exactly at the grid frequency at all times. Once synchronized, such a machine has inherent tendency to remain

TABLE 9-2

Solar II Design Features

Site

- Mojave Desert in California
- 1,949 feet above sea level
- 7.5 kWh/m^2-day annual average daily insolation
- 95 acres of land

Tower

- Reused from Solar I plant
- 277 feet to top of the receiver
- 211 feet to top of BCS deck

Heliostats

- 1,818 Solar I heliostats, 39.1 m^2, 91 percent reflectivity
- 108 new Lug heliostats, 95.1 m^2, 93 percent reflectivity
- 81,000 m^2 total reflective surface
- Can operate in winds up to 35 mph

Receiver

- New for Solar II plant
- Supplier Rockwell
- 42.2 MW thermal power rating
- Average flux 429 suns (429 kW/m^2)
- Peak flux 800 suns
- 24 panels, 32 tubes per panel
- 20 feet tall and 16.6 feet diameter
- 0.8125 inch tube OD
- 0.049 inch tube wall thickness
- Tubes 316H stainless steel

Thermal Storage System

- Supplier Pitt Des Moines
- Two new 231,000 gallon storage tanks, 38 ft ID
- Cold tank carbon steel, 25.8 ft high, 9 inch insulation
- Hot tank 304 stainless steel, 27.5 ft high, 18 inches insulation
- 3 hours of storage at rated turbine output

Nitrate salt — Chilean Nitrate

- 60% $NaNO_3$, 40% KNO_3
- Melting temperature 430°F
- Decomposing temperature 1,100°F
- Energy storage density two thirds of water
- Density two times that of water
- Salt inventory 3.3 million pounds

Steam Generator

- Supplier ABB Lummus
- New salt-in-shell superheater
- New slat-in-tube kettle boiler
- New salt-in-shell preheater

Turbine-Generator

- Supplier General Electric Company
- Refurbished from Solar I plant
- 10 MWe net
- 12 MWe gross

(Source: U.S. Department of Energy and Southern California Edison Company.[2])

TABLE 9-3

Solar II Operating Features

Thermodynamic Cycle	Electrical Power Generator
Hot salt temperature 1,050°F	Capacity 10 MWe
Cold salt temperature 550°F	Capacity factor 20%
Steam temperature 1,000°F	Overall solar-electric efficiency 16%
Steam pressure 1,450 psi	Cost of conversion from Solar I $40 M
Receiver salt flow rate 800,000 lbs/hour	
Steam generator flow rate 660,000 lbs/hour	

(Source: U.S. Department of Energy and Southern California Edison Company.)

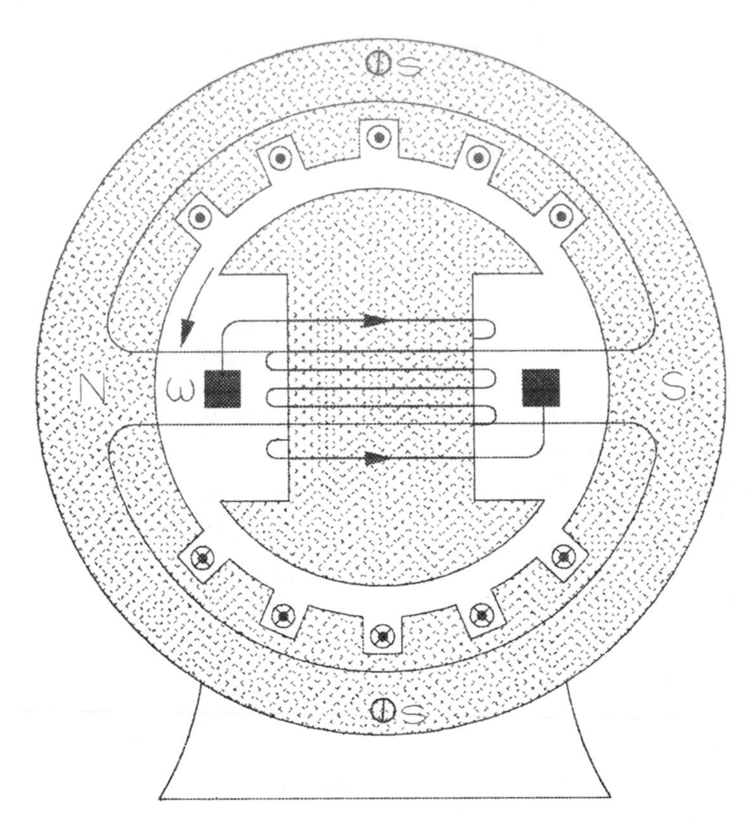

FIGURE 9-5
Cross section view of the synchronous generator.

in synchronism. However, a large sudden disturbance such as a step load can force the machine out of the synchronism, as discussed in Section 9.3.4.

9.3.1 Equivalent Electrical Circuit

The equivalent electricity circuit of the synchronous machine can be represented by a source of alternating voltage E and an internal series resistance R_s and reactance X_s representing the stator winding. The resistance, being much smaller than the reactance, can be ignored to reduce the equivalent circuit to a simple form shown in Figure 9-6. If the machine is supplying the load current I lagging the terminal voltage V by phase angle ϕ, it must internally generate the voltage E, which is the phasor sum of the terminal voltage and the internal voltage drop IX_s. The phase angle between the V and E is called the power angle δ. At zero power output, load current is zero and so is the IX_s vector, making V and E in phase having zero power angle. Physically, the power angle represents the angle by which the rotor position lags the stator-induced rotating magnetic field. The output power can be

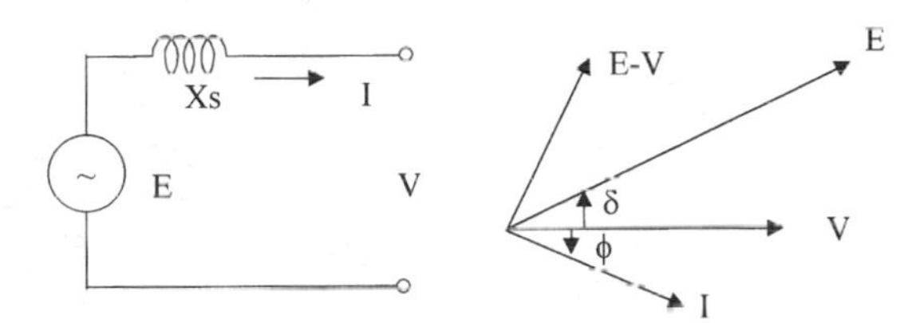

FIGURE 9-6
Equivalent electrical circuit and phasor diagram of synchronous machine.

increased by increasing the power angle up to a certain limit, beyond which the rotor would no longer follow the stator field and will step out of the synchronous mode of operation. In the nonsynchronous mode, it cannot produce steady power.

9.3.2 Excitation Methods

The synchronous machine excitation system must be designed to produce the required magnetic field which is controllable to control the voltage and the reactive power of the system. In modern high power machines, X_s can be around 1.5 units of the base impedance of the machine. With reactance of this order, the phasor diagram of Figure 9-6 can show that the rotor filed excitation required at rated load (100 percent load at 0.8 lagging power factor) is more than twice that at no load with the same terminal voltage. The excitation system has the corresponding current and voltage ratings, with capability of varying the voltage over a wide range of 1 to 3 or even more without undue saturation in the magnetic circuit. The excitation power, primarily to overcome the rotor winding I^2R loss, ranges from ½ to 1 percent of the generator rating. Most excitation systems operate at 200 to 1,000 Vdc.

For large machines, three types of excitation systems — DC, AC and static — are possible. In the DC system, a suitably designed DC generator supplies the main field winding excitation through conventional slip rings and brushes. Due to low reliability and high maintenance requirement, the conventional DC machine is seldom used in the synchronous machine excitation system.

Most utility scale generators use the AC excitation system shown in Figure 9-7. The main exciter is excited by a pilot exciter. The AC output of a permanent magnet pilot exciter is converted into DC by a floor standing rectifier and supplied to the main exciter through slip rings. The main exciter's AC output is converted into DC by means of phase controlled rectifiers, whose firing angle is changed in response to the terminal voltage variations. After filtering the ripples, this direct current is fed to the synchronous generator field winding.

An alternative scheme is the static excitation, as opposed to the dynamic excitation described in the preceding paragraph. In the static excitation

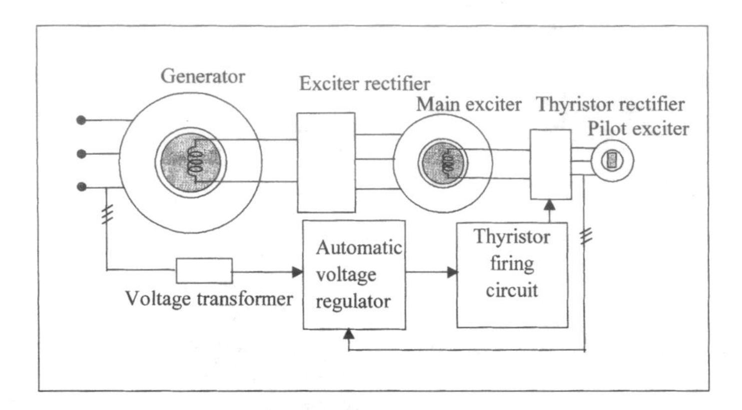

FIGURE 9-7
AC excitation system for synchronous generator.

scheme, the controlled DC voltage is obtained from a suitable stationary AC source rectified and filtered. The DC voltage is then fed to the main field winding through slip rings. This excitation scheme has a fast dynamic response and is more reliable because it has no rotating exciters.

The excitation control system modeling for analytical studies must be carefully done as it forms a multiple feedback control system that can become unstable. The IEEE has developed industry standards for modeling the excitation systems. The model enters nonlinearly due to magnetic saturation present in all practical designs. The stability can be improved by supplementing the main control signal by auxiliary signals such as speed and power.

9.3.3 Electrical Power Output

The electrical power output of the synchronous machine is as follows:

$$P = VI \cos \phi \tag{9-3}$$

Using the phasor diagram of Figure 9-6, the current can be expressed as follows:

$$I = \frac{E - V}{jXs} = \frac{E\angle\delta - V\angle o}{jXs} = \frac{E(\cos\delta + j\sin\delta) - V}{jXs} \tag{9-4}$$

The real part of this current is $I_{real} = \dfrac{E \cdot sin\delta}{X_s}$.

This part, when multiplied with the terminal voltage V gives the output power:

$$P = \frac{VE}{Xs}\sin\delta \tag{9-5}$$

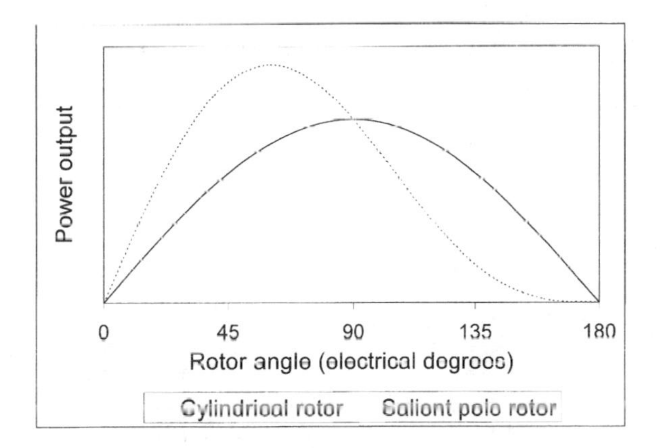

FIGURE 9-8
Power versus power angle of round rotor and salient pole synchronous machine.

The output power versus the power angle is a sine curve shown by the solid line in Figure 9-8, having the maximum value at $\delta = 90°$. The maximum power that can be generated by the machine is therefore:

$$P_{max} = \frac{VE}{Xs} \tag{9-6}$$

Some synchronous machine rotors have magnetic saliency in the pole structure. The saliency produces a small reluctance torque superimposed on the main torque, modifying the power angle curve as shown by the dotted line.

The electromechanical torque required at the shaft to produce this power is the power divided by the angular velocity of the rotor. That is as follows:

$$T_e = \frac{VE}{\omega Xs}\sin\delta \tag{9-7}$$

The torque also has the maximum limit corresponding to the maximum power limit, and is given as follows:

$$T_{max} = \frac{VE}{\omega Xs} \tag{9-8}$$

9.3.4 Transient Stability Limit

The maximum power limit just described is called the steady state stability limit. Any loading beyond this value will cause the rotor to lose synchronism, and hence, the power generation capability. The steady state limit must not

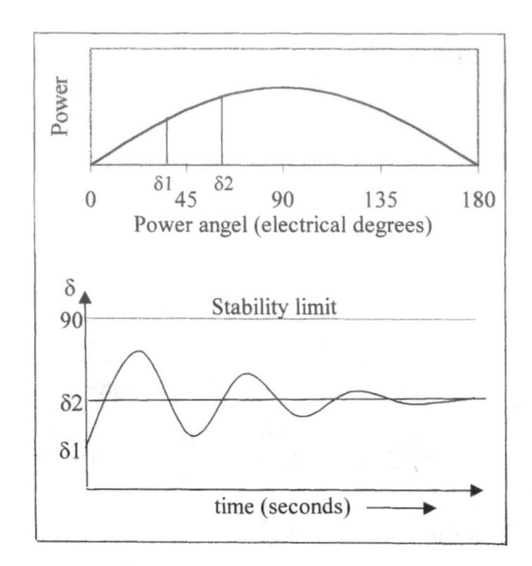

FIGURE 9-9
Load step transient and stability limit of synchronous machine.

be exceeded under any condition, including those that can be encountered during transients. For example, if a sudden load step is applied to the machine initially operating in a steady state at load power angle δ_1 (Figure 9-9), the rotor power angle would increase from δ_1 to δ_2 corresponding to the new load that it must supply. This takes some time depending on the electromechanical inertia of the machine. No matter how short or long it takes, the rotor inertia and the electromagnetic restraining torque will set the rotor in a mass-spring type of oscillatory mode, swinging the rotor power angle beyond its new steady state value. If the power angle exceeds 90° during this swing, the machine stability and the power generation are lost. For this reason, the machine can be loaded only to the extent that even under the worst-case step load, planned or accidental, or during all possible faults, the power angle swing will remain below 90° with sufficient margin. This limit on loading the machine is called the transient stability limit.

Equation 9-5 shows that the stability limit at given voltages can be increased by designing the machine with low synchronous reactance X_s, which is largely made of the stator armature reaction component.

9.4 Commercial Power Plants

The commercial power plants using the solar thermal system are being explored in a few hundred MWe capacity. Based on the Solar II power plant operating experience, the design studies made by the National Renewable Energy Laboratory for the U.S. Department of Energy have estimated the

TABLE 9-4

Comparison of 10 MWe Solar II and 100 MWe Prototype Design

Performance Parameter	Solar II Plant 10 MWe	Commercial Plant 100 MWe
Mirror reflectivity	90%	94%
Field efficiency	73%	73%
Mirror cleanliness	95%	95%
Receiver efficiency	87%	87%
Storage efficiency	99%	99%
Electromechanical conversion efficiency of generator	34%	43%
Auxiliary components efficiency	90%	93%
Overall solar to electric conversion efficiency	16%	23%

(Source: U.S. Department of Energy and Southern California Edison Company.)

performance parameters that are achievable for a 100 MWe commercial plant. Table 9-4 summarizes these estimates and compares with those achieved in the 10 MWe experimental Solar II power plant. The 100 MWe prototype design shows that the overall solar to electric efficiency of 23 percent can be achieved in a commercial plant using the existing technology. For comparison, the conventional coal thermal plants typically operate at 40 percent conversion efficiency, and the photovoltaic power systems have the overall solar-to-electricity conversion efficiency of 8 to 10 percent with amorphous silicon and 15 to 20 percent with crystalline silicon technologies.

Major conclusions of the studies to date are the following:

1. First plants as large as 100–200 MWe are possible to design and build based on the demonstrated technology to date. Future plants could be larger.
2. The plant capacity factors up to 65 percent are possible, including outage.
3. Fifteen percent annual average solar-to-electric conversion efficiency is achievable.
4. The energy storage feature of the technology makes possible to meet the peak demand on the utility lines.
5. Leveled energy cost is estimated to be 6 to 7 cents per kWh.
6. A 100 MWe plant with a capacity factor of 40 percent requires 1.5 square miles of land.
7. The capital cost of $2,000 per kWe capacity for first few commercial plants and less for future plants.
8. A comparable combined cycle gas turbine plant would cost $1,000 kWe, which includes no fuel cost.
9. Solar-fossil hybrids are the next step in development of this technology.

Compared to the pv and wind power, the solar thermal power technology is less modular. Its economical size is estimated to be in the 100 to 300 MWe range. The cost studies at the National Renewable Energy Laboratory have shown that a commercially designed utility-scale power plant using the central receiver power tower can produce electricity at a cost of 6 to 11 cents per kWh.

In other countries, the low temperature solar thermal power finds growing applications. The high temperature applications, however, have yet to develop on a large scale. An integrated combined cycle 40 MWe solar thermal with 100 MWe gas turbine power plant is proposed at Jodhpur in Rajasthan, India. According to the Indian Renewable Energy Development Agency, India has the target to install 150 MWe solar thermal and photovoltaic capacity by the year 2002, and 1,500 MWe by 2012.

References

1. Mancini, T. 1994. "Solar thermal power today and tomorrow," *Mechanical Engineering, publication of the Institution of Mechanical Engineers,* London, August 1994.
2. Southern California Edison Company. 1995. *Technology Report No. 14,* Irwindale, CA, Fall 1995.

10

Energy Storage

Electricity is more versatile in use because it is a highly ordered form of energy that can be converted efficiently into other forms. For example, it can be converted into mechanical form with efficiency near 100 percent or into heat with 100 percent efficiency. The heat energy, on the other hand, cannot be converted into electricity with high efficiency, because it is a disordered form of energy in atoms. For this reason, the overall thermal to electrical conversion efficiency of a typical fossil thermal power plant is under 40 percent.

A disadvantage of electricity is that it cannot be easily stored on a large scale. Almost all electrical energy used today is consumed as it is generated. This poses no hardship in conventional power plants, where the fuel consumption is varied with the load requirements. The photovoltaic and wind, being intermittent sources of power, cannot meet the load demand all of the time, 24 hours a day, 365 days of the year. The energy storage, therefore, is a desired feature to incorporate with renewable power systems, particularly in stand-alone plants. It can significantly improve the load availability, a key requirement for any power system.

The present and future energy storage technology that may be considered for stand-alone photovoltaic or wind power systems falls in the following broad categories:

- electrochemical battery.
- flywheel.
- compressed air.
- superconducting coil.

10.1 Battery

The battery stores energy in the electrochemical form, and is the most widely used device for energy storage in a variety of applications. The electrochemical energy is a semi-ordered form of energy, which is in between the electrical and thermal forms. It has one-way conversion efficiency of 85 to 90 percent.

There are two basic types of electrochemical batteries:

- the primary battery, which converts the chemical energy into the electrical energy. The electrochemical reaction in the primary battery is nonreversible, and the battery after discharge is discarded. For this reason, it finds applications where high energy density for one time use is needed.
- the secondary battery, which is also known as the rechargeable battery. The electrochemical reaction in the secondary battery is reversible. After a discharge, it can be recharged by injecting direct current from an external source. This type of battery converts the chemical energy into electrical energy in the discharge mode. In the charge mode, it converts the electrical energy into chemical energy. In both the charge and the discharge modes, a small fraction of energy is converted into heat, which is dissipated to the surrounding medium. The round trip conversion efficiency is between 70 and 80 percent.

The internal construction of a typical electrochemical cell is shown in Figure 10-1. It has positive and negative electrode plates with insulating separators and a chemical electrolyte in-between. The two groups of electrode plates are connected to two external terminals mounted on the casing. The

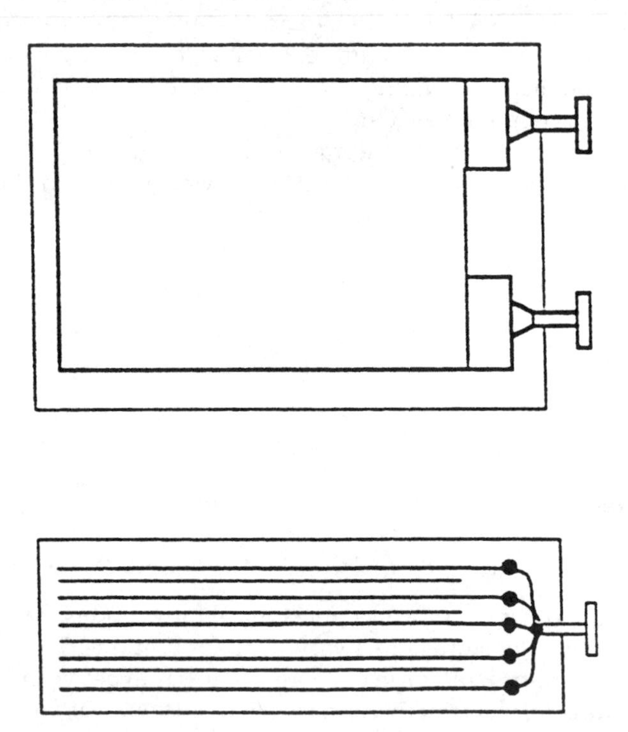

FIGURE 10-1
Electrochemical energy storage cell construction.

cell stores electrochemical energy at low electrical potentials, typically a few volts. The cell capacity, denoted by C, is measured in Ampere-hours (Ah), meaning it can deliver C amperes for one hour or C/n amperes for n hours.

The battery is made of numerous electrochemical cells connected in a series-parallel combination to obtain the desired operating voltage and current. The higher the battery voltage, the higher the number of cells required in series. The battery rating is stated in terms of the average voltage during discharge and the Ah capacity it can deliver before the voltage drops below the specified limit. The product of the voltage and the Ah forms the Wh energy rating it can deliver to a load from the fully-charged condition. The battery charge and discharge rates are stated in unit of its capacity in Ah. For example, charging a 100 Ah battery at C/10 rate means charging at 10 A rate. Discharging that battery at C/2 rate means draining 50 A, at which rate the battery will be fully discharged in 2 hours. The State of Charge (SOC) of the battery at any time is defined as the following:

$$SOC = \frac{\text{Ah capacity remaining in the battery}}{\text{Rated Ah capacity}}$$

10.2 Types of Batteries

There are at least six major rechargeable electrochemistries available today. They are as follows:

- lead-acid (Pb-acid).
- nickel-cadmium (NiCd).
- nickel-metal hydride (NiMH).
- lithium-ion (Li-ion).
- lithium-polymer (Li-poly).
- zinc-air.

New electrochemistries are being developed by the United States Advance Battery Consortium for a variety of applications, such as electric vehicles, spacecraft, utility load leveling and, of course, for renewable power systems.[1]

The average voltage during discharge depends on the electrochemistry, as listed in Table 10-1. The energy density of various batteries, as measured by Wh capacity per unit mass and per unit volume, are compared in Figure 10-2. The selection of the electrochemistry for a given application is a matter of performance and cost optimization.

Some construction and operating features of the above electrochemistries are presented in the proceeding sections.

TABLE 10-1

Average Cell Voltage During Discharge in Various Rechargeable Batteries

Electrochemistry	Cell Voltage	Remark
Lead-acid	2.0	Least cost technology
Nickel-cadmium	1.2	Exhibits memory effect
Nickel-metal hydride	1.2	Temperature sensitive
Lithium-ion	3.4	Safe, contains no metallic lithium
Lithium-polymer	3.0	Contains metallic lithium
Zinc-air	1.2	Requires good air management to limit self-discharge rate

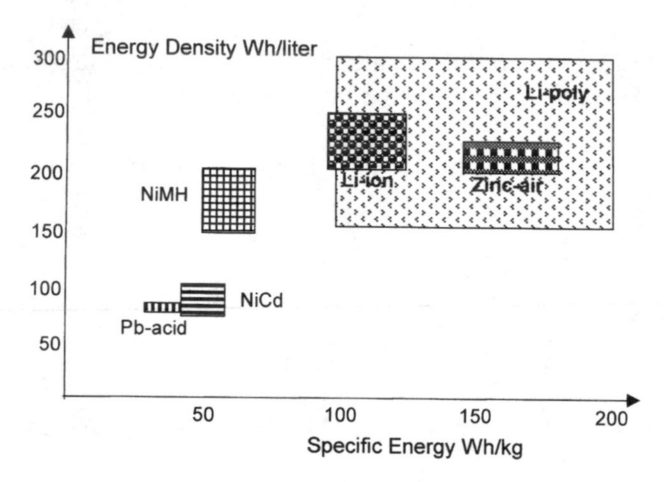

FIGURE 10-2
Specific energy and energy density of various electrochemistries.

10.2.1 Lead-Acid

This is the most common type of rechargeable battery used today because of its maturity and high performance over cost ratio, even though it has the least energy density by weight and volume. In the lead-acid battery under discharge, water and lead sulfate are formed, the water dilutes the sulfuric acid electrolyte, and the specific gravity of the electrolyte decreases with the decreasing state of charge. The recharging reverses the reaction in the lead and lead dioxide is formed at the negative and positive plates, respectively, restoring the battery into its originally charged state.

The lead-acid battery comes in various versions. The shallow-cycle version is used in automobiles where a short burst of energy is drawn from the battery when needed. The deep-cycle version, on the other hand, is suitable for repeated full charge and discharge cycles. Most energy storage applications require deep-cycle batteries. The lead-acid battery is also available in sealed 'gel-cell' version with additives which turns the electrolyte into a non-spillable gel. Such batteries can be mounted sideways or upside down. The high cost, however, limits its use in the military avionics.

10.2.2 Nickel Cadmium

The NiCd is a matured electrochemistry. The NiCd cell has positive electrodes made of cadmium and the negative electrodes of nickel hydroxide. Two electrodes are separated by Nylon® (registered trade name of E. I. Dupont Numerous and Company, Wilmington, Delaware) separators and potassium hydroxide electrolyte in stainless steel casing. With sealed cell and half the weight of conventional lead-acid, NiCd batteries have been used to power most rechargeable consumer applications. They have a longer deep cycle life, and are more temperature tolerant than the lead-acid batteries. However, this electrochemistry has a memory effect (explained later), which degrades the capacity if not used for a long time. Moreover, cadmium has recently come under environmental regulatory scrutiny. For these reasons, the NiCd is being replaced by NiMH and Li-ion batteries in laptop computers and other similar high-priced consumer electronics.

10.2.3 Nickel-Metal Hydride

The NiMH is an extension of NiCd technology, and offers an improvement in energy density over that in NiCd. The major construction difference is that the anode is made of a metal hydride. This eliminates the environmental concerns of cadmium. Another performance improvement is that it has negligible memory effect. The NiMH, however, is less capable of delivering high peak power, has high self-discharge rate, and is susceptible to damage due to overcharging. Compared to NiCd, NiMH is expensive at present, although the future price is expected to drop significantly. This expectation is based on current development programs targeted for large-scale application of this technology in electric vehicles.

10.2.4 Lithium-Ion

Lithium-ion technology is a new development, which offers three times the energy density over that of lead-acid. Such large improvement in the energy density comes from lithium's low atomic weight of 6.9 versus 207 for lead. Moreover, the lithium-ion has higher cell voltage of 3.5 versus 2.0 for lead-acid and 1.2 for other electrochemistries. This requires fewer cells in series for a given battery voltage, thus reducing the manufacturing cost.

On the negative side, the lithium electrode reacts with any liquid electrolyte, creating a sort of passivation film. Every time when the cell is discharged and then charged, the lithium is stripped away, a free metal surface is exposed to the electrolyte and a new film is formed. To compensate, the cell uses thick electrodes, adding into the cost. Or else, the life would be shortened. For this reason, it is more expensive than NiCd.

In operation, the lithium-ion electrochemistry is vulnerable to damage from overcharging or other shortcomings in the battery management. Therefore, it requires more elaborate charging circuitry with adequate protection against overcharging.

10.2.5 Lithium-Polymer

This is a lithium battery with solid polymer electrolytes. It is constructed with a film of metallic lithium bonded to a thin layer of solid polymer electrolyte. The solid polymer enhances the cell's specific energy by acting as both the electrolyte and the separator. Moreover, the metal in solid electrolyte reacts less than it does with liquid electrolyte.

10.2.6 Zinc-Air

The zinc-air battery has a zinc negative electrode, a potassium hydroxide electrolyte, and a carbon positive electrode, which is exposed to the air. During discharge, oxygen from the air is reduced at the carbon electrode (the so-called air cathode), and the zinc electrode is oxidized. During discharge, it absorbs oxygen from the air and converts into oxygen ions for transport to the zinc anode. During charge, it evolves oxygen. A good air management is essential for the performance of the zinc-air battery.

10.3 Equivalent Electrical Circuit

For steady-state electrical performance calculations, the battery is represented by an equivalent circuit shown in Figure 10-3. In its simplest form, the battery works as a constant voltage source with small internal resistance. The open-circuit (or electrochemical) voltage E_i of the battery decreases linearly with the Ah of discharge Q_d, and the internal resistance R_i increases linearly with Q_d. That is, the battery open-circuit voltage is lower and the internal resistance is higher in a partially discharged state as compared to the E_o and R_o values in the fully charged state. Quantitatively,

$$E_i = E_0 - K_1 Q_d$$
$$R_i = R_0 + K_2 Q_d \qquad (10\text{-}1)$$

where K_1 and K_2 are constants to be found by curve-fitting the test data.

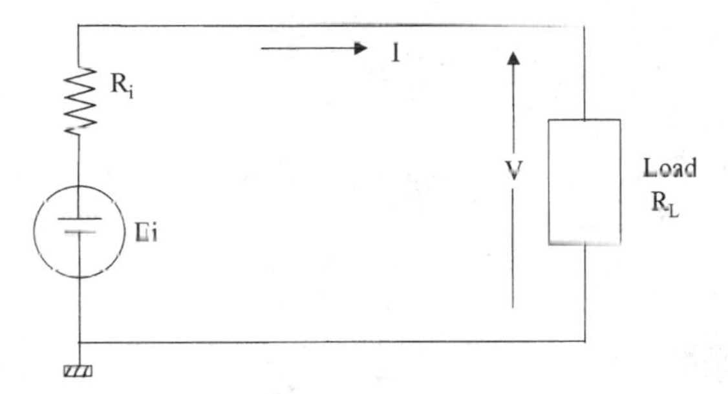

FIGURE 10-3
Equivalent electrical circuit of the battery showing internal voltage and resistance.

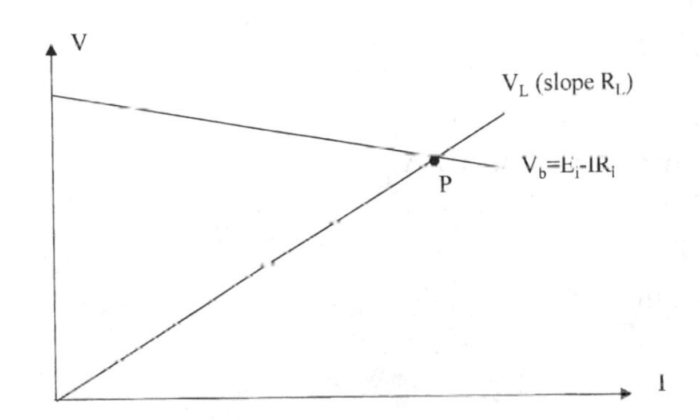

FIGURE 10-4
Battery source line intersecting with load line at the operating point.

The terminal voltage drops with increasing load as shown by the V_b line in Figure 10-4, and operating point is the intersection of the source line and the load line (point P). The power delivered to the external load resistance is I^2R_L.

In fast discharge applications, such as for starting a heavily loaded motor, the battery may be required to deliver the maximum possible power for a short time. The peak power it can deliver can be derived using the maximum power transfer theorem in electrical circuits. It states that the maximum power can be transferred from the source to the load when the internal impedance of the source equals the conjugate of the load impedance. The battery can deliver maximum power to a DC load when $R_L = R_i$. This gives the following:

$$P_{max} = \frac{E_i^2}{4R_i} \tag{10-2}$$

Since E_i and R_i vary with the state of charge, the P_{max} also vary accordingly. The internal loss is I^2R_i. The efficiency at any state of charge is therefore:

$$\eta = \frac{R_L}{R_L + R_i} \tag{10-3}$$

The efficiency decreases as the battery is discharged, thus generating more heat.

10.4 Performance Characteristics

The basic performance characteristics of the battery which influence the design are as follows:

- charge/discharge voltages.
- charge/discharge ratio (c/d ratio).
- round trip energy efficiency.
- charge efficiency.
- internal impedance.
- temperature rise.
- life in number of c/d cycles.

10.4.1 Charge/Discharge Voltages

The cell voltage variation during a typical charge/discharge cycle is shown in Figure 10-5 for cell with nominal voltage of 1.2 V, such as NiMH and NiCd. The voltage is maximum when the cell is fully charged (state of

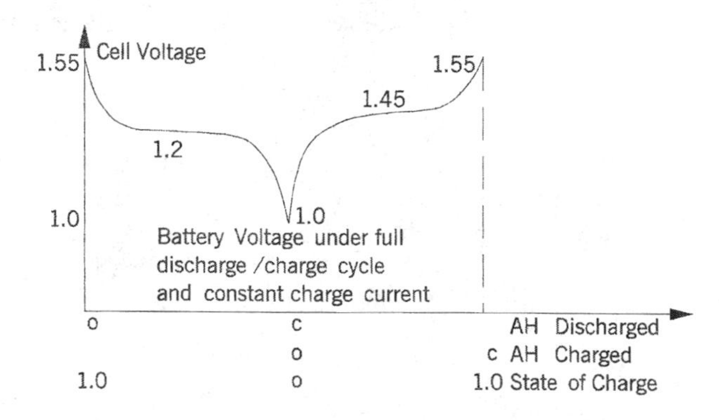

FIGURE 10-5
Voltage variation during charge/discharge cycle of nickel-cadmium cell with nominal voltage of 1.2 volt.

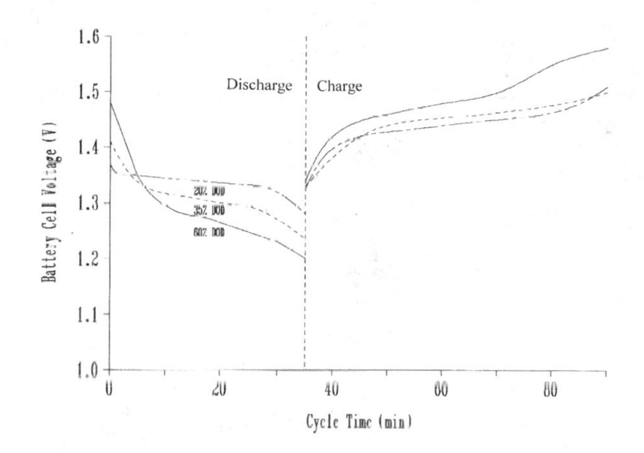

FIGURE 10-6
Cell voltage curves at different charge/discharge rates.

charge = 1.0, or Ah discharged – 0). As the cell is discharged, the cell voltage (Vc) drops quickly to a plateau value of 1.2, which holds for a long time before dropping to 1.0 at the end of its capacity (SOC = 0). In the reverse, when the cell is recharged, the voltage quickly rises to a plateau value of 1.45 and then reaches a maximum value of 1.55 volts. The charge/discharge characteristic also depends on how fast the battery is charged and discharged (Figure 10-6).

10.4.2 Charge/Discharge Ratio

After discharging certain Ah to load, the battery requires more Ah of charge to restore the full state of charge. The charge/discharge ratio is defined as the Ah input over the Ah output with no net change in the state of charge. This ratio depends on the charge and discharge rates and also on temperature, as shown in Figure 10-7. At 20°C, for example, the charge/discharge ratio is 1.1, meaning the battery needs 10 percent more Ah than what was discharged for restoring to fully charged state.

10.4.3 Energy Efficiency

The energy efficiency over a round trip of full charge and discharge cycle is defined as the ratio of the energy output over the energy input at the electrical terminals of the battery. For a typical battery of capacity C with an average discharge voltage of 1.2 V, average charge voltage of 1.45 V and the charge/discharge ratio of 1.1, the efficiency is calculated as follows:

$$\text{The energy output over the full discharge} = 1.2 \times C$$

$$\text{The energy input required to restore full charge} = 1.45 \times 1.1\ C$$

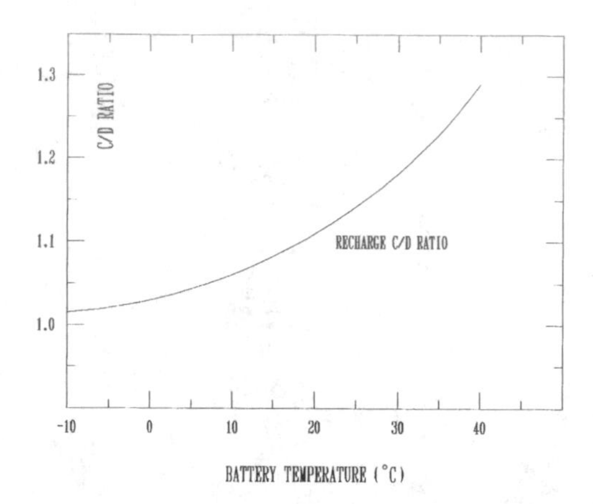

FIGURE 10-7
Temperature effect on charge/discharge ratio.

Therefore, the round trip energy efficiency is as follows:

$$\eta_{energy} = \frac{1.2 \cdot C}{1.45 \cdot 1.1 \cdot C} = 0.75 \text{ or } 75 \text{ percent}$$

10.4.4 Internal Resistance

The above efficiency calculations indicate that 25 percent energy is lost per charge/discharge cycle, which is converted into heat. This characteristic of the battery can be seen as having an internal resistance R_i. The value of R_i is a function of the battery capacity, operating temperature and the state of charge. The higher the cell capacity, the larger the electrodes and the lower the internal resistance. The R_i varies with the state of charge as per Equation 10-1. It also varies with temperature as shown in Figure 10-8, which is for a high quality 25 Ah NiCd cell.

10.4.5 Charge Efficiency

The charge efficiency is defined as the ratio of the Ah being deposited internally between the plates over that delivered to the external terminals during the charging process. It is different from the energy efficiency. The charge efficiency is almost 100 percent when the cell is empty of charge, the condition in which it converts all Ah received into useful electrochemical energy. As the state of charge approaches one, the charge efficiency tapers down to zero. The knee point where the charge efficiency starts tapering off depends on the charge rate (Figure 10-9). For example, at C/2 charge rate, the charge efficiency is 100 percent up to about 75 percent SOC. At a fast

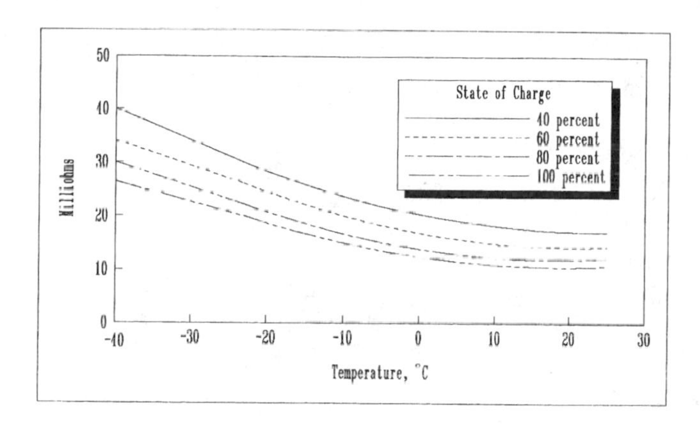

FIGURE 10-8
Temperature effect on internal resistance in 25 Ah nickel-cadmium cell.

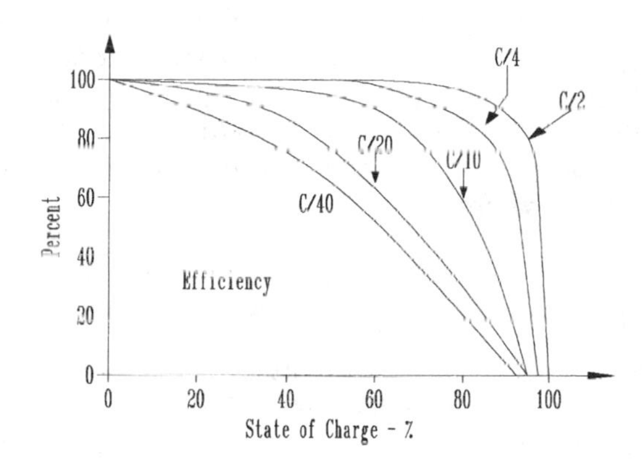

FIGURE 10-9
Charge efficiency versus state-of-charge at various charge rates.

charge rate of C/40, on the other hand, the charge efficiency at 60 percent SOC is only 50 percent.

10.4.6 Self-Discharge and Trickle Charge

Even with open-circuit terminals, the battery slowly self-discharges. In order to maintain the full state of charge, it is continuously trickle-charged to counter the self-discharge rate. This rate is usually less than one percent per day for most electrochemistries in normal working conditions.

After the battery is fully charged, the charge efficiency drops to zero. Any additional charge will be converted into heat. If overcharged at higher rate than the self-discharge rate for an extended period, the battery would overheat posing a safety hazard of potential explosion. For this reason, battery

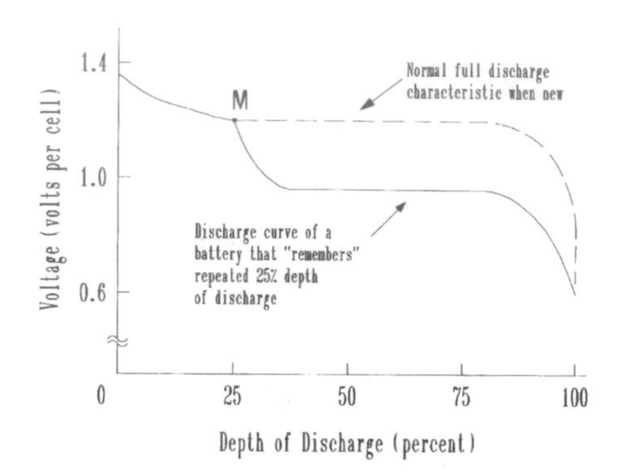

FIGURE 10-10
Memory effect degrades discharge voltage in nickel-cadmium cell.

charge should have a regulator, which cuts back the charge rate to the trickle-rate once the battery is fully charged. Any excessive overcharging will produce excessive gassing, which scrubs the electrode plates. Continuous scrubbing at high rate produces excessive heat, and wears out electrodes leading to shortened life. The trickle charging produces a controlled amount of internal gassing. It causes mixing action of the battery electrolyte, keeping it ready to deliver full charge.

10.4.7 Memory Effect

One major disadvantage of the NiCd battery is the memory effect. It is the tendency of the battery to remember the voltage at which it has delivered most of its capacity in the past. For example, if the NiCd battery is repeatedly charged and discharged 25 percent of its capacity to point M in Figure 10-10, it will remember point M. Subsequently, if the battery is discharged beyond point M, the cell voltage will drop much below its original normal value shown by the dotted line in Figure 10-10. The end result is the loss of full capacity after repeatedly using shallow discharge cycles. The phenomenon is like losing a muscle due to lack of use over a long time. A remedy for restoring the full capacity is "reconditioning", in which the battery is fully discharged to almost zero voltage once every few months and then fully charged to about 1.55 volts per cell. Other types of batteries do not have the memory effect.

10.4.8 Effects of Temperature

As seen in the preceding sections, the operating temperature significantly influences the battery performance as follows:

TABLE 10-2

Optimum Working Temperature Range for Nickel-Cadmium Battery

Operating Temperature °C	Charge Efficiency Percent	Discharge Efficiency Percent	Self-discharge Rate % Capacity/Day
–40	0	72	0.1
–35	0	80	0.1
–30	15	85	0.1
–25	40	90	0.2
–20	75	95	0.2
–15	85	97	0.2
–10	90	100	0.2
–5	92	100	0.2
0	93	100	0.2
5	94	100	0.2
10	94	100	0.2
15	94	100	0.3
20	93	100	0.4
25	92	100	0.6
30	91	100	1.0
35	90	100	1.4
40	88	100	2.0
45	85	100	2.7
50	82	100	3.6
55	79	100	5.1
60	75	100	8.0
65	70	100	12
70	60	100	20

- the capacity and the charge efficiency decrease with increasing temperature. The capacity drops at temperatures above or below certain range, and drops sharply at temperatures below freezing.
- the self discharge rate increases with temperature.
- the internal resistance increases as temperature decreases.

Table 10-2 shows the influence of temperature on the charge efficiency, the discharge efficiency, and the self-discharge rate in the NiCd battery. The process of determining the optimum operating temperature is also indicated in the table. It is seen that different attributes have different desirable operating temperature ranges. With all attributes jointly considered, the most optimum operating temperature is the intersection of all the desirable ranges. If we wish to limit the self discharge rate below 1 percent, and the charge efficiency above 90 percent, Table 10-2 indicates that the optimum working temperature range is between –10°C and 25°C.

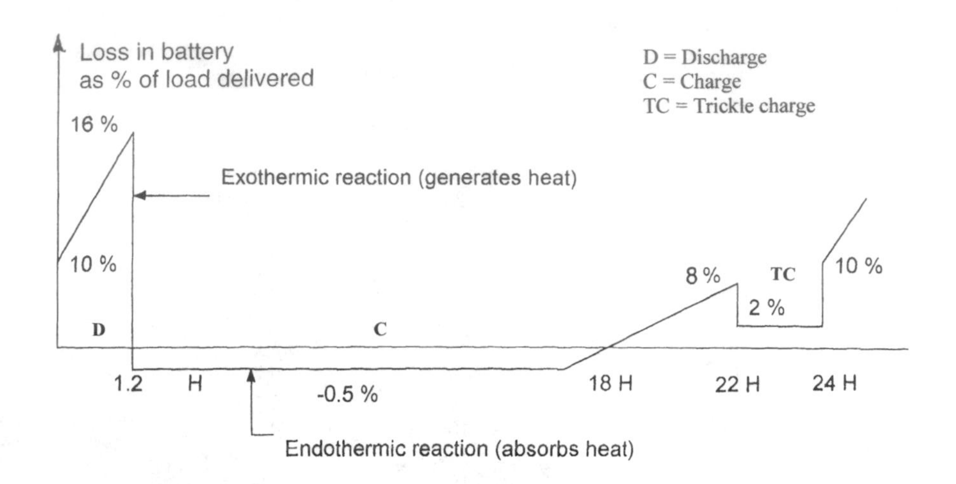

FIGURE 10-11
Internal energy loss in battery during charge/discharge cycle showing endothermic and exothermic periods.

10.4.9 Internal Loss and Temperature Rise

The battery temperature varies over the charge/discharge cycle. Taking NiCd as an example, the heat generated in one such cycle with 1.2 hours of discharge and 20.8 hours of charge every day is shown in Figure 10-11. Note that the heat generation increases as the depth of discharge increases because of the increased internal resistance. When the battery is put to charge, the heat generation is negative for a while, meaning that the electrochemical reaction during the initial charging period is endothermic (absorbing heat), as opposed to the exothermic reaction during other periods with positive heat generation. The temperature rise during the cycle depends on the cooling method used to dissipate the heat by conduction, convection and radiation.

Different electrochemistries, however, generate internal heat at different rates. The heat generation between various batteries can be meaningfully compared in terms of the adiabatic temperature rise during discharge, which is given by the following relation:

$$\Delta T = \frac{WH_d}{M \cdot C_p} \left[1 - \eta_v + \frac{E_d}{E_o} \right] \tag{10-4}$$

where ΔT = adiabatic temperature rise of the battery, °C
WH_d = watt-hour energy discharged
C_p = battery specific heat, Wh/kg.C
η_v = voltage efficiency factor on discharge

TABLE 10-3

Battery Characteristics Affecting Thermal Design

Electrochemistry	Operating temperature range °C	Overcharge tolerance	Heat capacity Wh/kg-K	Mass density kg/liter	Entropic heating on discharge W/A-cell
Lead-acid	−10 to 50	High	0.35	2.1	−0.06
Nickel-cadmium	−20 to 50	Medium	0.35	1.7	0.12
Nickel-metal hydride	−10 to 50	Low	0.35	2.3	0.07
Lithium-ion	10 to 45	Very low	0.38	1.35	0
Lithium-polymer	50 to 70	Very low	0.40	1.3	0

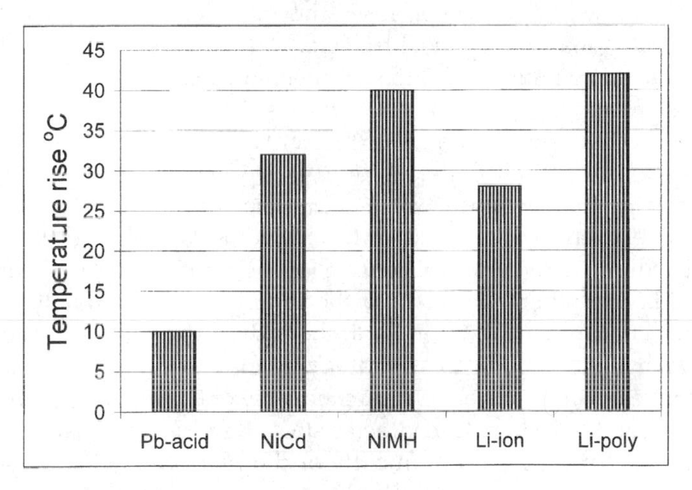

FIGURE 10-12
Adiabatic temperature rise for various electrochemistries.

E_d = average cell entropy energy per coulomb during discharge, i.e. average power loss per ampere of discharge, W/A

E_o = average cell open circuit voltage, volts

For full discharge, the WH_d/M ratio in the above expression becomes the specific energy. This indicates that higher specific energy cells would also tend to have higher temperature rise during discharge, requiring enhanced cooling design. Various battery characteristics affecting the thermal design are listed in Table 10-3. Figure 10-12 depicts the adiabatic temperature rise ΔT for various electrochemistries after a full discharge in short burst.

10.4.10 Random Failure

The battery fails when at least one cell fails. The cell failure is theoretically defined as the condition in which the cell voltage drops below certain low value before discharging rated capacity at room temperature. The low value is generally taken as a 1.0 volt in cells with nominal voltage of 1.2 V. This is a very conservative definition of a battery failure. In practice, if one cell shows less than 1.0 V, other cells can make up without detecting the failure at the battery level. Even if all cells show stable voltage below 1.0 V at full load, the load can be reduced to maintain the desired voltage for some time until the voltage degrades further.

The cell can fail open, short, or somewhere in between (a soft short). A short that starts soft eventually develops into a hard short. In a low voltage battery, any attempt to charge with a shorted cell may result in physical damage to the battery and/or the charge regulator. On the other hand, the shorted cell in a high voltage battery with numerous series connected cells may work forever. It, however, loses the voltage and Ah capacity, hence, would work as load on the healthy cells. An open cell, on the other hand, disables the entire battery of series-connected cells.

In a system having two parallel batteries, if one cell in one battery fails short, the two batteries would have different terminal characteristics. Charging or discharging such batteries as a group can result in highly uneven current sharing, subsequently overheating one of the batteries. Two remedies are available to avoid this. One is to charge and discharge both batteries with individual current controls such that they both draw their rated share of the load. The other is to replace the failed cell immediately, which can sometimes be impractical. In general, the individual charge/discharge control for each battery is the best strategy. It may also allow replacement of any one battery with different electrochemistry or different age, which would have different load sharing characteristics. Batteries are usually replaced several times during the economic life of the plant.

10.4.11 Wear-Out Failure

In addition to the random failure, the battery has the wear out failure mode. It is associated with the electrode wear due to repeated charge/discharge cycles. The number of times the battery can be discharged and recharged before the electrodes wear out depends on the electrochemistry. The battery life is measured in the number of charge/discharge cycles it can deliver before failure. The life depends strongly on the depth of discharge and the temperature as shown in Figure 10-13 for high quality NiCd batteries. The life also depends, to a lesser degree, on the electrolyte concentration and the electrode porosity and thickness. The former factors are application related, whereas the latter are construction related.

It is noteworthy from Figure 10-14 that the life at a given temperature is an inverse function of the depth of discharge. If the life is 100 units at 50 percent DoD, then it would be 200 units at 25 percent DoD. This makes the product of the cycles to fail and the DoD remain constant. This product decreases with increasing temperature. Such is true for most batteries. This

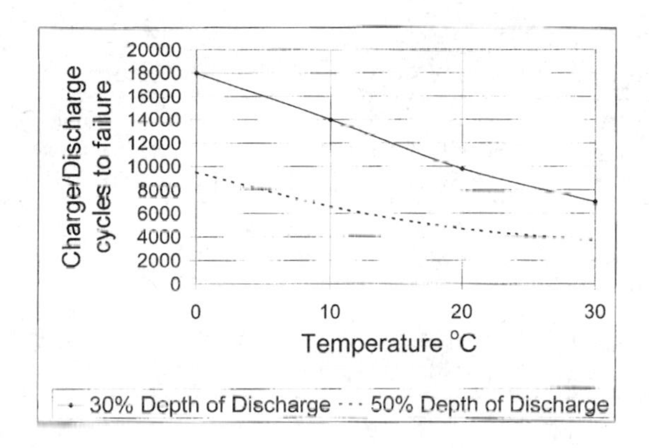

FIGURE 10-13
Charge/discharge cycle life of sealed nickel-cadmium battery versus temperature and depth of discharge.

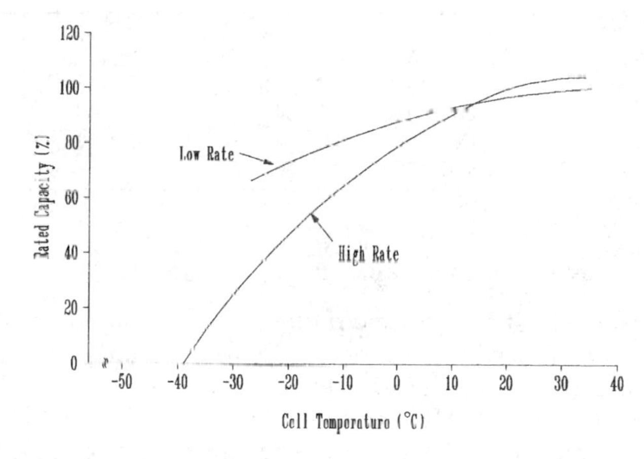

FIGURE 10-14
Lead-acid battery capacity variations with temperature.

means that the battery at a given temperature can deliver the same number of equivalent full cycles of energy regardless of the depth of discharge. The total Wh energy the battery can deliver over its life is approximately constant. Such observation is useful in comparing the costs of various batteries for a given application.

The life consideration is the dominant design driver in the battery sizing. Even when the load may be met with a smaller capacity, the battery is oversized to meet the life requirement as measured in number of charge/discharge cycles. For example, with the same Wh load, the battery that must charge/discharge twice as many cycles per period needs approximately double the capacity to have the same calendar life.

TABLE 10-4

Specific Energy and Energy Density of Various Batteries

Electrochemistry	Specific Energy Wh/kg	Energy Density Wh/liter	Specific Power W/kg	Power Density W/liter
Lead-acid	30–40	70–75	~200	~400
Nickel-cadmium	40–60	70–100	150–200	220–350
Nickel-metal hydride	50–65	140–200	~150	450–500
Lithium-ion	90–120	200–250	200–220	400–500
Lithium-polymer	100–200	150–300	>200	>350
Zinc-air	140–180	200–220	~150	~200

TABLE 10-5

Life and Cost Comparison of Various Batteries

Electrochemistry	Cycle life in full discharge cycles	Calendar life in years	Self discharge rate %/month at 25 °C	Relative cost $/kWh
Lead-acid	500–1000	5–8	3–5	200–500
Nickel-cadmium	1000–2000	10–15	20–30	1500
Nickel-metal hydride	1000–2000	8–10	20–30	2500
Lithium-ion	500–1000	—	5–10	3000
Lithium-polymer	500–1000	—	1-2	>3000
Zinc-air	200–300	—	4–6	—

10.4.12 Various Batteries Compared

The performance characteristics and properties of various electrochemistries presented in the preceding sections are summarized and compared in Tables 10-4 and 10-5. Note that the overall cost of the lead-acid battery is low compared to NiCd, NiMH and Li-ion batteries. Because of its least cost per Wh delivered over the life, the lead-acid battery has been the workhorse of the industry.[2]

10.5 More on Lead-Acid Battery

The lead-acid battery is available in small to large capacities in various terminal voltages, such as 6 V, 12 V and 24 V. Like in other batteries, the Ah capacity of the lead-acid battery is sensitive to temperature. Figure 10-14 shows the capacity variations with temperature for deep-cycle lead-acid batteries. At –20°F, for example, the high-rate battery capacity is about 20 percent of its capacity at 100°F. The car is hard to start in winter for this reason. On the other hand, the self-discharge rate decreases significantly at cold temperatures, as seen in Figure 10-15.

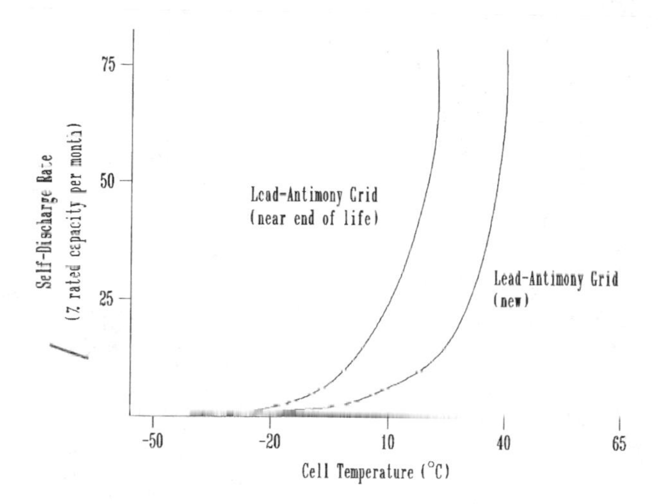

FIGURE 10-15
Lead-acid battery self-discharge rate versus temperature.

TABLE 10-6
Effects of SOC on Specific Gravity and Freezing Point of Lead-Acid Battery

State of Charge	Sp. Gravity	Freezing Point	120 Volts Battery Voltage
1 (Fully Charged)	1.27	-65°F	128
75 percent	1.23	-40°F	124
50 percent	1.19	-10°F	122
25 percent	1.15	+5°F	120
0 (Fully Discharged)	1.12	+15°F	118

Table 10-6 shows the effect of state-of-charge on the voltage, the specific gravity and the freezing point of the lead acid battery. The electrolyte in a fully charged battery has high specific gravity and freezes at -65°F. On the other hand, a fully discharged battery freezes at +15°F. The table shows the importance of keeping the battery fully charged in winters.

The cycle life versus DOD for the lead-acid battery is depicted in Figure 10-16, again showing half-life at double the depth of discharge.

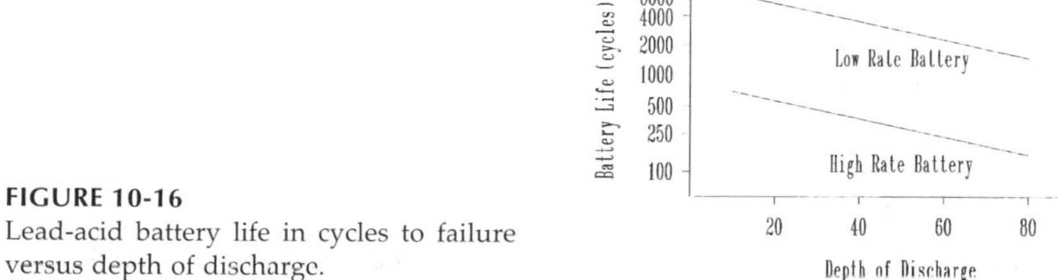

FIGURE 10-16
Lead-acid battery life in cycles to failure versus depth of discharge.

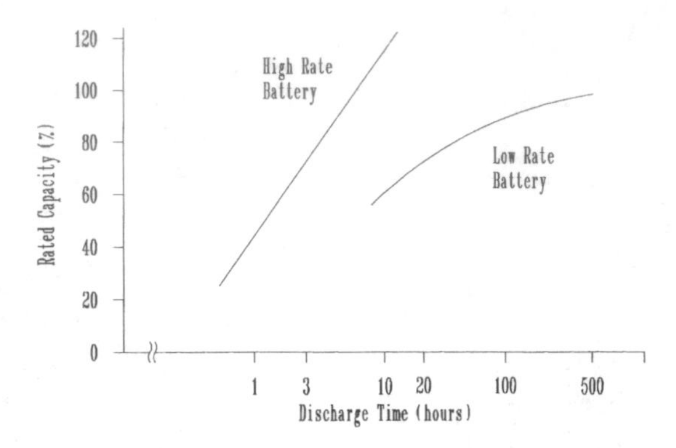

FIGURE 10-17
Lead-acid battery capacity with discharge time.

The discharge rate influences the lead-acid battery capacity, as shown in Figure 10-17. The shorter the discharge time (i.e., higher the discharge rate), the lower the Ah capacity the battery can deliver.

The lead-acid cell voltage is 2.0 volts nominal, and the internal resistance is around one milliohm per cell. The cycle life is 500-1,000 full charge discharge cycles for medium rate batteries. The operating temperature range is between –20 to 50°C and the survival temperature is –55 to 60°C.

10.6 Battery Design

The battery design for given application depends on the following system requirements:

- voltage and current.
- charge and discharge rates and duration.
- operating temperature during charge and discharge.
- life in number of charge and discharge cycles.
- cost, size, and weight constraints.

Once these system level design parameters are identified, the battery design proceeds in the following steps:

- select the electrochemistry suitable for the overall system requirements.
- determine the number of series cells required to meet the voltage requirement.
- determine the Ah discharge required to meet the load demand.

- for the required number of charge/discharge cycles, determine the maximum allowable depth of discharge.
- the total Ah capacity of the battery is then determined by dividing the Ah discharge required by the allowable depth of discharge calculated above.
- determine the number of battery packs required in parallel for the total Ah capacity.
- determine the temperature rise and the thermal controls required.
- provide the charge and discharge rate controls as needed.

Each cell in the battery pack is electrically insulated from each other and from the ground. The electrical insulation must be good conductor of heat to maintain low temperature gradient between the cells and also to the ground.

The battery performs better under slow charge and discharge rates. It accepts less energy when charged at a faster rate. Also the faster the discharge rate, the faster the voltage degradation and lower the available capacity to the load. For these reasons, high charge and discharge rate applications require different design considerations than the low rate applications.

10.7 Battery Charging

During the battery charging, the energy management software monitors the state-of-charge, the overall health, and the safe termination criteria. The operating parameters monitored are the voltage, the current, and the temperature. The charging timer is started after all initial checks are successfully completed. Charging may be suspended (but not reset) if it detects violation of critical safety criteria. The timer stops charging if the defect persists beyond a certain time limit.

The normal charging has the following three phases:

- bulk (fast) charge, which deposits 80 to 90 percent of the drained capacity.
- taper charge in which the charge rate is gradually cut back to top off the remaining capacity.
- trickle (float) charge after the battery is fully charged to counter the self-discharge rate.

The bulk charge and the taper charge termination criteria are preloaded in the battery management software to match with the battery electrochemistry and the system design parameters. For example, the NiCd and NiMH batteries are generally charged at constant current (Figure 10-18), terminating

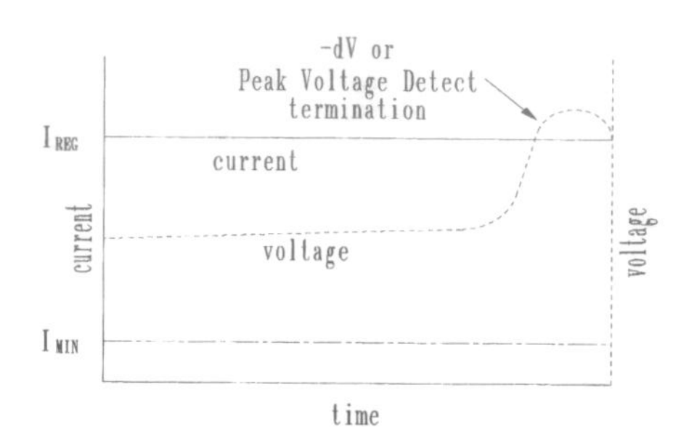

FIGURE 10-18
Constant current charging of nickel-cadmium and nickel-metal-hydride batteries.

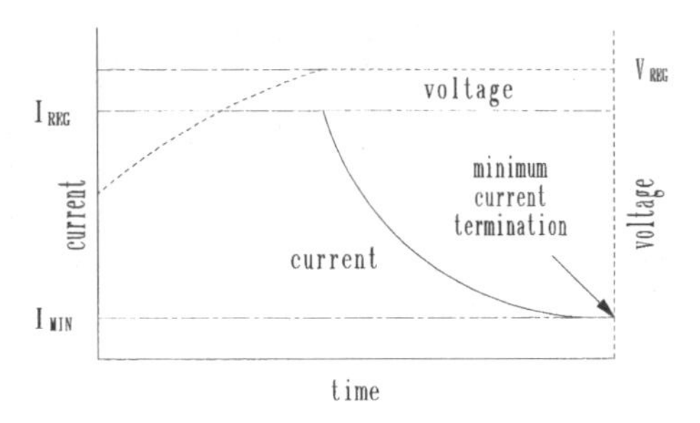

FIGURE 10-19
Constant voltage charging of lithium-ion battery.

the charging when the continuously monitored ΔV is detected negative. On the other hand, the Li-ion batteries, being sensitive to overcharging, are charged at constant voltage, tapering off the charge current as needed (Figure 10-19).

10.8 Charge Regulators

For safety reasons, it is extremely important that excessive charging of the battery is avoided at all times. Overcharging causes internal gassing, which causes loss of water in the lead-acid battery and premature aging. The charge regulator allows the maximum rate of charging until the gassing starts. Then the charge current is tapered off to trickle charge rate so that the full charge is approached gently.

The batteries are charged in the following three different manners.

10.8.1 Multiple Charge Rates

This is the best method, in which the battery is charged gently in multiple steps. First the battery is charged at full charge rate until 80 to 90 percent of the capacity is achieved. The charge current is then cut back, in steps, until the battery is fully charged. At this time, the charge current is further reduced to trickle-charge rate, keeping it fully charged until the next load demand comes on the battery. This method, therefore, needs at least three charge rates in the charge regulator design.

10.8.2 Single Charge Rate

This method uses a simple low cost regulator which is either on or off. The regulator is designed for only one charge rate. When the battery is fully charged, as measured by its terminal voltage, the charger is turned off by a relay. When the battery voltage drops below a preset value, the charger is again connected in full force. Since the charging is not gentle in this method, full charge is difficult to achieve and maintain. An alternate version of this charging method is the multiple pulse charging. Full current charges the battery up to the high preset voltage just below the gassing threshold. At this time, the charger is shut off for a short time to allow the battery chemicals to mix and voltage to fall. When the voltage falls below the low preset threshold, the charger is reconnected, again passing full current to the battery.

10.8.3 Unregulated Charging

This least cost method can be used in photovoltaic power systems. It uses no charge regulator. The battery is charged directly from a solar module dedicated just for charging. The charging module is properly designed for safe operation for the given number of cells in the battery. For example, in 12 volts lead-acid battery, the maximum photovoltaic module voltage is kept below 15 V, thus, making it difficult to overcharge. When the battery is fully charged, the array is fully shunted to ground by a shorting switch (transistor). The shunt transistor switch is open when the battery voltage drops below certain value. The isolation diode blocks the battery powering the array or the shunt at night, a discussed in Section 8.8.

10.9 Battery Management

Drawing the electrical power from the battery when needed, and charging it back when access power is available, requires a well-controlled charge and discharge process. Otherwise, the battery performance could suffer, the life shortened and the maintenance increased. Some common performance problems are as follows:

- low charge efficiency resulting in low state of charge.
- loss of capacity to hold the rated Ah charge.

- excessive gassing and heating, leading to a short life.
- unpredictable premature failure leading to loss of load availability.
- positive plate corrosion shortening the life.
- stratification and sulfation degrading the performance.

The following features incorporated in the battery management can avoid the above problems:

- controlled voltage charging, preferably at constant voltage.
- temperature-compensated charging, in that the charge termination occurs earlier if the battery temperature is higher than the reference temperature.
- individual charge control if two or more batteries are charged in parallel.
- accurate set points to start and to stop the charge and discharge modes.

10.9.1 Monitoring and Controls

The batteries in modern power systems are managed by dedicated computer software. The software monitors and controls the following performance parameters:

- voltage and current.
- temperature and pressure (if applicable).
- ampere-hour in and out of the battery.
- state of charge and discharge.
- rate of charge and discharge.
- depth of discharge.
- number of charge and discharge cycles

The Ah integrating meter is commercially available, which keeps track of the Ah in and out of the battery and sends required signals to the mode controller.

The temperature compensation on the maximum battery voltage and the state-of-charge can improve the battery management, particularly in extreme cold temperatures. It can allow additional charging during cold periods when the battery can accept more charge. The low voltage alarm is a good feature to have, as excessive discharging below the threshold low voltage can cause cell voltage reversal, leading to battery failure. The alarm can be used to shed noncritical loads from the battery to avoid battery damage.

Figure 10-20 depicts a commercially available battery management system incorporating a dedicated microprocessor with software.

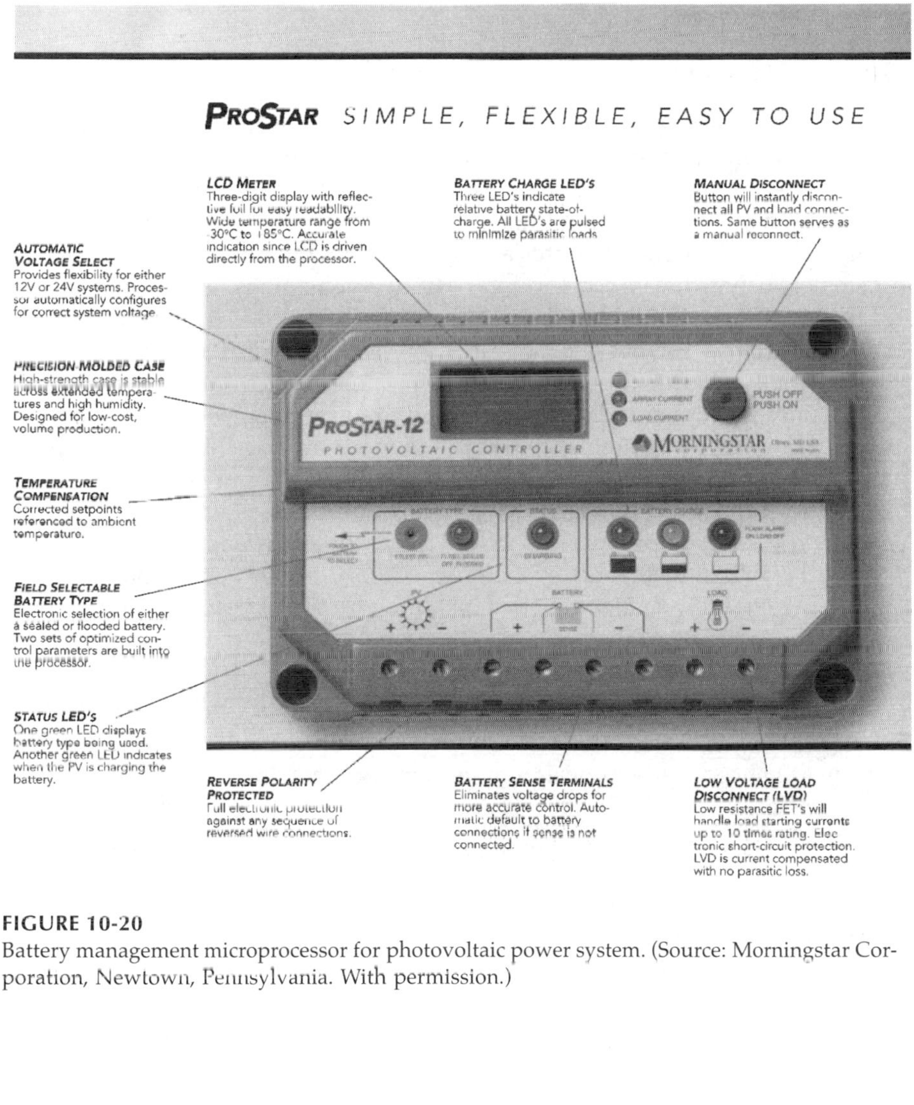

FIGURE 10-20
Battery management microprocessor for photovoltaic power system. (Source: Morningstar Corporation, Newtown, Pennsylvania. With permission.)

10.9.2 Safety

The battery operation requires certain safety considerations. The most important is not to overcharge the battery. Any overcharge above the trickle charge rate is converted into heat, which can explode the battery if allowed to build up beyond limit. This is particularly critical when the battery is charged directly from a photovoltaic module without a charge regulator. In such cases, the array is sized below a certain safe limit. As a rule of thumb, the photovoltaic array rating is kept below the continuous overcharge current that can be tolerated by the battery. This is typically below C/15 amperes for the lead-acid batteries.

10.10 Flywheel

The flywheel stores kinetic energy in a rotating inertia. This energy can be converted from and to electricity with high efficiency. The flywheel energy storage is an old concept, which is getting commercially viable due to advances made in high strength, light-weight fiber composite rotors, and the magnetic bearings that operates at high speeds. The flywheel energy storage system is being developed for a variety of applications, and is expected to make significant inroads in the near future. The round trip conversion efficiency of a large flywheel system can approach 90 percent, much higher than that in the battery.

The energy storage in the flywheel is limited by the mechanical stress due to centrifugal force at high speed. Small to medium-size flywheels have been in use for years. Considerable development efforts are underway around the world for high-speed flywheels to store large amounts of energy. The present goal of these developments is to achieve five times the energy density of the currently available secondary batteries. This goal is achievable with the following enabling technologies, which are already in place in their component forms:

- high-strength fibers having ultimate tensile strength of over one million psi.
- advances made in designing and manufacturing fiber-epoxy composites.
- high-speed, magnetic bearings which eliminate friction, vibrations, and noise.

The flywheel system is made of a composite fiber rotor supported on magnetic bearings rotating in a vacuum, and coupled with a brushless motor-generator machine. Two counter-rotating wheels are placed side by side in moving vehicles where gyroscopic effects must be eliminated.

10.10.1 Energy Relations

The energy stored in flywheel having the moment of inertia J rotating at the angular speed ω is given by the following:

$$E = \frac{1}{2} J \cdot \omega^2 \tag{10-5}$$

The centrifugal force in the rotor material of density ρ at radius r is given by $\rho(r\omega)^2$, which is supported by the hoop stress in the rotor rim. Since the linear velocity $V = 2\pi r\omega$, the maximum centrifugal stress in the rotor is proportional to the square of the outer tip velocity. The allowable stress in the material places an upper limit on the rotor tip speed. Therefore, smaller rotors can run at high speed, and vice versa. The thin rim type rotors have high inertia, and, hence, stores more energy per unit of weight. For this reason, most energy storage rotors have annular rim type construction. For such rotor with inner radius R_1 and the outer radius R_2, it can be shown that the maximum energy that can be stored for an allowable rotor tip velocity V is as follows:

$$E_{max} = K_1 \cdot V^2 \left[1 + \left(\frac{R_1}{R_2}\right)^2\right] \tag{10-6}$$

where K_1 is the proportionality constant. The thin rim flywheel with $R_1/R_2 \rightarrow 1$ results in high specific energy for given allowable stress. The higher the ultimate strength of the material, the higher is the specific energy. The lower the material density, the lower the centrifugal stress produced, and, therefore, higher the allowable speed and the specific energy. The E_{max} can therefore be expressed as follows:

$$E_{max} = K_2 \frac{\sigma_{max}}{\rho} \tag{10-7}$$

where K_2 is another proportionality constant, σ_{max} is the maximum allowable hoop stress and ρ is the mass density of the rotor material. A good flywheel design therefore has high σ_{max}/ρ ratio for high specific energy. It also has high E/ρ ratio for rigidity, where E is the Young's modulus of elasticity.

The metallic flywheel has low specific energy because of low σ_{max}/ρ ratio, whereas the high strength polymer fiber, such as graphite, silica and boron, having much higher σ_{max}/ρ ratio, store an order of magnitude higher energy per unit of weight. Table 10-7 compares the specific energy of various metallic and polymer fiber composite rotors. In addition to the high specific energy, the composite rotor has a safe mode of failure, as it disintegrates to fluff rather than fragmenting like the metal flywheel.

Figure 10-21 shows a rotor design recently developed at the Oakridge National Laboratory. The fiber/epoxy composite rim is made of two rings. The outer ring is made of high-strength graphite, and the inner ring of low-cost glass fibers. The hub is made of single piece aluminum in the radial spoke form. Such construction is cost effective because it uses costly material only where it is needed for strength. That is in the outer ring where the centrifugal stress is the maximum, resulting in high hoop stress.

Figure 10-22 shows a prototype 5 kWh flywheel weight and specific energy (Wh per unit weight) versus σ_{max}/ρ ratio of the material. It is noteworthy that the weight decreases inversely and the specific energy increases linearly with σ_{max}/ρ.

TABLE 10-7

Maximum Specific Energy Storable in a Thin Rim Flywheel with Various Rim Materials

Wheel Material	Maximum Specific Energy Storable Wh/kg
Aluminum alloy	25
Maraging steel	50
E-glass composite	200
Carbon fiber composite	220
S-glass composite	250
Polymer fiber composite	350
Fused silica fiber composite	1000
Lead-acid battery	30–40
Lithium-ion battery	90–120

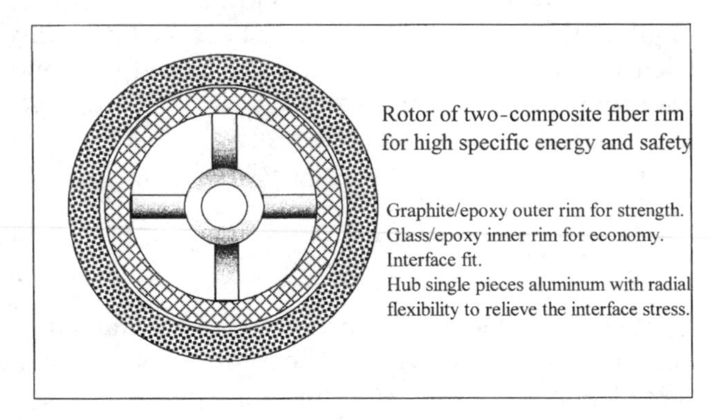

FIGURE 10-21
Flywheel rotor design using two composite rings. (Adapted from the DOE/Oakridge National Laboratory's prototype design.)

10.10.2 Flywheel System Components

The complete flywheel energy storage system requires the following components:

- high-speed rotor attached to the shaft via strong hub.
- bearings with good lubrication system or with magnetic suspension.
- electromechanical energy converter, usually one machine which can work as a motor during charging, and as a generator while discharging the energy.
- power electronics to drive the motor and to condition the generator power.
- control electronics for controlling the magnetic bearings and other functions.

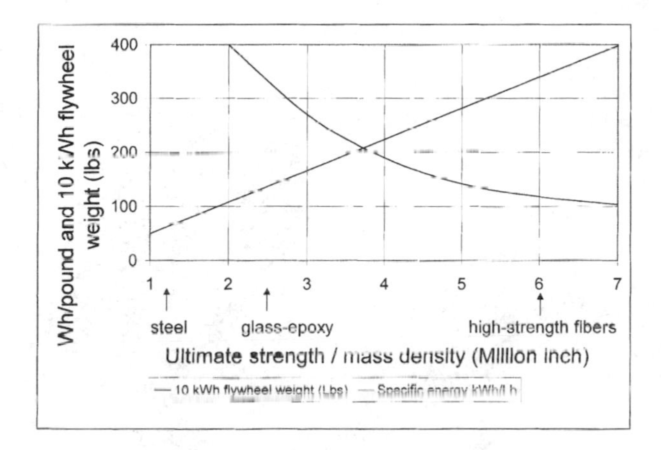

FIGURE 10-22
Specific energy versus specific strength in flywheel design.

Good bearings have low friction and vibration. Conventional bearings are used up to speeds in a few tens of thousands rpm. Speeds approaching 100,000 rpm are possible by using magnetic bearings, which support the rotor by magnetic repulsion and attraction. The mechanical contact is eliminated, thus eliminating the friction. The windage is eliminated by running the rotor in vacuum.

Magnetic bearings come in a variety of configurations using permanent magnets and dynamic current actuators to achieve the required restraints. A rigid body can have six degrees of freedom. The bearings retain the rotor in five degrees of freedom, leaving one for rotation. The homopolar configuration is depicted in Figure 10-23. The permanent magnets are used to provide free levitation support of the shaft. The permanent magnet bias also helps stabilize the shaft under a rotor drop. The electromagnet coils are used for stabilization and control. The control coils operate at low duty cycle, and only one servo controller loop is needed for each axis. The servo control coils provide active control to maintain the shaft stability by providing restoring forces as needed to maintain the shaft in the centered position. The position and velocity sensors are used in the active feedback loop. The electric current variation in the actuator coils compels the shaft to remain centered in the air gaps.

Small flux pulsation as the rotor rotates around the discrete actuator coils produce small electromagnetic loss in the metallic parts. This loss, however, is negligible compared to the loss due to friction in conventional bearings.

In the flywheel system configuration, the rotor can be located radially outward, as shown in Figure 10-24. It forms a volume-efficient packaging. The magnetic bearing has permanent magnets inside. The magnetic flux travels through the pole shoes on the stator and the magnetic feedback ring on the rotor. The reluctance lock between the pole shoes and the magnetic feedback ring provides the vertical restraint. The horizontal restraint is provided by two sets of the dynamic actuator coils. The currents in the coils are controlled in response to the feedback loop controlling the rotor position.

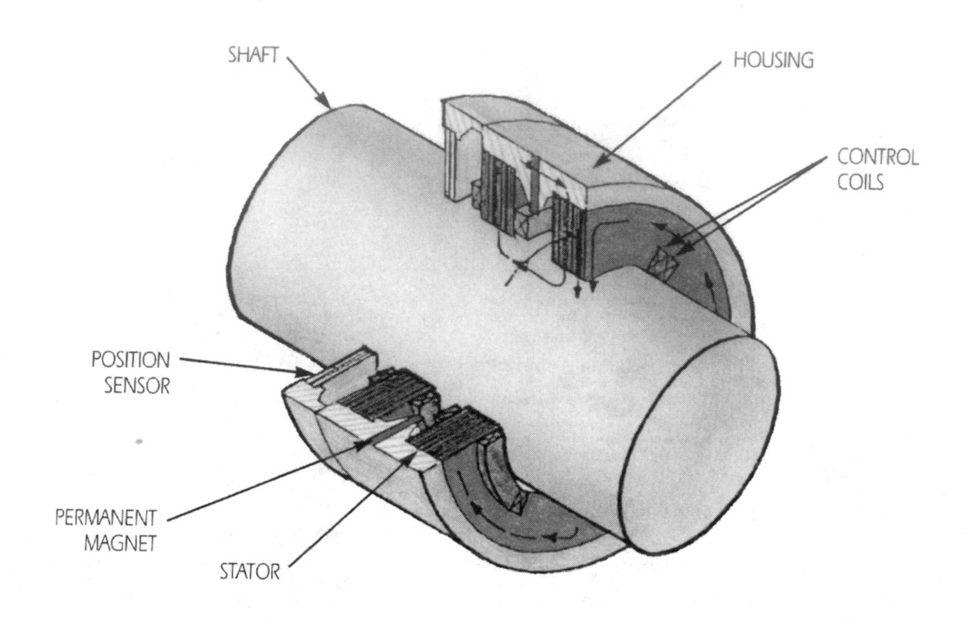

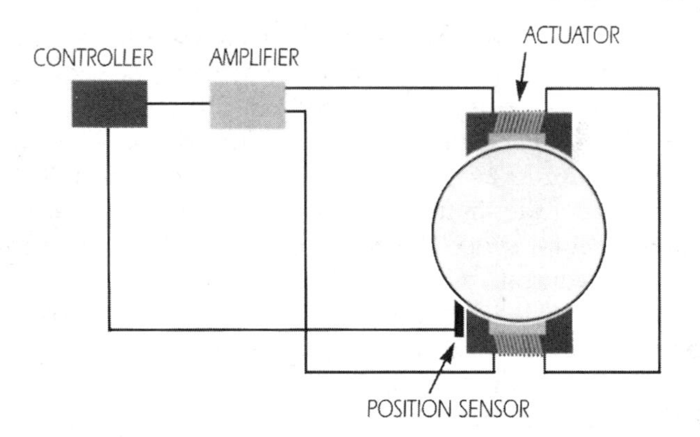

AVCON'S SERVO CONTROLLER

FIGURE 10-23
Avcon's patented homopolar permanent magnet active bearing. (Source: Avcon Inc., Woodland Hills, California. With permission.)

The electromechanical energy conversion is achieved with one machine, which works as a motor for spinning up the rotor for energy charge, and as a generator while decelerating the rotor for discharge. Again, two types of

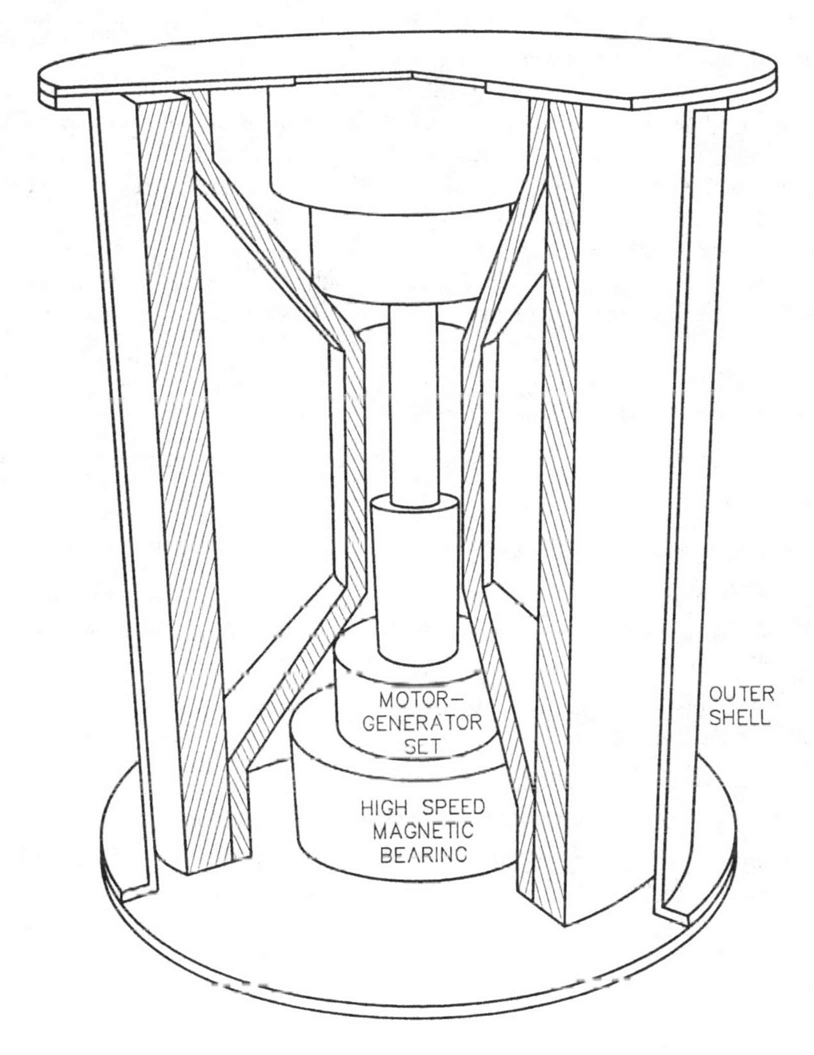

FIGURE 10-24
Flywheel configuration with rotor outside enclosing the motor generator and the bearing.

electrical machines can be used, the synchronous machine with variable frequency converter or the permanent magnet brushless DC machine.

The machine voltage varies over a wide range with speed. The power electronic converters provide interface between widely varying machine voltage and the fixed bus voltage. It is possible to design the discharge converter and the charge converter with input voltage varying over 1 to 3 ranges. This allows the machine speed to vary over the same range. That is, the low rotor speed can be one-third of the full speed. Since the energy storage is proportional to the speed squared, the flywheel state of charge at low speed can be as low as 0.10. That means 90 percent of the flywheel energy can be discharged with no hardship on the power electronics, or other components of the system.

As to the number of charge-discharge cycles the flywheel can withstand, the fatigue life of the composite rotor is the limiting factor. Experience indicates that the polymer fiber composites in general have a longer fatigue life than solid metals. The properly-designed flywheel, therefore, can last much longer than the battery and can discharge to a much deeper level. Flywheels made of composite rotors have been fabricated and tested to demonstrate more than 10,000 cycles full charge and discharge. This is an order of magnitude more than any battery presently available.

10.10.3 Flywheel Benefits Over Battery

The main advantages of the flywheel energy storage over the battery are as follows:

- high energy storage capacity per unit of weight and volume.
- high depth of discharge.
- long cycle life, which is insensitive to the depth of discharge.
- high peak power capability without overheating concerns.
- easy power management, as the state of charge is simply measured by the speed.
- high round trip energy efficiency.
- flexibility in designing for a given voltage and current.
- improved quality of power as the electrical machine is stiffer than the battery.

These benefits have the potential of making the flywheel the least-cost energy storage alternative per Wh delivered over the operating life.

10.11 Compressed Air

The compressed air energy storage system consists of:

- air compressor.
- expansion turbine.
- electric motor-generator.
- overhead storage tank or an underground cavern.

The energy stored in compressed air is given by the following derivations.

If P and V represent the air pressure and volume, respectively, and if the air compression from pressure P_1 to P_2 follows the gas law PV^n = constant, then the work required during this compression is the energy stored in the compression air. It is given by the following:

$$\text{Energy stored} = \frac{n(P_2V_2 - P_1V_1)}{n-1} \tag{10-8}$$

And the temperature at the end of the compression is given by the following:

$$\frac{T_2}{T_1} = \left(\frac{P_2}{P_1}\right)^{\frac{n-1}{n}} \tag{10-9}$$

When the elevated temperature air at the end of the constant volume compression cools down, a part of the pressure is lost with the corresponding decrease in the stored energy.

The smaller the value of n, the smaller the energy stored. The isentropic value of n for air is 1.4. Under normal working conditions, n is about 1.3.

The electrical power is generated by venting the compressed air through an expansion turbine which drives the generator. The compressed air system may work under constant-volume or constant-pressure.

In the constant-volume compression, the compressed air is stored in pressure tanks, mine caverns, depleted oil or gas fields, or abandoned mines. One million cubic feet of air storage at 600 psi provide an energy storage capacity enough to supply about ¼ million kWh_e. This system, however, has a disadvantage. The air pressure reduces as the compressed air is depleted from the storage, and the electrical power output decreases with the decreasing pressure.

In the constant-pressure compression, the air storage may be in an above ground variable-volume tank or an underground aquifer. One million cubic feet of air storage at 600 psi provide an energy storage capacity enough to supply about 0.07 million kWh_e. A variable-volume tank maintains a constant pressure by the weight on the tank cover. If aquifer is used, the pressure remains approximately constant while the storage volume increases because of water displacement in the surrounding rock formation. During electric generation, the water displacement of the compressed air causes a decrease of only a few percentages in the storage pressure, keeping the electrical generation rate essentially constant.

The energy storage efficiency of the compressed air-storage system is a function of a series of component efficiencies, such as the compressor efficiency, the motor-generator efficiency, heat losses, and the compressed air leakage. The overall round trip efficiency of about 50 percent has been estimated.

10.12 Superconducting Coil

The development efforts to use the superconducting technology for storing electrical energy has started yielding highly promising results. In its working principle, the energy is stored in the magnetic field of a coil, and is given by the following expression:

$$\begin{aligned}\text{Energy stored in a coil} &= \tfrac{1}{2} B^2/\mu \ \text{Joules/m}^3 \\ &= \tfrac{1}{2} I^2 \cdot L \ \text{Joules}\end{aligned} \tag{10-10}$$

where B = the magnetic field density produced by the coil, Tesla
μ = magnetic permeability of air = $4\ \pi\ 10^{-7}$ henry/meter
L = inductance of the coil, henry

The coil must carry current in order to produce the required magnetic field. The current requires a voltage to be applied to the coil terminals. The relation between the coil current I and the voltage V is as follows:

$$V = RI + L\frac{di}{dt} \tag{10-11}$$

where R and L are the resistance and inductance of the coil, respectively. For storing energy in a steady state, the second term in Equation 10-11 must be zero. Then the voltage required to circulate the needed current is simply $V = R \cdot I$.

The resistance of the coil is temperature dependent. For most conducting materials, it is higher at higher temperature. If the temperature of the coil is reduced, the resistance drops as shown in Figure 10-25. In certain materials, the resistance abruptly drops to a precise zero at some critical temperature. In the figure, this point is shown as T_c. Below this temperature, no voltage is required to circulate current in the coil, and the coil terminals can be shorted. The current will continue to flow in the coil indefinitely, with the corresponding energy stored in the coil also indefinitely. The coil is said to have attained the superconducting state, one which has attained precisely zero resistance. The energy in the coil then freezes.

Although the superconducting phenomenon was discovered decades ago, industry interest in developing practical applications started in the early 1970s. In the U.S.A., the pioneering work has been done in this field by the General Electric Company, the Westinghouse Research Center and the University of Wisconsin. During the 1980s, a grid-connected 8 kWh superconducting energy storage system was built with funding from the Department of Energy, and was operated by the Bonneville Power Administration in Portland, Oregon.

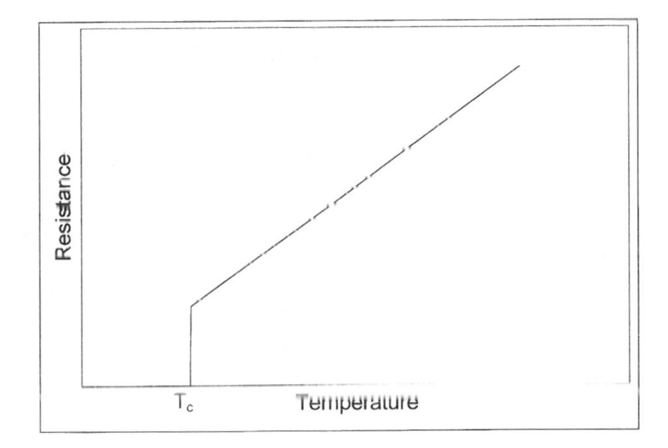

FIGURE 10-25
Resistance versus temperature with abrupt loss of resistance at the critical superconducting temperature.

The system demonstrated over one million charge-discharge cycles, meeting its electrical, magnetic and structural performance goals. Conceptual designs of large superconducting energy storage systems up to 5,000 MWh energy for utility applications have been developed in the past.

The main components in a typical superconducting energy storage system are shown in Figure 10-26. The coil is charged by an AC to DC converter in the magnet power supply. Once fully charged, the converter continues providing small voltage needed to overcome losses in the room temperature parts of the circuit components. This keeps the constant DC current flowing (frozen) in the superconducting coil. In the storage mode, the current is circulated through the normally closed switch.

The system controller has three main functions:

- control the solid state isolation switch.
- monitor the load voltage and current.
- interface with the voltage regulator that controls the DC power flow to and from the coil.

If the system controller senses the line voltage dropping, it interprets that the system is incapable of meeting the load demand. The switch in the voltage regulator opens in less than one millisecond. The current from coil now flows into the capacitor bank until the system voltage recovers the rated level. The capacitor power is inverted into 60 or 50 Hz AC and is fed to the load. As the capacitor energy is depleted, the bus voltage drops. The switch opens again, and the process continues to supply energy to the load continually. The system is sized to store sufficient energy to power the load for specified duration.

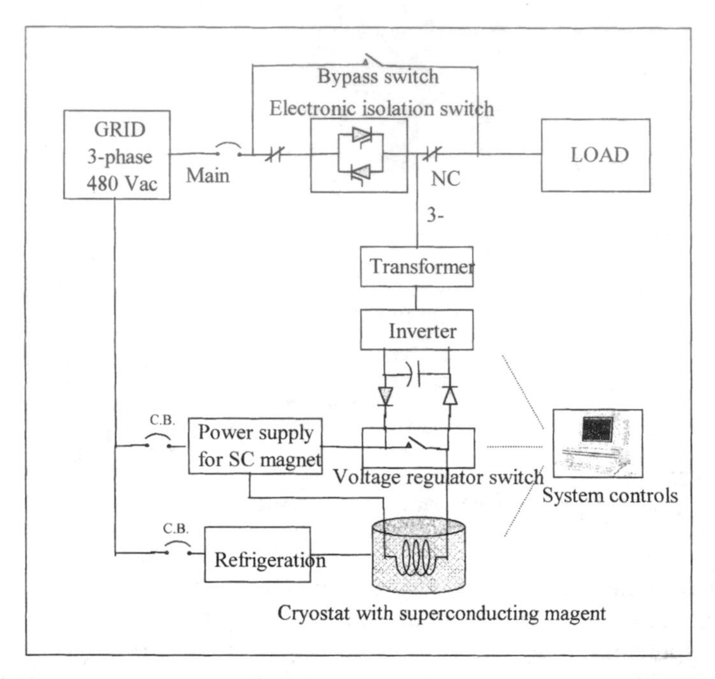

FIGURE 10-26
Superconducting energy storage schematic.

The superconducting energy storage has several advantages over other technologies:

- the round trip efficiency of the charge-discharge cycle is high at 95 percent. This is higher than that attainable by any other technology.
- much longer life, up to about 30 years.
- the charge and discharge times can be extremely short, making it attractive for supplying large power for a short time if needed.
- has no moving parts in the main system, except in the refrigeration components.

In the superconducting energy storage system, the main cost is to keep the coil below the critical superconducting temperature. Until now the niobium-titanium alloy has been extensively used, which has the critical temperature of about 9°K. This requires liquid helium as coolant at around 4°K. The 1986 discovery of high temperature superconductors has accelerated the industry interest in this technology. Three types of high temperature superconducting materials are available now, all made from bismuth or yttrium-cuprate compounds. These superconductors have the critical temperature around 100°K. Therefore, they can be cooled by liquid nitrogen, which needs orders of magnitude less refrigeration power. As a result, numerous programs around the world have started to develop commercial applications. Toshiba Corporation of Japan and GEC-Alsthom along with Electricite de France are actively perusing development in this field.[3-4]

References

1. Riezenman, M. J. 1995. "In search of better batteries," *IEEE Spectrum*, p. 51-56, May 1995.
2. Wicks, F. and Halls, S. 1995. "Evaluating performance enhancement of lead-acid batteries by force circulation of the electrolytic," *1995 Proceedings of the Intersociety Engineering Conference on Energy Conversion, Paper No. 180, 1995.*
3. DeWinkel, C. C., and Lamopree J. D. 1993. "Storing power for critical loads," *IEEE Spectrum*, p. 38-42, June 1993.
4. Balachandran, U. 1997. "Super Power, Progress in developing the new superconductors," *IEEE Spectrum*, p. 18-25. July 1997.

11

Power Electronics

The power electronic circuits in wind and photovoltaic power systems basically perform the following functions:

- convert AC into DC.
- convert DC into AC.
- control voltage.
- control frequency.
- convert DC into DC.

These functions are performed by solid state semiconductor devices periodically switched on and off at desired frequency. In terms of applications, no other technology has brought greater change in power engineering, or holds greater potential of bringing improvements in the future, than the power electronic devices and circuits. In this chapter, we review the power electronic circuits used in modern wind and photovoltaic power systems.

11.1 Basic Switching Devices

A great variety of solid state devices is available in the market. Some of the more commonly used devices are as follows:

- bipolar junction transistor (BJT).
- metal-oxide semiconducting field effect transistor (MOSFET).
- insulated gate bipolar transistor (IGBT).
- silicon controlled rectifier (SCR), also known as the thyristor.
- gate turn off thyristor (GTO).

For specific application, the choice depends on the power, voltage, current, and the frequency requirement of the system. A common feature among these devices is that all are three-terminal devices as shown in their generally

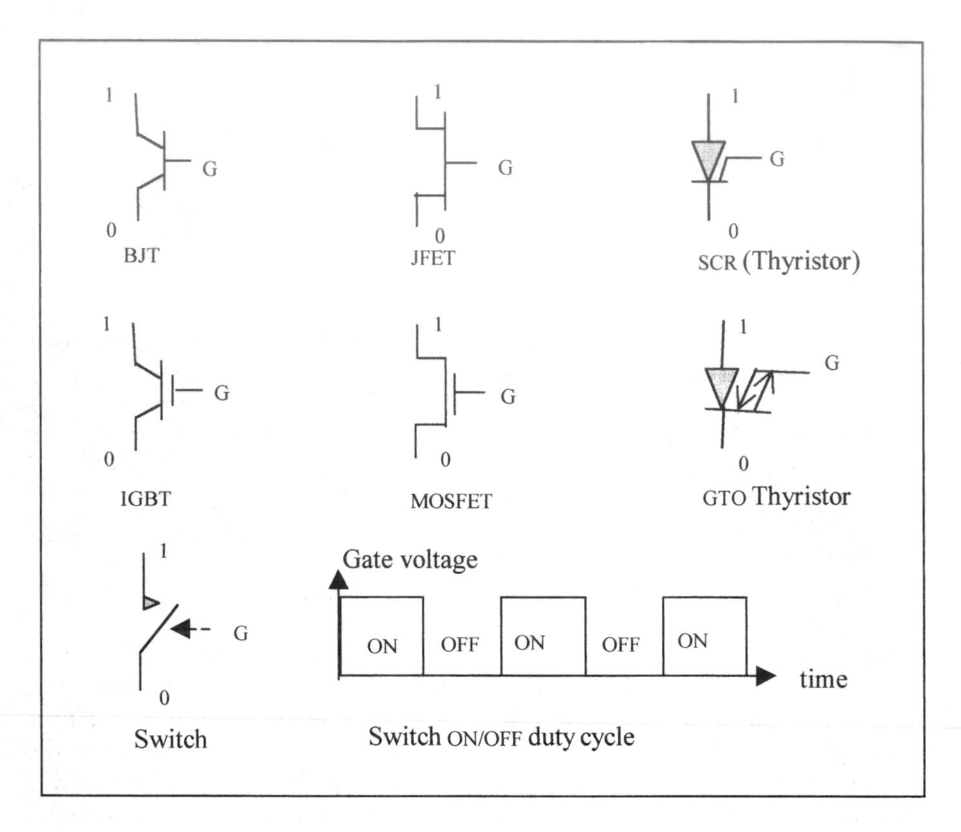

FIGURE 11-1
Basic semiconductor switching devices.

used circuit symbols in Figure 11-1. The two power terminals 1 and 0 are connected in the main power circuit, and one control terminal G. In normal conducting operation, terminal 1 is generally at higher voltage than terminal 0. Terminal G, known as the gate terminal, is connected to the auxiliary control circuit.

Since the devices are primarily used for switching power on and off as required, they are functionally represented by the gate-controlled switch shown in (f). In absence of the control signal at the gate, the device resistance between the power terminals is large, with the functional equivalence of an open switch. When the control signal is applied at the gate, the device resistance approaches zero, making the device behave like a closed switch. The device in this state lets the current flow freely through its body.

The voltage and current ratings of the switching devices available in the market vary. The presently available ratings are listed in Table 11-1.

The switch is triggered periodically on and off by a train of gate signals of suitable frequency. The gate signal may be of rectangular or other wave shape, and is generated by a separate triggering circuit, which is often called

TABLE 11-1

Maximum Voltage and Current Ratings of Power Electronic Switching Devices

Device	Voltage Rating, Volts	Current Rating, Amperes	Remark
BJT	1500	200	Requires large current signal to turn on
IGBT	1200	100	Combines the advantages of BJT, MOSFET and GTO
MOSFET	1000	100	Higher switching speed
SCR	6000	3000	Once turned on, requires heavy turn-off circuit

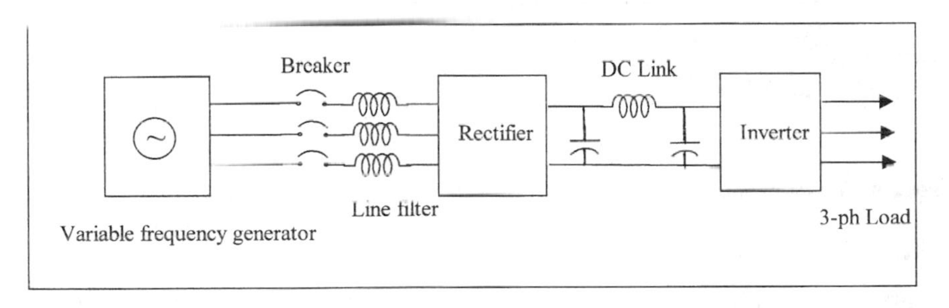

FIGURE 11-2
Variable-speed constant-frequency wind power system schematic.

the firing circuit. Although it has a distinct identity and many different design features, it is generally incorporated in the main power electronic component assembly.

In wind power system operating at variable speed for maximum annual energy production, the output frequency and voltage of the induction generator vary with the wind speed. The variable-frequency, variable-voltage output is converted into fixed voltage 60 Hz or 50 Hz terminal output to match with utility requirement. In modern plants, this is accomplished by power electronics scheme shown in Figure 11-2. The variable frequency is first rectified into DC, and the DC is then inverted back into the fixed frequency AC. The increase in the energy production from the variable speed wind turbine over the plant life more than offsets the added cost of the power electronics.

In photovoltaic power systems, the DC power produced by the pv modules is inverted into 60 or 50 Hz AC power using the inverter. The inverter circuit in the pv system is essentially the same that is used in the variable speed wind power system.

The main power electronic components of the wind and pv power systems are, therefore, the rectifier and the inverter. Their circuits and the AC and DC voltage and current relationships are presented in the following sections.

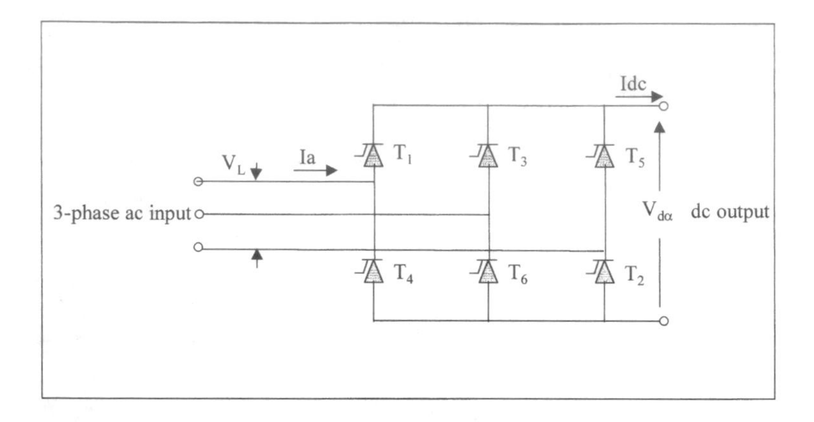

FIGURE 11-3
Three-phase full bridge AC to DC silicon-controlled rectifier circuit.

11.2 AC to DC Rectifier

The circuit diagram of the full-bridge, three-phase, AC to DC rectifier is shown Figure 11-3. The power switch generally used in the rectifier is the silicon-controlled rectifier. The average DC output voltage in this circuit is given by the following:

$$V_{dc} = \frac{3\sqrt{2}}{\pi} V_L \cos\alpha \tag{11-1}$$

where V_L = line-to-line voltage on three-phase AC side of the rectifier
α = angle of firing delay in the switching.

The delay angle is measured from the zero crossing in the positive half of the AC voltage wave. Equation 11-1 shows that the output DC voltage is controllable by varying the delay angle α, which in turn controls the conduction (on-time) of the switch.

The load determines the DC side current:

$$I_{DC} = \frac{DC\ load\ power}{V_{DC}}$$

In the steady state operation, the balance of power must be maintained on both AC and DC sides. That is, the power on the AC side must be equal to the sum of the DC load power and the losses in the rectifier circuit. The AC side power is therefore:

$$P_{AC} = \frac{DC\ load\ power}{rectifier\ efficiency} \tag{11-2}$$

Moreover, as we will see in Chapter 14, the three-phase AC power is given by the following:

$$P_{AC} = \sqrt{3} \cdot V_L \cdot I_L \cdot \cos\phi \tag{11-3}$$

where $\cos\phi$ is the power factor on the AC side. With a well-designed power electronic converter, the power factor on the AC side is approximately equal to that of the load.

From Equation 11-2 with 11-3, we obtain the AC-side line current I_L

11.3 DC to AC Inverter

The power electronic circuit used to convert DC into AC is known as the inverter. The term "converter" is often used to mean either the rectifier or the inverter. The DC input to the inverter can be from any of the following sources:

- rectified DC output of the variable speed wind power system.
- DC output of the photovoltaic power modules.
- DC output of the battery used in the wind or photovoltaic power system.

Figure 11-4 shows the DC to three-phase AC inverter circuit diagram. The DC source current is switched successively in a 60 Hz three-phase time sequence such as to power the three-phase load. The AC current contains significant harmonics as discussed in Section 14-7. The fundamental frequency (60 or 50 Hz) phase-to-neutral voltage is as follows:

$$V_{ph} = \frac{2\sqrt{2}}{\pi}\cos\left(\frac{\pi}{6}\right) \cdot V_{DC} \tag{11-4}$$

The line-to-line AC voltage, as will be seen in the next chapter, is given by $\sqrt{3} \cdot Vph$.

Unlike in BJT, MOSFET, and IGBT, the thyristor current, once switched on, must be forcefully switched off (commutated) to cease conduction. If the thyristor is used as the switching device, the circuit must incorporate additional commutating circuit to perform this function. The commutating circuit is a significant part of the inverter circuit. There are two main types of inverters, the line commutated and the forced commutated.

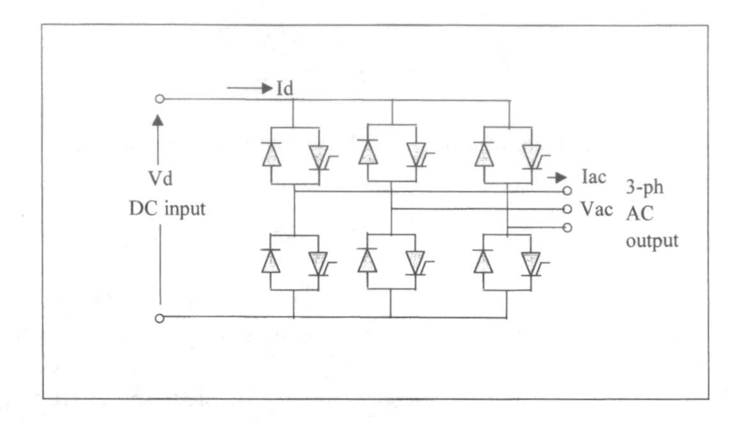

FIGURE 11-4
DC to three-phase AC inverter circuit.

The line-commutated inverter must be connected to the AC system into which they feed power. The design method is matured and has been extensively used in the high-voltage DC transmission line inverters. Such inverters are simple and inexpensive and can be designed in any size. The disadvantage is that they act as a sink of reactive power and generate high content of harmonics.

Poor power factor and high harmonic content in line commutated inverters significantly degrade the quality of power at the utility interface. This problem has been recently addressed by a series of design changes in the inverters. Among them is the 12-pulse inverter circuit and increased harmonic filtering. These new design features have resulted in today's inverters operating at near unity power factor and less than 3 to 5 percent total harmonic distortion. The quality of power at the utility interface at many modern wind power plants exceeds that of the grid they interface.

The force-commutated inverter does not have to be supplying load and can be free-running as an independent voltage source. The design is relatively complex and expensive. The advantage is that they can be a source of reactive power and the harmonics content is low.

11.4 Grid Interface Controls

At the utility interface, the power flow direction and magnitude depend on the voltage magnitude and the phase relation of the site voltage with respect to the grid voltage. The grid voltage being fixed, the site voltage must be controlled both in magnitude and in phase in order to feed power to the grid when available, and to draw from the grid when needed. If the inverter is already included in the system for frequency conversion, the magnitude

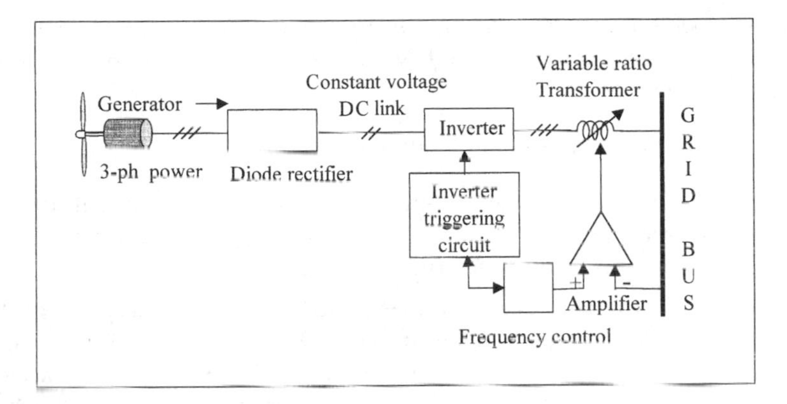

FIGURE 11-5
Voltage control by means of uncontrolled rectifier and variable ratio tap-changing transformer.

and phase control of the site voltage is done with the same inverter with no additional hardware cost. The controls are accomplished as follows:

11.4.1 Voltage Control

For interfacing with the utility grid lines, the renewable power system output voltage at the inverter terminals must be adjustable. The voltage is controlled by using one of the following two methods:

1. By controlling the alternating voltage output of the inverter using tap-changing autotransformer at the inverter output (Figure 11-5). The tap changing is automatically obtained in a closed-loop control system. If the transformer has a phase-changing winding also, a complete control on the magnitude and phase of the site voltage can be achieved. The advantages of this scheme are that the site output voltage waveshape does not vary over a wide range, and high input power factor is achieved by using uncontrolled diode rectifiers for the DC link voltage. The added cost of the transformer, however, can be avoided by using the method discussed below.
2. Since the magnitude of the alternating voltage output from the static inverter is proportional to the direct voltage input from the rectifier, the voltage control can be achieved by operating the inverter with the variable DC link voltage. Such a system also maintains the same output voltage, frequency and wave shape over a wide range. However, in circuits deriving the load current from the commutating capacitor voltage from the DC link, the commutating capability decreases when the output voltage is reduced. This could lead to an operational difficulty when the DC link voltage varies over a wide range, such as in motor drives controlling

the speed in ratio exceeding four to one. In renewable power applications, such commutation difficulty is unlikely as the speed varies over a narrow range.

The variable DC link voltage is obtained two ways:

- one way is to connect a variable ratio transformer on the input side of the rectifier. The secondary tap changing is automatically obtained in a closed-loop control system.
- the other way is to use the phase-controlled rectifier in place of the uncontrolled rectifier in Figure 11-5. At reduced output voltage, this method gives poor power factor and high harmonic content, and requires filtering the DC voltage before feeding to the inverter.

11.4.2 Frequency Control

The output frequency of the inverter solely depends on the rate at which the switching thyristors or transistors are triggered into conduction. The triggering rate is determined by the reference oscillator producing a continuous train of timing pulses, which are directed by logic circuits to the thyristor gating circuits. The timing pulse train is also used to control the turn-off circuits. The frequency stability and accuracy requirements of the inverter dictate the selection of the reference oscillator. A simple temperature compensated R-C relaxation oscillator gives the frequency stability within 0.02 percent. When better stability is needed, a crystal-controlled oscillator and digital counters may be used, which can provide stability of .001 percent or better. The frequency control in a stand-alone power system is an open-loop system. The steady state or transient load changes do not affect the frequency. This is one of the major advantages of the power electronics inverter over the old electromechanical means of frequency controls.

11.5 Battery Charge/Discharge Converters

The stand-alone photovoltaic power system uses the DC to DC converter for battery charging and discharging.

11.5.1 Battery Charge Converter

Figure 11-6 is the most widely used DC-DC battery charge converter circuit, also called the buck converter. The switching device used in such converters may be the BJT, MOSFET, or the IGBT. The buck converter steps down the input bus voltage to the battery voltage during battery charging. The transistor switch is turned on and off at high frequency (in tens of kHz). The duty ratio D of the switch is defined as the following:

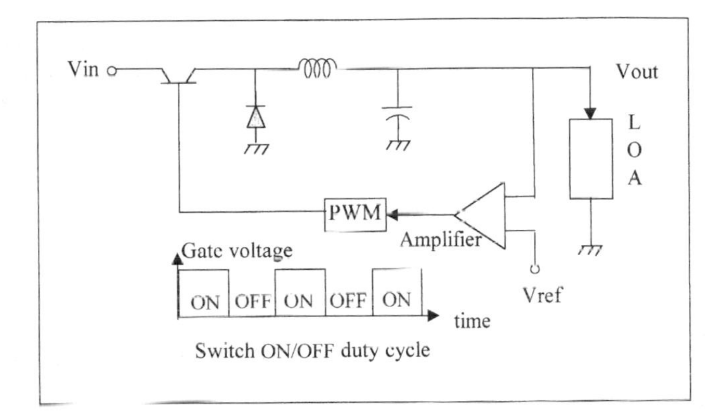

FIGURE 11-6
Battery charge converter for pv systems (DC to DC buck converter).

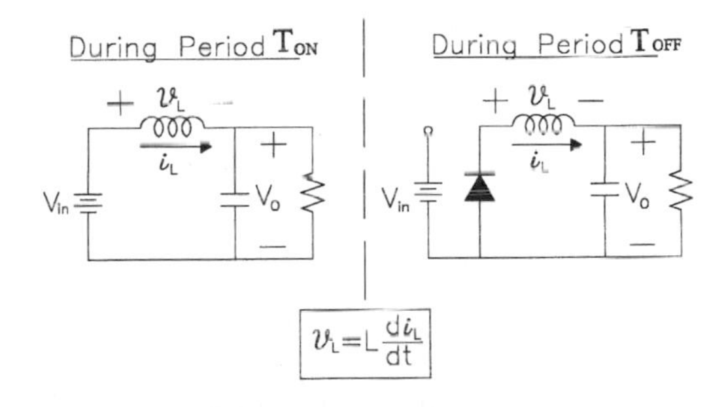

FIGURE 11-7
Charge converter operation during switch on-time and off-time.

$$\text{Duty ratio } D = \frac{\text{Time on}}{\text{Period}} = \frac{T_{on}}{T} = T_{on} \cdot \text{Switching Frequency} \tag{11-5}$$

The charge converter operation during one complete cycle of the triggering signal is shown in Figure 11-7. During the on time, the switch is closed and the circuit operates as on the left. The DC source charges the capacitor and supplies power to the load via the inductor. During the off time, the switch is open and the circuit operates as on the right. The power drawn from the DC source is zero. However full load power is supplied by the energy stored in the inductor and the capacitor, with the diode providing the return circuit. Thus, the inductor and the capacitor provide short-time energy storage to ride through the off period of the switch.

The simple analyses of this circuit is shown below. It illustrates the basic method of analyzing all power electronic circuits.

The power electronic circuit analysis is based on the energy balance over one period of the switching signal. That is as follows:

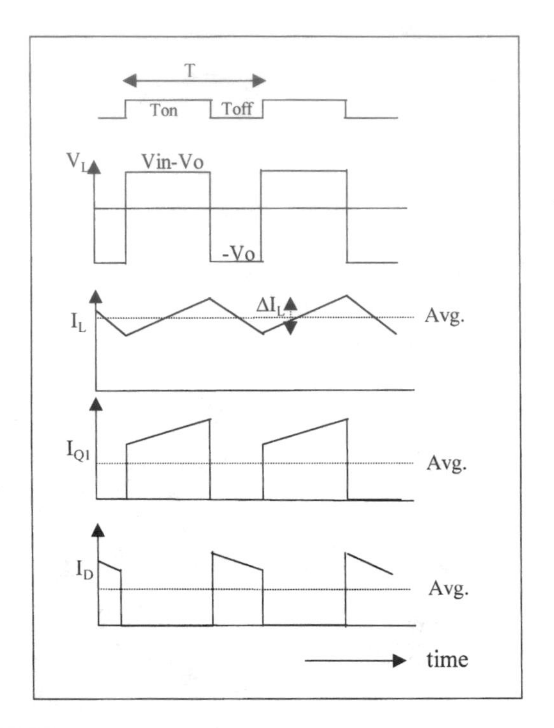

FIGURE 11-8
Current and voltage waveforms in the buck converter.

Energy supplied to the load over the total period of repetition.
= Energy drawn from the source during the on-time, and
Energy supplied to the load during the off-time
= Energy drawn from the inductor and the capacitor during off-time

Alternatively, the volt-second balance method is used, which in essence gives the energy balance.

The voltage and current waveforms are displayed in Figure 11-8. In the steady state condition, the inductor volt-second balance during the on and off periods must be maintained. Since the voltage across the inductor must equal L dI_L/dt,

During on-time, $$\Delta I_L \cdot L = \left(V_{in} - V_{out}\right) \cdot T_{on} \tag{11-6}$$

And during off-time $$\Delta I_L \cdot L = V_{out} \cdot T_{off} \tag{11-7}$$

If the inductor is large enough, as usually is the case in practical circuits, the change in the inductor current is small, and the peak value of the inductor current is given by the following:

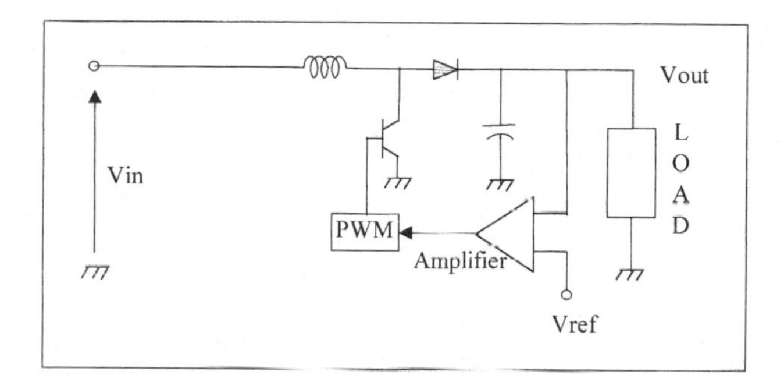

FIGURE 11-9
Battery discharge converter circuit for pv systems (DC to DC boost converter).

$$I_{peak} = I_o + \tfrac{1}{2}\Delta I_L \quad (11\text{-}8)$$

where the load current $I_o = V_{out}/R_{load}$
= average value of the inductor current.

The algebraic manipulation of the above equations leads to the following:

$$V_{out} = V_{in} \cdot D \quad (11\text{-}9)$$

It is seen from Equation 11-9 that the output voltage is controlled by varying the duty ratio D. This is done in a feedback control loop with the required battery charge current as the reference. The duty ratio is controlled by modulating the pulse width of T_{on}. Such a converter is, therefore, also known as the Pulse Width Modulated (PWM) converter.

11.5.2 Battery Discharge Converter

The battery discharge converter circuit is shown in Figure 11-9. It steps up the sagging battery voltage during discharge to the required output voltage. When the transistor switch is on, the inductor is connected to the DC source. When the switch is off, the inductor current is forced to flow through the diode and the load. The output voltage of the boost converter is derived again from the volt-second balance in the inductor. With duty ratio D of the switch, the output voltage is given by the following:

$$V_{out} = \frac{V_{in}}{1 - D} \quad (11\text{-}10)$$

For all values of D < 1, the output voltage is always greater than the input voltage. Therefore, the boost converter can only step up the voltage. On the

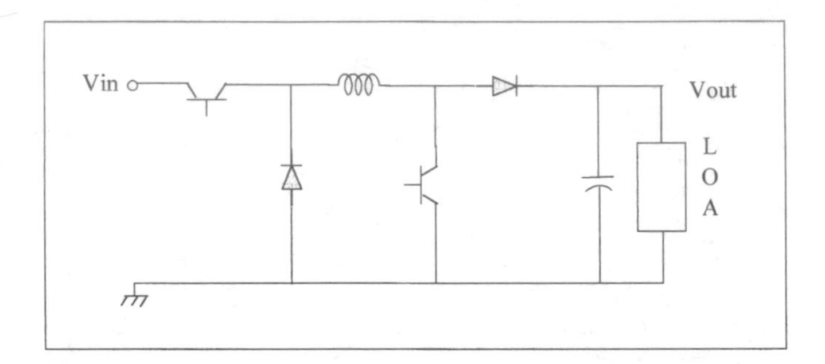

FIGURE 11-10
Buck-boost converter circuit (general DC to DC converter for pv systems).

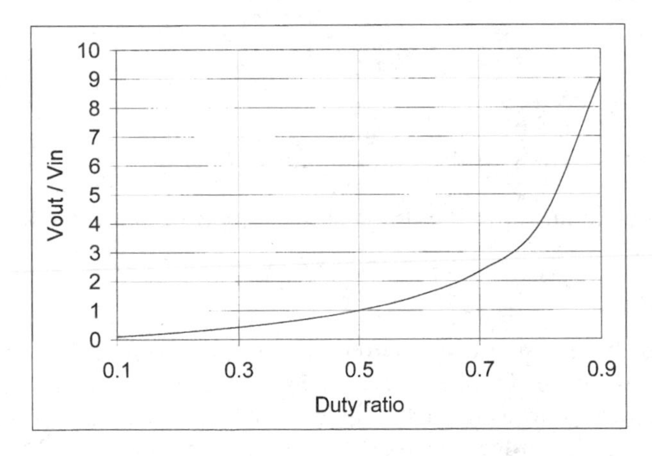

FIGURE 11-11
Buck-boost converter output to input ratio versus duty ratio.

other hand, the buck converter presented in the preceding section can only step down the input voltage. Combining the two converters in cascade, therefore, give a buck-boost converter, which can step down or step up the input voltage. A modified buck-boost converter often used for this purpose is shown in Figure 11-10. The voltage relation is obtained by cascading the buck and boost converter voltage relations. That is as follows:

$$V_{out} = \frac{V_{in} \cdot D}{1 - D} \tag{11-11}$$

Equation 11-11 for the buck-boost converter shows that the output voltage can be higher or lower than the input voltage depending on the duty ratio D (Figure 11-11).

In addition to its use in the battery charging and discharging, the buck-boost circuit is capable of four-quadrant operation with the DC machine

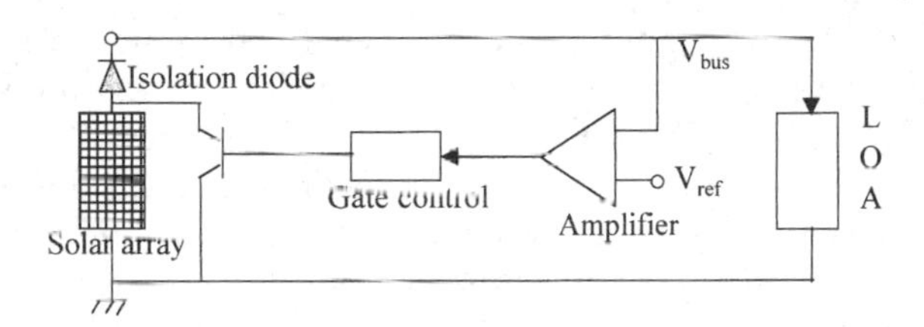

FIGURE 11-12
Power shunt circuit for shorting the module.

when used in the variable speed wind power systems. The converter is in the step-up mode during the generating operation, and in the step down mode during the motoring operation.

11.6 Power Shunts

In stand-alone photovoltaic systems, the power generation in excess of the load and battery charging requirements must be dissipated in dump load in order to control the output bus voltage. The dump load may be resistance-heaters. However when the heaters cannot be accommodated in the system operation, the dissipation of the excess power can pose a problem. In such situations, shorting (shunting) the photovoltaic module to ground forces the module to operate under the short circuit condition, delivering I_{sc} at zero voltage. No power is delivered to the load. The solar power remains on the photovoltaic modules, raising the module temperature and ultimately dissipating the excess power in the air. The photovoltaic module area is essentially used here as the dissipater.

The circuit used for shunting the pv module is shown in Figure 11-12. A transistor is used as the switch. When the excess power is available, the bus voltage will rise above the rated value. This is taken as the signal to turn on the shunt circuits across the required number of photovoltaic modules. The shunt circuit is generally turned on or off by the switch controlled by the bus voltage reference. Relays can perform this function, but the moving contacts have a much shorter life than the solid state power electronic switches. Therefore, relays are seldom used except in small systems.

Another application of the power shunt circuit is in the pv module dedicated to directly charge the battery without the charge regulator. When the battery is fully charged, the module is shunted to ground by shorting the switch. This way, the battery is protected from overcharging.

For array with several modules in parallel, basic configuration shown in Figure 11-12 is used for each module separately, but the same gate signal is supplied to all modules simultaneously. For shunting large power, multiple shunt circuits can be switched on and off in sequence to minimize switching transients and electromagnetic interference (EMI) with the neighboring equipment. For fine power control, one segment can be operated in the pulse width modulation (PWM) mode.

The power electronics is a very wide and developing subject. The intent of this chapter is to present an overview of the basic circuits used in the wind and photovoltaic power systems. Further details can be obtained from many excellent books on power electronics.[1-3]

References

1. Bose, B. "Power Electronics and Variable Frequency Drives," *IEEE Press,* 1996.
2. Dewan, S. B. and Straughen, A. 1975. "Power Semiconductor Circuits," New York, John Wiley & Sons, 1975.
3. Mohan, N. and Undeland, T. 1995. "Power Electronics Converters, Applications and Design," IEEE and John Wiley & Sons, 1995.

12

Stand-Alone System

The major application of the stand-alone power system is in remote areas where utility lines are uneconomical to install due to terrain, the right-of-way difficulties or the environmental concerns. Even without these constraints, building new transmission lines is expensive. A 230 kV line costs about $1 million per mile. For remote villages farther than two miles from the nearest transmission line, a stand-alone wind system could be more economical. The break-even distance for pv systems, however, is longer because the pv energy is four to five times more expensive than wind energy at present.

The solar and wind power outputs can fluctuate on an hourly or daily basis. The stand-alone system must, therefore, have some means of storing energy, which can be used later to supply the load during the periods of low or no power output. Alternatively, the wind or pv or both can also be used in a hybrid configuration with diesel engine generator in remote areas or with fuel cells in urban areas.

According to the World Bank, more than 2 billion people live in villages that are not yet connected to utility lines. These villages are the largest potential market of the hybrid stand-alone systems using diesel generator with wind or pv for meeting their energy needs. Additionally, the wind and pv systems create more jobs per dollar invested, which help minimize the migration to already strained cities.

Because power sources having differing performance characteristics must be used in parallel, the stand-alone hybrid system is technically more challenging and expensive to design than the grid-connected system that simply augments the existing utility system.

12.1 pv Stand-Alone

The typical pv stand-alone system consists of a solar array and a battery connection as shown in Figure 12-1. The array powers the load and charges the battery during daytime. The battery powers the load after dark. The

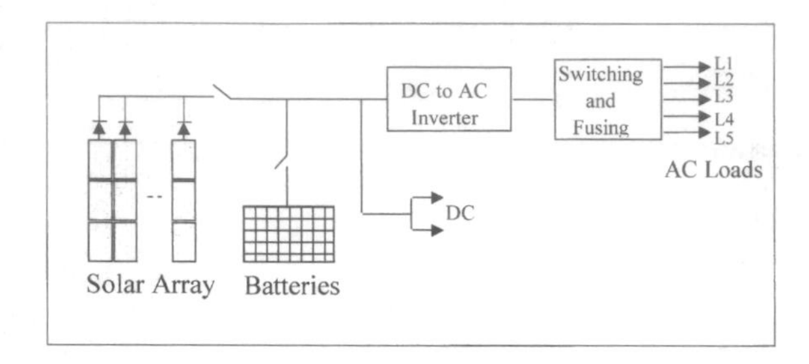

FIGURE 12-1
Photovoltaic stand-alone power system with battery.

inverter converts the DC power of the array and the battery into 60 or 50 Hz power. Inverters are available in a wide range of power ratings with efficiency ranging from 85 to 95 percent. The array is segmented with isolation diodes for improving the reliability. In such designs, if one string of the solar array fails, it does not load or short the remaining strings. Multiple inverters, such as three inverters each with 35 percent rating rather than one with 105 percent rating, are preferred. If one such inverter fails, the remaining two can continue supplying essential loads until the failed one is repaired or replaced. The same design approach also extends in using multiple batteries.

Most of the stand-alone pv systems installed in developing countries provide basic necessities, such as lighting and pumping water. Others go a step further (Figure 12-2).

12.2 Electric Vehicle

The solar electric car developed in the U.S.A. and in many other countries is an example of the stand-alone (rather the move-alone) pv power system. The first solar car was built in 1981 and driven across the Australian Outback by Hans Tholstrup of Australia. The solar car has been developed and is commercially available, although more expensive than the conventional car at present. However, the continuing development is closing the price-gap every year.

A new sport at the American universities these days is the annual solar car race. The DOE and several car manufacturers sponsor the race every two years. It is open to all engineering and business students, who design, build, and run their car across the heartland of America. The first U.S. solar car race was organized in 1990. Figure 12-3 shows one such car built by the University of Michigan. In 1993, it finished first in the 1,100-mile "Sunrayce" that began from Arlington, Texas, and cruised through Oklahoma, Kansas, Missouri, Iowa and ended in Minnesota. It covered 1,102 miles in six and a

FIGURE 12-2
A traveling medical clinic uses photovoltaic electricity to keep vaccines refrigerated in the African desert area. (Source: Siemens Solar Industries, Camarillo, California.)

FIGURE 12-3
The University of Michigan solar car raced 1,100 miles in the U.S.A. and 1,900 miles in Australia. This DOE-sponsored "Sunrayce" taps the bright young brains and displays the new technology across the country.

TABLE 12-1

Design Characteristic Range of Solar Cars Built by U.S. University Students for the Biennial 1,100-Mile 'Sunrayce'

Design Parameters	Parameter Range
Solar array power capability (silicon crystalline, gallium arsenide)	750–1500 watts
Battery (Lead-acid, Silver-Zinc)	3.5–7 kWh
Electric motor (DC brushless with permanent magnet)	4–8 horsepower
Car weight	500–1000 pounds
Car dimensions (approximate overall)	≈20′ long × 7′ wide × 3.5′ high

half days. In 1994, the Michigan car finished the 1,900-mile long World Solar Challenge from Darwin to Adelaide across the Australian Outback. Several dozens of teams participate in the race. Much more than just a race, the goal of this DOE program is to provide the hands-on, minds-on experience to young students in renewable power sources, and display it on the wheels across the country.

The solar cars entering the "Sunrayce" every year generally have a wide range of the design characteristics as listed in Table 12-1. The Michigan car was designed with silicon cells and lead-acid battery, which was changed to the silver-zinc battery later on. With a permanent magnet brushless DC motor, it reached the peak speed of 50 miles per hour. The Gallium arsenide pv cells are often used in the Australian race, but are not allowed in the U.S. race. A few race cars have been designed with a sterling engine driven by helium heated by the solar energy instead of using the photovoltaic cells. The design considerations include trading hundreds of technical parameters covered throughout this book and making decisions to meet the constraints. But certain key elements are essential for winning. They are as follows:

- photovoltaic cells with high conversion efficiency.
- peak power tracking design.
- lightweight battery with high specific energy.
- energy efficient battery charging and discharging.
- low aerodynamic drag.
- high reliability without adding weight.

The zinc-air battery discussed in Chapter 10 is an example of a lightweight battery. Chapter 8 covered the use of isolation diodes in solar array for achieving the reliability without adding weight, and the peak power tracking principle for extracting the maximum power output under the given solar radiation.

After the car is designed and tested, the strategy to optimize the solar energy capture and to use it efficiently while maintaining the energy balance for the terrain and the weather on the day of the race becomes the final test

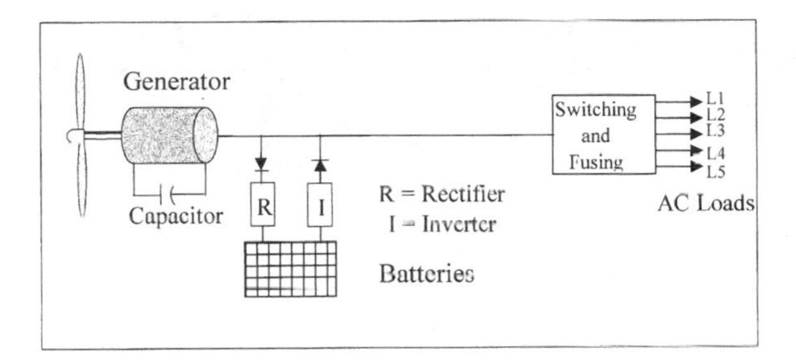

FIGURE 12-4
Stand-alone wind power system with battery.

for winning. The energy balance analysis method for sizing the solar array and battery is described later in this chapter.

12.3 Wind Stand-Alone

A simple stand-alone wind system using a constant speed generator is shown in Figure 12-4. It has many features similar to the pv stand-alone system. For a small wind system supplying local loads, a permanent magnet DC generator makes a wind system simple and easier to operate. The induction generator, on the other hand, gives AC power. The generator is self-excited by shunt capacitors connected to the output terminals. The frequency is controlled by controlling the turbine speed. The battery is charged by an AC to DC rectifier and discharged through a DC to AC inverter.

The wind stand-alone power system is often used for powering farms (Figure 12-5). In Germany, nearly half the wind systems installed on the farms are owned either by individual farmers or by an association. The performance of turbines under the "250 MW program" is monitored and published by ISET, the Institute of Solar Energy and Technology at the University of Kassel.[1] The performance reports are also available from the German Wind Energy Institute. The reports list all installations, their performance, and any technical problems.

The steady state performance of the electrical generator is determined by the theory and analyses presented in Chapter 6. For example, Equations 6-13 and 6-14 determine the rating of the capacitor needed to self-excite the generator for the desired voltage and frequency. The power factor of the load has a great effect on both the steady state and the transient performance of the induction generator. The load power factor can be unity, lagging, or leading depending on the load being resistive, inductive, or capacitive, respectively. Most loads in the aggregate are inductive with a power factor of about 0.9 lagging. Unlike in the synchronous generator, the induction generator output current and the power factor for a given load are determined

FIGURE 12-5
Remote farms are a major market for the stand-alone power systems. (Source: World Power Technologies, Duluth, Minnesota. With permission.)

by the generator parameters. Therefore, when the induction generator delivers a certain load, it also supplies a certain in-phase current and a certain quadrature current. The quadrature current is supplied by the capacitor bank connected to the terminal. Therefore, the induction generator is not suitable for supplying low power factor loads.

The transient performance of the stand-alone, self-excited induction generator, on the other hand, is more involved. The generalized d-q axis model of the generator is required. The computer simulation using the d-q axis model shows the following general transient characteristics:[2]

- under sudden loss of the self-excitation due to tripping-off the capacitor bank, the resistive and inductive loads cause the terminal voltage quickly reach the steady state zero. The capacitive load takes a longer time before the terminal voltage decays to zero.
- under sudden loading of the generator, the resistive and inductive loads result in sudden voltage drop, while the capacitive load has little effect on the terminal voltage.
- under sudden loss of resistive and inductive loads, the terminal voltage quickly rises to its steady state value.
- at light load, the magnetizing reactance will change to its unsaturated value, which is large. This makes the machine performance unstable, resulting in the terminal voltage to collapse.

To remedy the instability problem indicated above, the stand-alone induction generator must always have some minimum load, dummy if necessary, permanently connected to its terminals.

12.4 Hybrid System

12.4.1 Hybrid with Diesel

The certainty of meeting load demands at all times is greatly enhanced by the hybrid system using more than one power source. Most hybrids use diesel generator with pv or wind, since diesel provides more predictable power on demand. In some hybrids, batteries are used in addition to the diesel generator. The batteries meet the daily load fluctuation, and the diesel generator takes care of the long-term fluctuations. For example, the diesel generator is used in the worst case weather condition, such as extended overcasts or windless days or weeks.

Figure 12-6 is one of the largest pv-diesel hybrid systems installed in California. The project was part of the Environmental Protection Agency's PV-Diesel Program.

Figure 12-7 is a schematic layout of the wind/diesel/battery hybrid system. The power connection and control unit (PCCU) provides a central place to make organized connections of most system components. In addition, the PCCU houses the following components:

- battery charge and discharge regulators.
- transfer switches and protection circuit breakers.
- power flow meters.
- mode controller.

Figure 12-8 is a commercially available PCCU for hybrid power systems.

The transient analysis of the integrated wind-pv-diesel requires an extensive model that takes the necessary input data and event definitions for computer simulation.[3]

12.4.2 Hybrid with Fuel Cell

In stand-alone renewable power systems of hybrid designs, the fuel cell has the potential to replace the diesel engine in urban areas. In these applications, the diesel engine would be undesirables due to its environmental negatives. The airborne emission of fuel cell power plant is 25 grams per MWh delivered.

FIGURE 12-6
A 300 kW photovoltaic-diesel hybrid system in Superior Valley, California. (Source: ASE Americas Inc., Billerica, MA. With permission.)

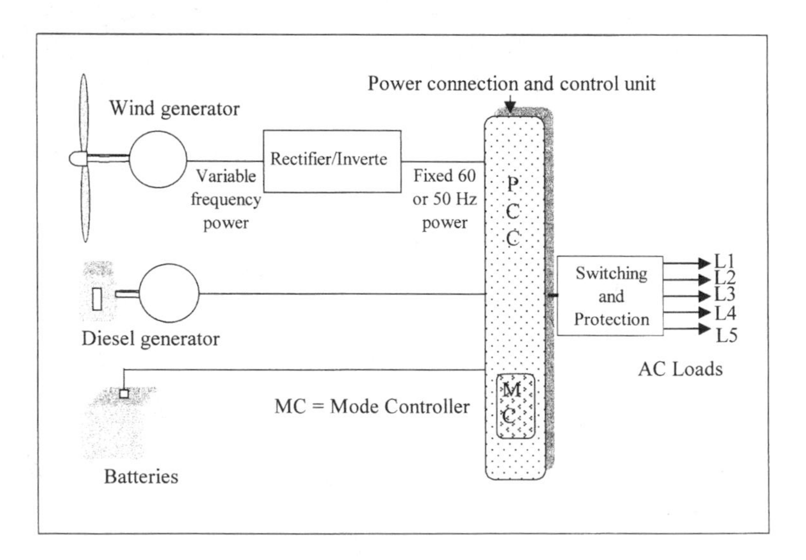

FIGURE 12-7
Wind-diesel-battery hybrid system.

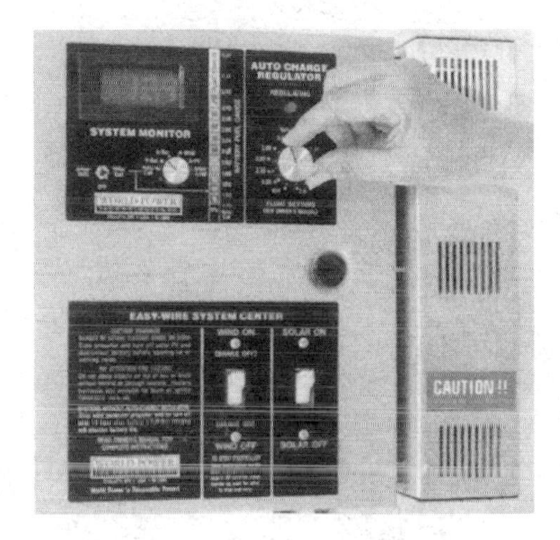

FIGURE 12-8
Integrated power connection and control unit for wind-pv-battery hybrid system. (Source: World Power Technologies, Duluth, Minnesota. With permission.)

Factories and hospitals have considered the fuel cell to replace the diesel generator in the uninterruptible power system. Electric utility companies are considering the fuel cell for meeting the peak demand and for load leveling between the day and night and during the week.

The fuel cell is an electrochemical device that generates electricity by chemical reaction without altering the electrodes or the electrolyte materials. This distinguishes the fuel call from the electrochemical batteries. The concept of the fuel cell is the reverse of the electrolysis of water, in which the hydrogen and oxygen are combined to produce electricity and water. The fuel cell is a static device that converts the chemical energy directly into electrical energy. Since fuel cell bypasses the thermal-to-mechanical conversion, and since its operation is isothermal, the conversion efficiency is not Carnot-limited. This way, it differs from the diesel engine.

The fuel cell, developed as an intermediate-term power source for space applications, was first used in a moon buggy and continues to be used to power NASA's space shuttles. It also finds other niche applications at present. Providing electrical power for a few days or a few weeks is not practical using the battery, but is easily done with the fuel cell.

The basic constructional features of the fuel cell are shown in Figure 12-9. The hydrogen 'fuel' is combined with oxygen of the air to produce electricity. The hydrogen, however, does not burn as in the internal combustion engine, rather it produces electrically by an electrochemical reaction. Water and heat are the byproducts of this reaction if the fuel is pure hydrogen. With the natural gas, ethanol or methanol as the source of hydrogen, the byproducts include carbon dioxide, and traces of carbon monoxide, hydrocarbons and nitrogen oxides. However, they are less than 1 percent of those emitted by the diesel engine. The superior reliability with no moving parts is an additional benefit of the fuel cell over the diesel generator. Multiple fuel cells

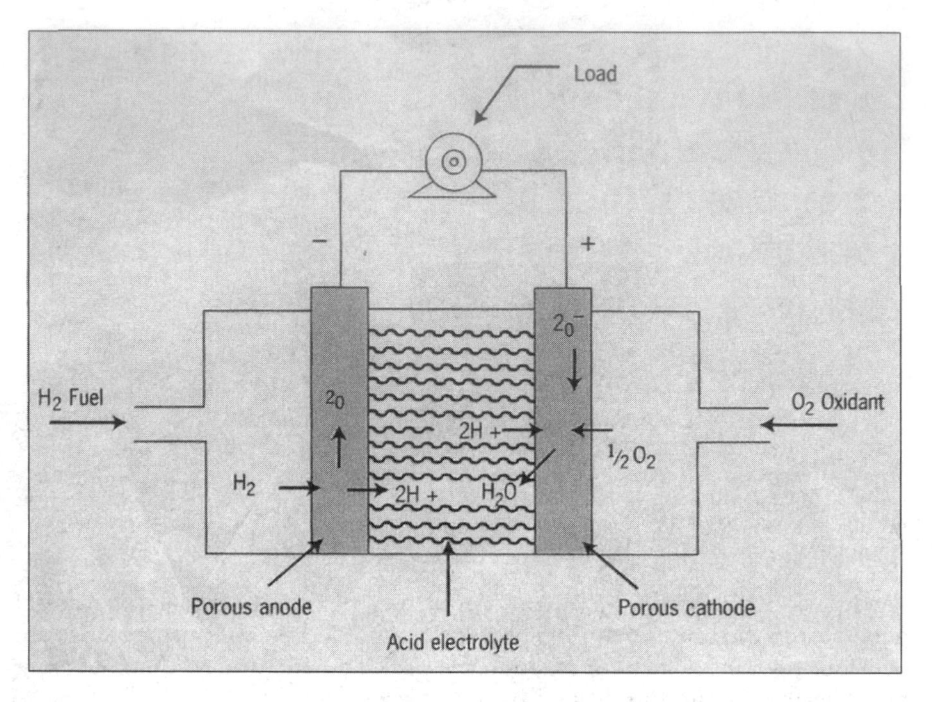

FIGURE 12-9
Fuel cell principle: hydrogen and oxygen in, electrical power and water out.

stack up in series-parallel combinations for the required voltage and current, just as the electrochemical cells do in the battery.

The low temperature (250°C) fuel cell is now commercially available from several sources. It uses phosphoric acid as the electrolytic solution between the electrode plates. A typical low temperature fuel cell with a peak power rating of 200 kW costs under $1,800 per kW at present, which is over twice the cost of the diesel engine. The fuel cell price, however, is falling with new developments being implemented every year.

The high temperature fuel cell has a higher power generation capacity per kilogram at a relatively high cost, limiting the use in special applications at present. Solid oxide, solid polymer, molten carbonate, and proton membrane exchange fuel cells in this category are being developed. The industry interest in such cells is in large capacity for use in a utility power plant. The Fuel Cell Commercialization Group in the U.S.A. recently field-tested molten carbonate direct-fuel cells for 2 MW utility-scale power plants. The test results were a qualified success. Based on the results, a commercial plant is being designed for a target date of operation by the year 2000.

Solid oxide fuel cells of several different designs, consisting of essentially similar materials for the electrolyte, the electrodes, and the interconnections, are being investigated worldwide. Most success to date has been achieved with the tubular geometry being developed by the Westinghouse Electric

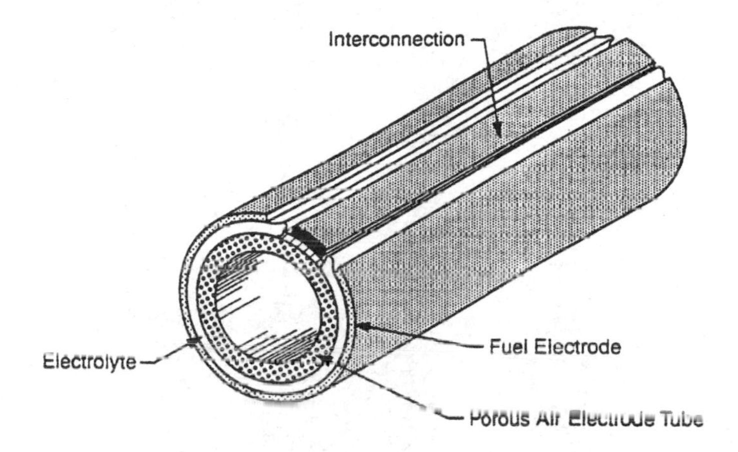

FIGURE 12-10
Air electrode supported type tubular solid oxide fuel cell design. (Courtesy of Westinghouse Electric Company, A Division of CBS Corporation, Pittsburgh, PA. Reprinted with permission.)

Corporation in the U.S.A. and Mitsubishi Heavy Industries in Japan. The cell element in this geometry consists of two porous electrodes separated by a dense oxygen ion-conducting electrolyte as depicted in Figure 12-10.[4] It uses ceramic tube operating at 1,000°C. The fuel cell is an assembly of such tubes. SureCELL™ (Trademark of Westinghouse Electric Corporation, Pittsburgh, Pennsylvania) is a solid oxide high temperature tubular fuel cell shown in Figure 12-11. It is being developed for multi-megawatt combined cycle gas turbine and fuel cell plants and targeted for distributed power generation and cogeneration plants of up to 60 MW capacity. It fits well for utility scale wind and photovoltaic power plants. Inside SureCELL, natural gas or other fuels are converted to hydrogen and carbon monoxide by internal reformation. No external heat or stream is needed. Oxygen ions produced from an air stream react with the hydrogen and carbon monoxide to generate electric power and high temperature exhaust gas.

Because of the closed-end tubular configuration, no seals are required and relative cell movement due to differential thermal expansions is not restricted. This enhances the thermal cycle capability. The tubular configuration solves many of the design problems facing other high temperature fuel cells. The target for the SureCELL development is to attain 75 percent overall efficiency, compared to 60 percent possible using only the gas turbine (Figure 12-12). Environmentally, the solid oxide fuel cell produces much lower CO_2, NOx and virtually zero SOx compared with other fuel cell technologies.

During the eight years of failure-free steady state operation of early prototypes, these cells were able to maintain the output voltage within 0.5 percent per 1,000 hours of operation. The second generation of the Westinghouse fuel cell shows voltage degradation of less then 0.1 percent per 1,000 hours of operation, with life in tens of thousands of hours of operation. The SureCELL prototype has been tested for over 1,000 thermal cycles with

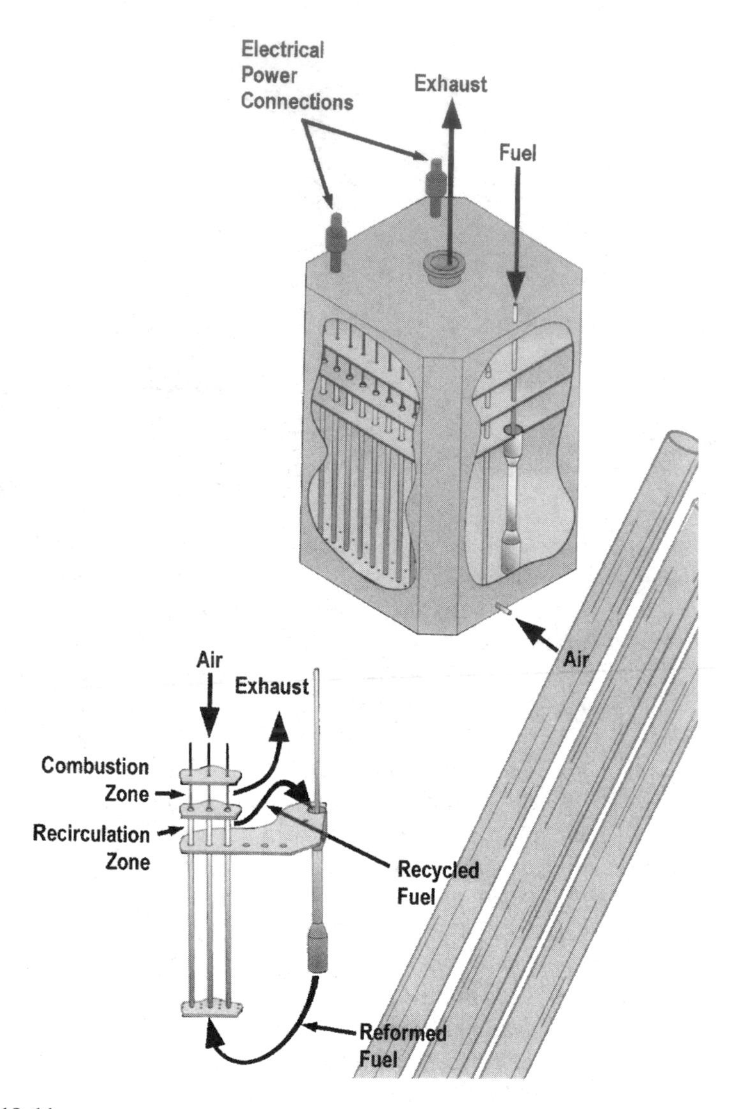

FIGURE 12-11
Seal-less solid oxide fuel cell power generator. (Courtesy of Westinghouse Electric Company, A Division of CBS Corporation, Pittsburgh, PA. Reprinted with permission.)

zero performance degradation, and 12,000 hours of operation with less then 1 percent performance degradation.

The transient electrical performance model of the fuel cell includes electrochemical, thermal, and mass flow elements that affects the electrical output.[5] Of primary interest is the electrical response of the cell to a load change. To design for the worst case, the performance is calculated under both the constant reactant flow and the constant inlet temperature.

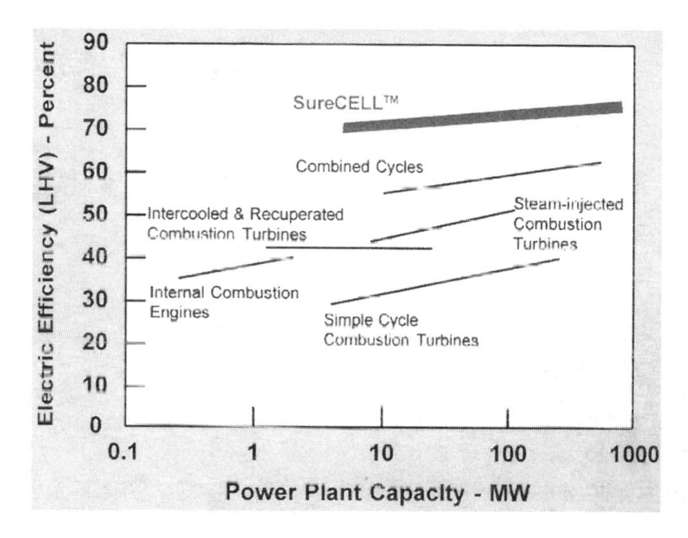

FIGURE 12-12
Natural gas power generation system efficiency comparison. (Courtesy of Westinghouse Electric Company, A Division of CBS Corporation, Pittsburgh, PA. Reprinted with permission.)

The German-American car-manufacturer Daimler Chrysler and Ballard Power Systems of Canada are developing the solid polymer fuel cell for automobiles as an alternative to the battery-powered vehicles. Their target is to sell the first commercial fuel cell powered car by the year 2004.

12.4.3 Mode Controller

The overall system must be designed for a wide performance range to accommodate the characteristics of the diesel generator (or fuel cell), the wind generator, and the battery. As and when needed, switching to the desired mode of generation is done by the mode controller. Thus, the mode controller is the central monitor and controller of the hybrid systems. It houses the microcomputer and software for the source selection, the battery management, and load shedding strategy. The mode controller performs the following functions:

- monitors and controls the health and state of the system.
- monitors and controls the battery state-of-charge.
- brings up the diesel generator when needed, and shuts off when not needed.
- sheds low priority loads in accordance with the set priorities.

The battery comes on-line by automatic transfer switch, which takes about 5 ms to connect to the load. The diesel, on the other hand, is generally brought

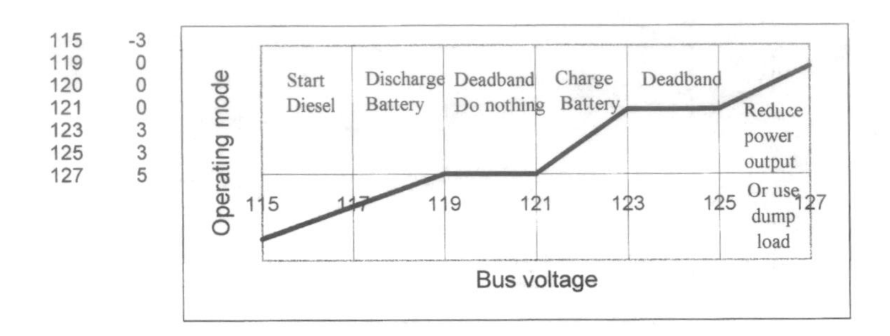

FIGURE 12-13
Mode controller deadbands eliminates the system chatter.

on-line, manually or automatically after going through the preplanned strategy algorithm. Even with automatic transfer switches, the diesel generator takes a long time to come on-line. Typically, this delay time is approximately 20 seconds.

The mode controller is designed and programmed with deadbands to avoid change over of the sources for correcting small variation on the bus voltage and frequency. The deadbands avoid chatters in the system. Figure 12-13 is an example of 120 volts hybrid system voltage-control regions. The deadbands are along the horizontal segments of the control line.

As a part of the overall system controller, the mode controller may incorporate the maximum power extraction algorithm. The dynamic behaviors of the closed-loop system, following common disturbances such as insolation changes due to cloud, wind fluctuation, sudden load changes and short circuit faults, are taken into account in a comprehensive design.[6]

12.4.4 Load Sharing

Since the wind, pv, battery, and diesel (or fuel cell) in various combinations, operate in parallel, the load sharing between them is one of the key design aspects of the hybrid system. For example, in the wind / diesel hybrid system (Figure 12-14), the electrical properties of the two systems must match so that they share load in proportion to their rated capacities.

For determining the load sharing, the two systems are first reduced to their respective Thevenin equivalent circuit model, in which each system is represented by its internal voltage and the series impedance. This is shown in Figure 12-14. The terminal characteristics of the two generators are then given by the following:

$$
\begin{aligned}
E_1 &= E_{01} - I_1 \cdot Z_1 \\
E_2 &= E_{02} - I_2 \cdot Z_2
\end{aligned}
\qquad (12\text{-}1)
$$

where subscripts 1 and 2 represent system 1 and 2 respectively, and

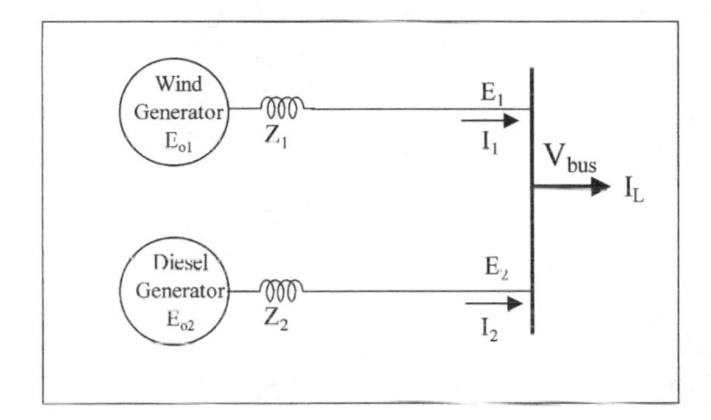

FIGURE 12-14
Thevenin's equivalent model of two sources in hybrid power system.

E_o = internally generated voltage
Z = internal series impedance
E = terminal voltage of each system

If the two generators are connected together, their terminal voltages E_1 and E_2 must be equal to the bus voltage V_{bus}. Additionally, the sum of the component loads I_1 and I_2 must be equal to the total load current I_L. Thus, the conditions imposed by the terminal connection are as follows:

$$E_1 = E_2 = V_{bus}$$

and

$$I_1 + I_2 = I_L \tag{12-2}$$

These imposed conditions, along with the machines internal characteristics E_o and Z, would determine the load sharing I_1 and I_2. The loading on individual generators is determined algebraically by solving the two simultaneous equations for the two unknowns, I_1 and I_2. Alternatively, the solution is found graphically as shown in Figure 12-15. In this method, E versus I characteristics of the two power systems are first individually plotted on the two sides of the current axis (horizontal). The distance between the two voltage axes (vertical) is kept equal to the total load current I_L. The electrical generators will share the load such that their terminal voltages are exactly equal, the condition imposed by connecting them together at the bus. This condition is met at the point of intersection of the two load lines. The point P in the figure, therefore, settles the bus voltage and the load sharing. The current I_1 and I_2 in the two generators are then read from the graph.

Controlling the load sharing requires controlling the E versus I characteristic of the machines. This may be easy in case of the separately excited DC or the synchronous generator used with the diesel engine. It is, however, difficult in case of the induction machine. Usually the internal impedance Z is fixed once the machine is built. Care must be exercised in the hybrid design

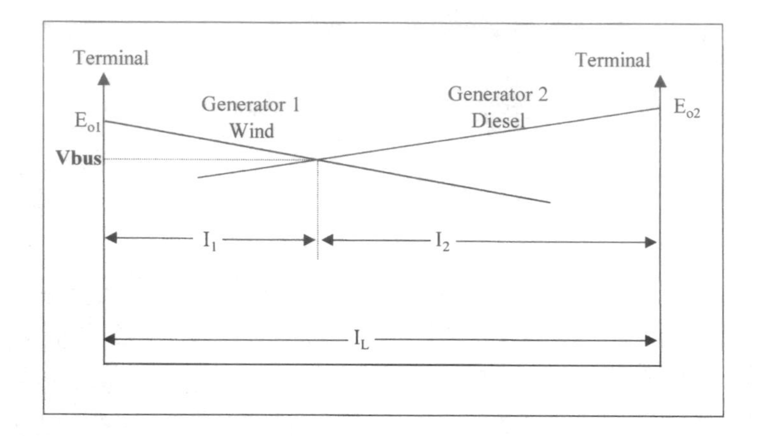

FIGURE 12-15
Graphical determination of load shared by two sources in hybrid power system.

to make sure that sufficient excitation control is built in for the desired load sharing between the two sources.

The load sharing strategy can vary depending on the priority of loads and the cost of electricity from alternative sources. In a wind-diesel system, for example, diesel electricity is generally more expensive than wind (~25 versus 5 cents per kWh). Therefore, all priority-1 (essential) loads are met first by wind as far as possible and then by diesel. If the available wind power is more than priority-1 loads, wind supplies part of priority-2 loads and the diesel is not run. If the wind power now fluctuates on the down side, the lower priority loads are shed to avoid running the diesel. If wind power drops further to cut into the priority-1 load, the diesel is brought on-line again. Water pumping and heater loads are examples of priority-2 loads.

12.5 System Sizing

For determining the required capacity of the stand-alone power system, estimating the peak load demand is only one aspect of the design. Estimating the energy required over the duration selected for the design is the first requirement for the system sizing.

12.5.1 Power and Energy Estimates

The system sizing starts with compiling a list of all loads that are to be served. Not all loads are constant. Time-varying loads are expressed in peak watts they consume and the duty ratio. The peak power consumption is used in determining the wire size for making a connection to the source.

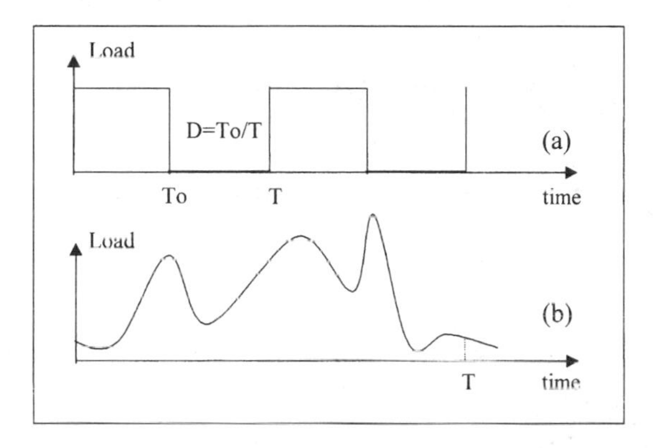

FIGURE 12-16
Duty ratio and peak power of intermittent loads.

The duty ratio is used in determining the contribution of individual load in the total energy demand. If the load has clean on-off periods as shown in Figure 12-16(a), then the duty ratio D is defined as D = To/T, where To is the time the load is on and T is the period of repetition. For irregularly varying loads shown in (b), the duty ratio is defined as the actual energy consumed in one period over the peak power times the period, i.e.:

$$D = \frac{\textit{Energy in watt} \cdot \textit{hours consumed in one repetition period}}{\textit{Peak power in watts} \cdot \textit{Repetition period in hours}} \tag{12-3}$$

The peak power consumption and the duty ratio of all loads are compiled, the product of the two is the actual share of the energy requirement of that load on the system during one repetition period. If there are distinct intervals in the period, say between the battery discharge and charge intervals, then the peak power and the duty ratio of each load are computed over the two intervals separately. As a simple example of this in a solar power system, one interval may be from 8 A.M. to 6 P.M. and the other from 6 P.M. to 8 A.M. The power table is then prepared as shown in Table 12-2.

TABLE 12-2

Power and Energy Compilation Table for Energy Balance Analysis

	8 A.M. to 6 P.M. (Interval A) (battery on charge)			**6 P.M. to 8 A.M. (Interval B) (battery on discharge)**		
Load	**Peak watts**	**Duty ratio**	**Energy per period Wh**	**Peak watts**	**Duty ratio**	**Energy per period Wh**
Load $_1$	P_{1a}	D_{1a}	E_{1a}	P_{1b}	D_{1b}	E_{1b}
Load $_2$	P_{2a}	D_{2a}	E_{2a}	P_{2b}	D_{2b}	E_{2b}
..	..	..	..	..	..	..
Load $_n$	P_{na}	D_{na}	E_{na}	P_{nb}	D_{nb}	E_{nb}
TOTAL	ΣP_a		ΣE_a	ΣP_b		ΣE_b

Total Battery Discharge Required = ΣE_b watts-hours

TABLE 12-3

NEC® Demand Factors (Adapted from National Electrical Code® Handbook, 7th Edition, 1996, Table 220-32)

Number of Dwellings	Demand Factor
3	0.45
10	0.43
15	0.40
20	0.38
25	0.35
30	0.33
40	0.28
50	0.26
>62	0.23

In a community of homes and businesses, not all connected loads draw power simultaneously. The statistical time staggering in their use times results in the average power capacity requirement of the plant significantly lower than the sum of the individually connected loads. The National Electrical Code® provides factors for determining the average community load in normal residential and commercial areas (Table 12-3). The average plant capacity is then determined as follows:

Required power system capacity = NEC® factor from Table 12-3
× Sum of connected loads.

12.5.2 Battery Sizing

The battery Ah capacity required to support the load energy requirement of E_{bat} as determined using a method of Table 12-2 or equivalent:

$$Ah = \frac{E_{bat}}{\eta_{disch}\left[N_{cell} \cdot V_{disch}\right] \cdot DoD_{allowed} \cdot N_{bat}} \tag{12-4}$$

where E_{bat} = energy required from the battery per discharge
η_{disch} = efficiency of discharge path, including inverters, diodes, wires, etc.
N_{cell} = number of series cells in one battery
V_{disch} = average cell voltage during discharge
$DOD_{allowed}$ = maximum DOD allowed for the required cycle life
N_{bat} = number of batteries in parallel

The following example illustrates the use of this formula to size the battery. Suppose we want to design a battery for a stand-alone power system, which charges and discharges the battery from 110 volts DC solar array. For the

DC-DC buck converter that charges the battery, the maximum available battery-side voltage is 70 volts for it to work efficiently in the PWM mode. For the DC-DC boost converter discharging the battery, the minimum required battery voltage is 45 volts. Assuming that we are using NiMH battery, the cell voltage can vary from 1.55 when fully charged to 1.1 when drained to the maximum allowable DOD. Then, the number of cells needed in the battery is less than 70/1.55 = 45 cells and more than 45/1.1 = 41. Thus, the number of cells required in the battery from the voltage considerations is between 41 and 45. It is generally more economical to use fewer cells of higher capacity than more lower-capacity cells. We, therefore, select 41 cells in the battery design.

Now again for an example, let us assume that the battery is required to discharge 2 kW load for 14 hours (28,000 Wh) every night for five years before replacement. The life requirement is, therefore, 5 × 365 = 1,825 cycles of deep discharge. For the NiMH battery, the cycle life at full depth of discharge is 2,000. Since this is greater than the 1,825 cycles required, we can fully discharge the battery every night for five years. If the discharge efficiency is 80 percent, the average cell discharge voltage is 1.2 V, and we desire three batteries in parallel for reliability, each battery Ah capacity calculated from the above equation is as follows:

$$Ah = \frac{28000}{0.80 \cdot [41 \cdot 1.2] \cdot 1.0 \cdot 3} = 237 \qquad (12\text{-}5)$$

Three batteries, each having 41 series cells of 237 Ah capacity, therefore, will meet the system requirement. Margin must be allowed to account for the uncertainty in estimating the loads.

12.5.3 pv Array Sizing

The basic tenet in sizing the stand-alone "power system" is to remember that it is really the stand-alone "energy system." It must, therefore, maintain the energy balance over the specified period. The energy drained during lean times must be made up by the positive balance during the remaining time of the period. A simple case of a constant load on the pv system using solar arrays perfectly pointing toward the sun normally for 10 hours of the day is shown in Figure 12-17 to illustrate the point. The solar array is sized such that the two shaded areas on two sides of the load line must be equal. That is, the area **oagd** must be equal to the area **gefb**. The system losses in the round trip energy transfers, e.g., from and to the battery, adjust the available load to a lower value as shown by the dotted line.

In general, the stand-alone system must be sized so as to satisfy the following energy balance equation over one period of repetition.

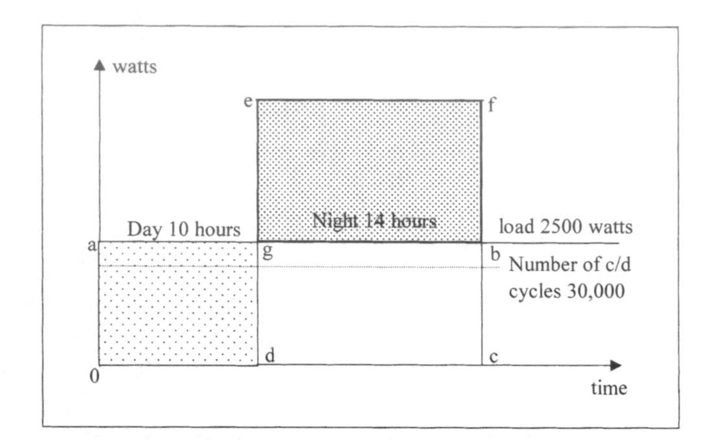

FIGURE 12-17
Energy balance analysis over one load cycle.

$$\int_{8\ A.M.}^{6\ P.M.} (solar\ radiation \cdot conversion\ efficiency) \cdot dt =$$

$$\int_{8\ A.M.}^{6\ P.M.} (loads + losses + charge\ power + shunt\ power) \cdot dt + \quad (12\text{-}6)$$

$$\int_{6\ P.M.}^{8\ A.M.} (loads + losses) \cdot dt$$

Or, in discrete time intervals of constant load and source power:

$$\sum_{8\ A.M.}^{6\ P.M.} (solar\ radiation \cdot conversion\ efficiency) \cdot \Delta t = \quad (12\text{-}7)$$

$$\sum_{8\ A.M.}^{6\ P.M.} (loads + losses + charge\ power + shunt\ power) \cdot \Delta t + \sum_{6\ P.M.}^{8\ A.M.} (loads + losses) \cdot \Delta t$$

12.6 Wind Farm Sizing

In a stand-alone wind farm, selecting the number of towers and the battery size depend on the load power availability requirement. A probabilistic

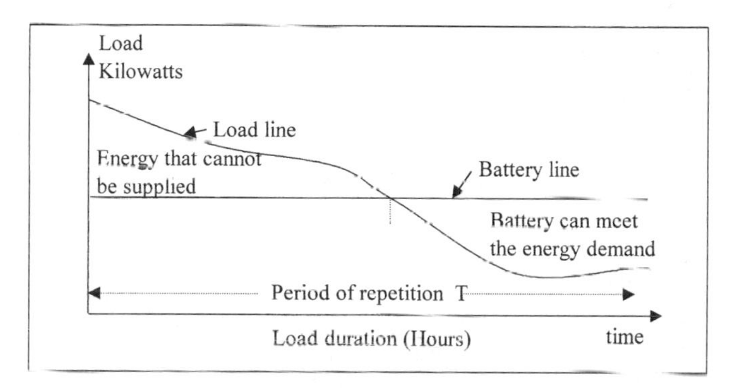

FIGURE 12-18
Effect of battery size on load availability for given load duration curve.

model can determine the number of towers and the size of the battery storage required for meeting the load with required certainty. Such a model can also be used to determine the energy to be purchased from or injected into the grid if the wind power plant was connected to the grid. In the probabilistic model, the wind speed is taken as the random variable. The load is treated as an independent variable. The number of wind turbines and the number of batteries are also the variables. Each turbine in a wind farm may or may not have the same rated capacity and the same outage rate. In any case the hardware failure rate is independent of each other. The resulting model has the joint distribution of the available wind power (wind speed variations), and the operating mode (each turbine working or not working). The events of these two distributions are independent. For a given load duration curve over a period of repetition, the expected energy not supplied to the load by the hybrid system clearly depends on the size of the battery as shown in Figure 12-18. The larger the battery, the higher the horizontal line, thus decreasing the duration of the load not supplied by the systems.

With such a probabilistic model, the expected product of power and time during which the power is not available is termed as the Expected Energy Not Supplied (EENS). This is given by the shaded area on the left hand side. The Energy Index of Reliability (EIR) is then given by the following:

$$EIR = 1 - \frac{EENS}{E_o} \qquad (12\text{-}8)$$

where E_o is energy demand on the system over the period under consideration, which is the total area under the load duration curve. The results of such probabilistic study[6-7] are shown in Figure 12-19, which indicates the following:

- the higher the number of wind turbines, the higher the EIR.
- the larger the battery size, the higher the EIR.
- the higher the requirement on EIR, the higher the number of required towers and batteries increasing the capital cost of the project.

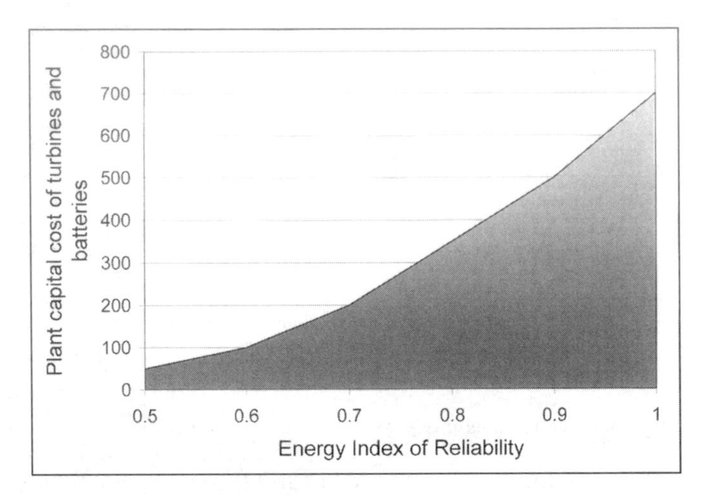

FIGURE 12-19
Relative capital cost versus EIR with different numbers of wind turbines and battery sizes.

Setting unnecessarily high EIR requirement can make the project uneconomical. For that reason, the Energy Index of Reliability must be set after a careful optimization of the cost and the consequences of not meeting the load requirement during some portion of the time period.

References

1. "Institute of Solar Energy and Technology Annual Report," Kassel, Germany, 1997.
2. Wang, L. and Su, J. 1997. "Dynamic performance of an isolated self-excited induction generator under various load conditions," *IEEE Power Engineering Paper No. PE-230-EC-1-09,* 1997.
3. Bonarino, F., Consoli, A., Baciti, A., Morgana, B. and Nocera, U. 1998. "Transient analysis of integrated diesel-wind-photovoltaic generation systems," *IEEE Paper No. PE-425-EC-104,* July 1998.
4. Singhal, S. C. 1996. "Status of solid oxide fuel cell technology," *Proceedings of the 17th Risø International Symposium on Material Science, High Temperature Electrochemistry, Ceramics and Metals, Roskilde,* Denmark, Sept. 1996.
5. Hall, D. J. and Colclaser, R. G. 1998. "Transient modeling and simulation of a tubular solid oxide fuel cell," *IEEE Paper No. PE-100-EC-004,* July 1998.
6. Abdin, E. S., Asheiba, A. M., and Khatee, M. M. 1998. "Modeling and optimum controllers design for a stand-alone photovoltaic-diesel generating unit," *IEEE Paper No. PE-1150-0-2,* 1998.
7. Baring-Gould, E. I. 1996. "Hybrid2, The hybrid system simulation model user manual," *NREL Report No. TP-440-21272,* June 1996.

13

Grid-Connected System

The wind and photovoltaic power systems have made a successful transition from small stand-alone sites to large grid-connected systems. The utility interconnection brings a new dimension in the renewable power economy by pooling the temporal excess or the shortfall in the renewable power with the connecting grid. This improves the overall economy and the load availability of the renewable plant; the two important factors of any power system. The grid supplies power to the site loads when needed, or absorbs the excess power from the site when available. One kWh meter is used to record the power delivered to the grid, and another kWh meter is used to record the power drawn from the grid. The two meters are generally priced differently.

Figure 13-1 is a typical circuit diagram of the grid-connected photovoltaic power system. It interfaces with the local utility lines at the output side of the inverter as shown. A battery is often added to meet short term load peaks. In the United States, the Environmental Protection Agency sponsors grid-connected pv programs in urban areas where wind towers would be impractical. In recent years, large building-integrated photovoltaic installations have made significant advances by adding the grid-interconnection in the system design. Figure 13-2 shows the building-integrated pv system on the roof of the Northeastern University Student Center in Boston, MA. The project was part of the EPA PV DSP Program. The system produces 18 kW pv power and is connected to the grid. In addition, it collects sufficient research data using numerous instruments and computer data loggers. The vital data are sampled every 10 seconds, and then are averaged and stored every 10 minutes. The incoming data includes information about the air temperature and wind speed. The performance parameters include the DC voltage and current generated by the pv roof, and the AC power on the inverter output side.

In the United Kingdom, a 390 square meter building-integrated pv system has been in operation since 1995 at the University of Northumbria, Newcastle (Figure 13-3). The system produces 33,000 kWh electricity per year and is connected to the grid. The pv panels are made of monocrystalline cells with the photoconversion efficiency of 14.5 percent.

On the wind side, most grid-connected systems are large utility-scale power plants. A typical equipment layout in such plants is shown in Figure 13-4. The wind generator output is at 480 volts AC, which is raised

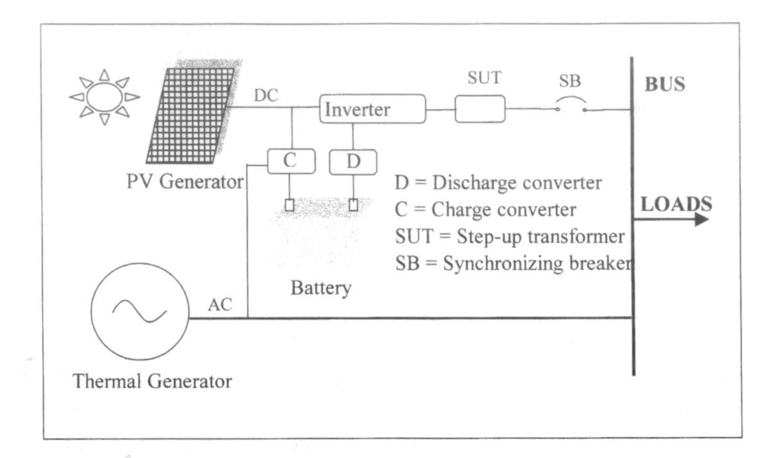

FIGURE 13-1
Electrical schematic of the grid-connected photovoltaic system.

FIGURE 13-2
18 kW grid-connected pv system on the Northeastern University Student Center in Boston, MA. (Source: ASE Americas, Billerica, Massachusetts. With permission.)

to an intermediate level of 21 kV by a pad-mounted transformers. An overhead transmission line provides the link to the site substation, where the voltage is raised again to the grid level. The site computer, sometimes using mulitplexer and remote radio links, controls the wind turbines in response to the wind conditions and the load demand.

project	University of Northumbria
location	Newcastle
consulting engineer	Ove Arup and Partners
date completed	October 1995
area of solar facade	390m^2
electricity generated	33,000 kWh/year
cell material	monocrystalline
efficiency	14.5%
number of panels	465
orientation	16° east of south
angle of orientation	25°
grid connected	yes

FIGURE 13-3
Grid-connected pv system at the University of Northumbria, Newcastle, U.K. The 390 square meter monocrystalline modules produce 33,000 kWh per year. (Source: Professional Engineer, Publication of the Institution of Mechanical Engineer's, London. With permission.)

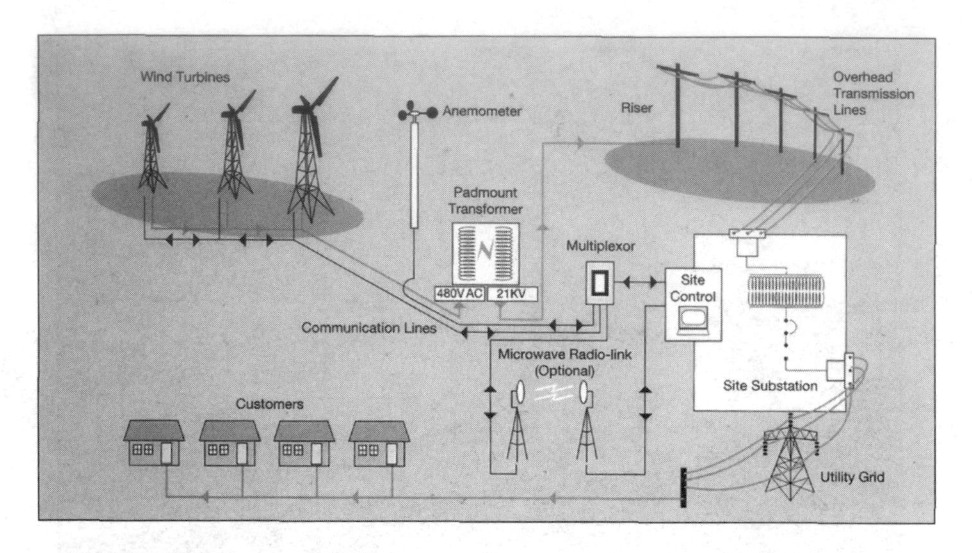

FIGURE 13-4
Electrical component layout of the grid-connected wind power system. (Source: AWEA/IEA/CADDET Technicla Brochure, 1995.)

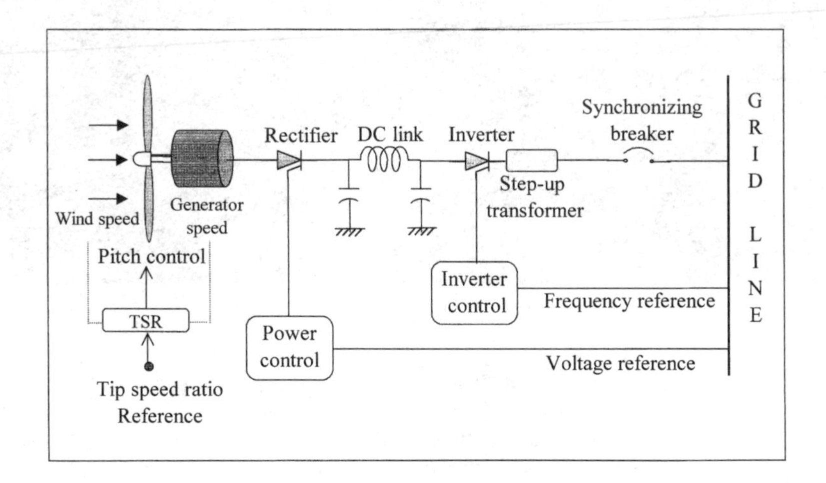

FIGURE 13-5
Electrical schematic of the grid-connected variable speed wind power system.

Large wind systems being installed now tend to have the variable-speed design. The power schematic of such a system is shown in Figure 13-5. The variable-frequency generator output is first rectified into DC, and then inverted into a fixed-frequency AC. Before the inversion, the rectifier harmonics are filtered out from the DC by the inductor and capacitors. The frequency reference for the inverter firing and the voltage reference for the rectifier phase-angle control are taken from the grid lines. The optimum reference value of the tip-speed ratio is stored and continuously compared with the value computed from the measured speeds of the wind and the rotor. The turbine speed is accordingly changed to assure maximum power production at all times.

13.1 Interface Requirements

Both the wind and the pv systems interface the grid at the output terminals of the synchronizing breaker at the output end of the inverter. The power flows in either direction depending on the site voltage at the breaker terminals. The fundamental requirements on the site voltage for interfacing with the grid are as follows:

- the voltage magnitude and phase must equal to that required for the desired magnitude and direction of the power flow. The voltage is controlled by the transformer turn ratio and/or the rectifier/inverter firing angle in a closed-loop control system.
- the frequency must be exactly equal to that of the grid, or else the system will not work. To meet the exacting frequency requirement, the only effective means is to use the utility frequency as a reference for the inverter switching frequency.
- in the wind system, the synchronous generators of the grid system supply magnetizing current for the induction generator.

The interface and control issues are similar in many ways between both the pv and the wind systems. The wind system, however, is more involved since the electrical generator and the turbine with large inertia introduce certain dynamic issues not applicable in the static pv system. Moreover, wind plants generally have much greater power capacity than the pv plants. For example, many wind plants that have been already installed around the world have capacity in tens of MW each. The newer wind plants in the hundreds of MW capacity are being installed and more are planned.

13.2 Synchronizing with Grid

The synchronizing breaker in Figures 13-1 and 13-5 has internal voltage and phase angle sensors to monitor the site and grid voltages and signal the correct instant for closing the breaker. As a part of the automatic protection circuit, any attempt to close the breaker at an incorrect instant is rejected by the breaker. Four conditions which must be satisfied before the synchronizing switch will permit the closure are as follows:

- the frequency must be as close as possible with the grid frequency, preferably about one-third of a hertz higher.
- the terminal voltage magnitude must match with that of the grid, preferably a few percent higher.
- the phase sequence of the two three-phase voltages must be the same.
- the phase angle between the two voltages must be within 5 degrees.

Taking the wind power system as an example, the synchronizing process specifically runs as follows:

1. With the synchronizing breaker open, the wind power generator is brought up to speed using the machine in the motoring mode.
2. Change the machine into the generating mode, and adjust the controls such that the site and grid voltages match to meet the above requirements as close as possible.
3. The match is monitored by the synchroscope or three synchronizing lamps, one in each phase (Figure 13-6). The voltage across the

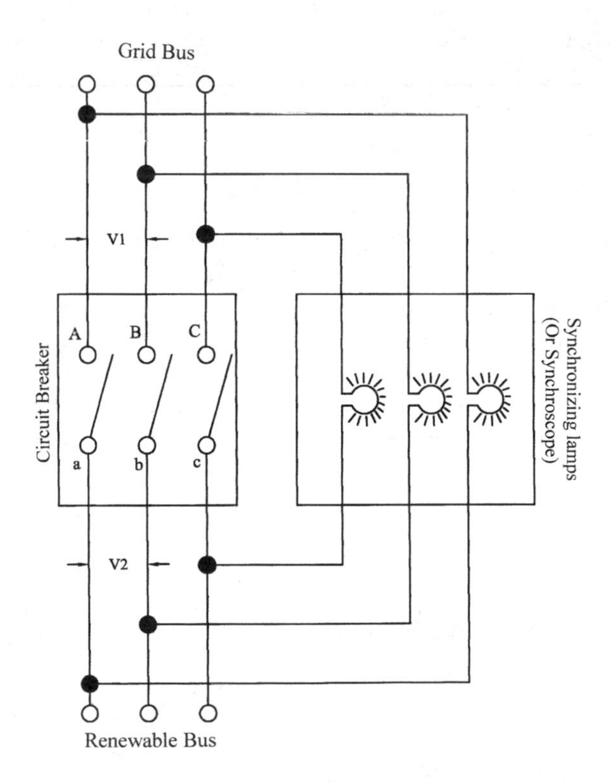

FIGURE 13-6
Synchronizing circuit using three synchronizing lamps or the synchroscope.

lamp in each phase is the difference between the renewable site voltage and the grid voltage at any instant. When the site and the grid voltages are exactly equal in all three phases, all three lamps will be dark. However, it is not enough for the lamps to be dark at any one instant. They must remain dark for a long time. This condition which will be met only if the generator and the grid voltages have nearly the same frequency. If not, one set of the two three-phase voltages will rotate faster relative to the other, and the phase difference between the two voltages will light the lamps.

4. The synchronizing breaker is closed if the lamps remain dark for ¼ to ½ second.

Following the closure, any small mismatch between the site voltage and the grid voltage will circulate the inrush current between the two such that the two systems will come to perfect synchronous operation.

13.2.1 Inrush Current

The small unavoidable difference between the site and the grid voltages will result in an inrush current to flow between the site and the grid. The inrush current eventually decays to zero at an exponential rate that depends on the internal resistance and inductance. The initial magnitude of this current in the instant the circuit breaker is closed depends on the degree of mismatch between the two voltages. It is not all bad, as it produces the synchronizing power which acts to bring the two systems in synchronous lock. However, it produces a mechanical torque step, setting up the electromechanical oscillations before the two machines come into synchronism and get locked with each other. The magnitude of the inrush current is calculated as follows:

Let ΔV be the difference between the site voltage and the grid voltage at the closing instant due to any reason. Since this voltage is suddenly applied on the system, the resulting inrush current is determined by the subtransient reactance of the machine X_d''. That is as follows:

$$I_{inrush} = \frac{\Delta V}{X_d''} \tag{13-1}$$

The inrush current is primarily reactive, as is solely determined by X_d''. Its magnitude is kept within the allowable limit, else the thermal or mechanical damage may result.

The synchronizing power produced by the inrush current brings the wind system and the grid in synchronism after the oscillations decay out. Once synchronized, the generator has a natural tendency to remain in synchronism with the grid, although it can fall out of synchronization if excessive load is extracted, large load steps are applied, or during system faults. Small perturbation swings in the load angle decay out over a time, restoring the

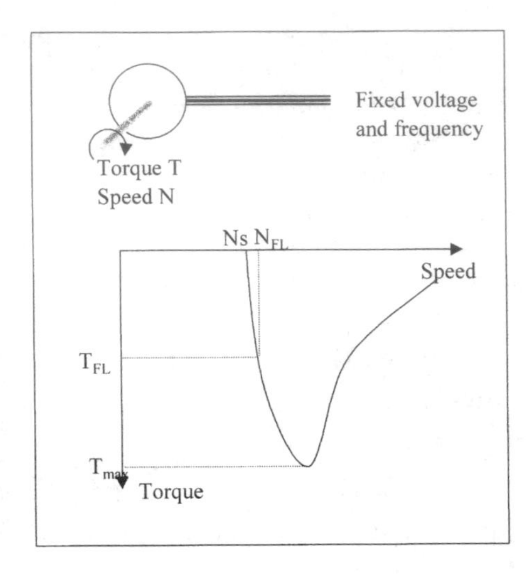

FIGURE 13-7
Resilience in the torque versus speed characteristic of induction generator.

synchronous condition. The magnitude of the restoring power, also known as the synchronizing power, is highest if the machine is running at no load, and is zero if it is running at its steady state stability limit.

13.2.2 Synchronous Operation

Once synchronized, the voltage and frequency of the wind system need to be controlled. When the induction generator is directly connected to the grid, the grid serves as the frequency reference for the generator output frequency. The grid also acts as the excitation source supplying the reactive power. Since the torque versus speed characteristic of the induction generator has a steep slope near zero slip (Figure 13-7), the speed of the wind turbine remains approximately constant within a few percentages. Higher load torque is met by increased slip up to a certain point (Q_m), beyond which the generator becomes unstable. If the load torque is immediately reduced, the generator will return to the stable operation. From the operating point of view, the induction generator is softer, as opposed to the relatively stiff operation of the synchronous generator, which works at an exact constant speed or falls out of stability.

If the synchronous generator is used, as in wind farms installed in California in the 1980s, the voltage is controlled by controlling the rotor field excitation current. The frequency control, however, is not required on a continuous basis. Once synchronized and connected with the lines, the synchronous generator has an inherent tendency to remain in synchronous lock with the grid. Only during transients and system faults, the synchronism can be lost. In such cases the generator must be resynchronized.

In the variable-speed induction generator system using the inverter at the interface, the inverter gate signal is derived from the grid voltage to assure synchronism. The inverter stability depends a great deal on the design. For example, with line commutated inverter, there is no stability limit. The power limit in this case is the steady state load limit of the inverter with any short-term overload limit.

13.2.3 Load Transient

During steady state operation, if the renewable power system output is fully or partially lost, the grid will pick up the area load. The effect of this will be felt in two ways:

- the grid generators slow down slightly to increase their power angle needed to make up for the lost power. This will result in a momentary drop in frequency.
- small voltage drop results throughout the system, as the grid conductors carry more load.

The same effects are felt if a large load is suddenly switched in at the green power site, starting the wind turbine as the induction motor draws a large current. This will result in the above effect. Such load transients are minimized by soft-starting large generators. In wind farms consisting of many generators, individual generators are started in sequence, one after another.

13.2.4 Safety

Safety is a concern when renewable power is connected to the utility grid lines. The interconnection may endanger the utility repair crew working on the lines by continuing to feed power into the grid even when the grid itself went down. This issue has been addressed by including an internal circuit that takes the inverter off line immediately if the system detects grid outage. Since this circuit is critical for human safety, it has a built-in redundancy.

The site-grid interface breaker can get suddenly disconnected, accidentally or to meet an emergency situation. The high wind speed cut out is a usual condition when the power is cut off to protect the generator from overloading. In systems where large capacitors are connected at the wind site for power factor improvement, the site generator would still be in the self-excitation mode, drawing excitation power from the capacitors and generating terminal voltage. In absence of such capacitors, one would assume that the voltage at the generator terminals would come down to zero. The line capacitance, however, can keep the generator self excited. The protection circuit is designed to avoid both of these situations, which are potential safety hazards to unsuspecting site crew.

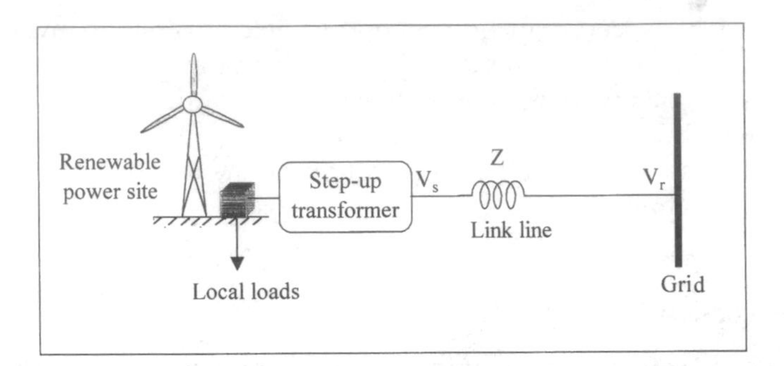

FIGURE 13-8
Equivalent circuit of renewable power plant connected to grid via transmission line link.

When the grid is disconnected for any reason, the generator will experience a loss of frequency regulation, as the frequency synchronizing signal derived from the grid lines is now lost. When a change in frequency is detected beyond a certain limit, the automatic control can shut down the system, cutting off all possible sources of excitation.

13.3 Operating Limit

The link line connecting the renewable power site with the utility grid introduces the operating limit in two ways, the voltage regulation and the stability limit. In most cases, the line can be considered as an electrically short transmission line. The ground capacitance and the ground leakage resistance are generally negligible and are ignored. The equivalent circuit of such a line, therefore, reduces to a series resistance R and reactance L (Figure 13-8). Such an approximation is valid in lines up to 50 miles long. The line carries power from the renewable site to the utility grid, or from the grid to the renewable site to meet local peak demand. There are two major effects of the transmission line impedance, one on the voltage regulation and the other on the maximum power transfer capability of the link line.

13.3.1 Voltage Regulation

The phasor diagram of the voltage and current at the sending and receiving ends are shown in Figure 13-9. Since the shunt impedance is negligible, the sending end current I_s is the same as the receiving end current I_r, i.e., $I_s = I_r = I$. The voltage at the receiving end is the vector sum of the sending end voltage plus the impedance voltage drop I·Z in the line, i.e.:

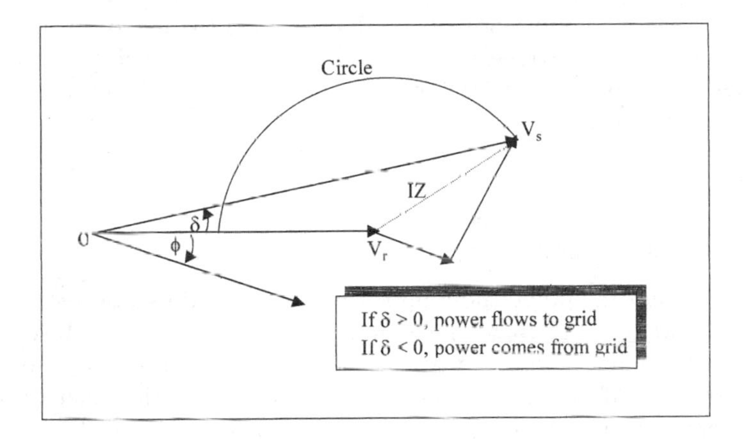

FIGURE 13-9
Phasor diagram of the link line carrying rated current.

$$V_s = V_r + I(R + jX) \tag{13-2}$$

The voltage regulation is defined as the rise in the receiving end voltage, expressed in percent of the full load voltage, when full load at a specified power factor is removed, holding the sending end voltage constant. That is as follows:

$$\textit{percent voltage regulation} = \frac{V_{nl} - V_{fl}}{V_{fl}} \times 100 \tag{13-3}$$

where V_{nl} = magnitude of receiving end voltage at no load = V_s
V_{fl} = magnitude of receiving end voltage at full load = V_r

With reference to the phasor diagram of Figure 13-9, $V_{nl} = V_s$ and $V_{fl} = V_r$.

The voltage regulation is a strong function of the load power factor. For the same load current at different power factors, the voltage drop in the line is the same, but is added to the sending end voltage at different phase angles to derive the receiving end voltage. For this reason, the voltage regulation is greater for lagging power factor, and the least or even negative for leading power factor.

In Figure 13-9, suppose the magnitude of V_r and I are held constant and the power factor of the load is varied from zero lagging to zero leading. The vector V_s will vary such that its end point will lie on a semicircle since the magnitude $I \cdot (R + jX)$ is constant.[1] Such a circle diagram is useful for plotting the sending end voltage versus load power factor for the given load voltage and KVA.

If the voltages at both ends of the lines are held constant in magnitude, the receiving end real power and reactive power points plotted for several loads would lie on a circle known as the power circle diagram. The reader is referred to Stevenson[1] for further reading on the transmission line circle diagrams.

13.3.2 Stability Limit

The direction of the power flow depends on the sending and receiving end voltages, and the electrical phase angle between the two. However, the maximum power the line can transfer while maintaining stable operation has a limit. We derive below the stability limit assuming that the power flows from the renewable power site to the grid, although the same limit applies in the reverse direction as well. The series resistance in most lines is negligible, hence, is ignored here.

The power transferred to the grid by the transmission line is as follows:

$$P = V_r I \cos \phi \qquad (13\text{-}4)$$

Using the phasor diagram of Figure 13-9, the current I can be expressed as follows:

$$I = \frac{V_s - V_r}{jX} = \frac{V_s \angle\delta - V_r \angle o}{jX} = \frac{V_s(\cos\delta + j\sin\delta) - V_r}{jX} \qquad (13\text{-}5)$$

The real part of this current is as follows:

$$I_{real} = \frac{V_s \cdot \sin\delta}{X} \qquad (13\text{-}6)$$

This, when multiplied with the receiving end voltage Vr, gives the following power:

$$P = \frac{V_s \cdot V_r}{X} \sin\delta \qquad (13\text{-}7)$$

Thus, the magnitude of the real power transferred by the line depends on the power angle δ. If $\delta > 0$, the power flows from the site to the grid. On the other hand, if $\delta < 0$, the site draws power from the grid.

The reactive power depends on (Vs–Vr). If Vs > Vr, the reactive power flows from the site to the grid. If Vs < Vr, the reactive power flows from the grid to the site.

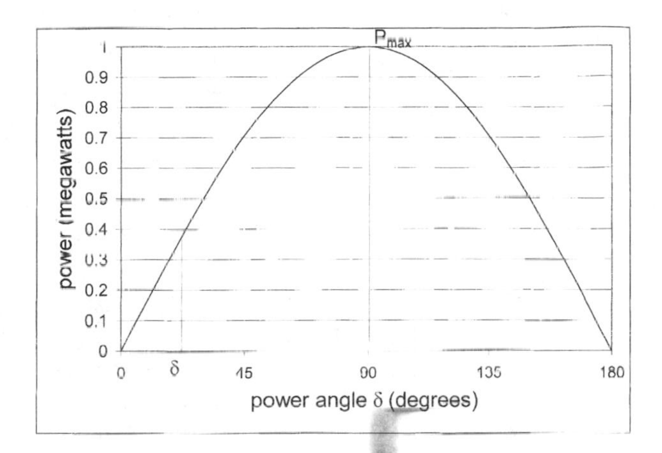

FIGURE 13-10
Power versus power angle showing static and dynamic stability limits of the link line.

Obviously, the power flow in either direction is maximum when $\delta = 90°$ (Figure 13-10). Beyond P_{max}, the link line becomes unstable and will fall out of synchronous operation. That is, it will lose its ability to synchronously transfer power from the renewable power plant to the utility grid. This is referred to as the steady state stability limit. In practice, the line loading must be kept well below this limit to allow for transients such as sudden load steps and system faults. The maximum power the line can transfer without losing the stability even during system transients is referred to as the dynamic stability limit. In typical systems, the power angle must be kept below 10° to 20° to assure dynamic stability.

Since the generator and the link line are in series, the internal impedance of the generator is added in the line impedance for determining the maximum power transfer capability of the link line, the dynamic stability and the steady state performance.

13.4 Energy Storage and Load Scheduling

For large wind and pv plants on grid, it may be economical to store some energy locally in the battery or other energy storage systems. The short-term peak demand is met by the battery without drawing from the grid and paying the demand charge. For formulating the operating strategy for scheduling and optimization, the system constraints are first identified. The usual constraints are then battery size, the minimum on/off times and ramp rates for the thermal units, the battery charge and discharge rates, and the renewable capacity limits. The optimization problem is formulated to minimize

TABLE 13-1

Production Cost of 300 MW Thermal-pv-Battery System

System Configuration	Battery Depletion MWh/day	Production cost $/day	Savings $/day
Thermal only	—	750,000	—
Thermal + pv	—	710,000	40,000
Thermal + pv + battery	344	696,000	54,000

the cost of all thermal and renewable units combined subject to the constraints by arriving at the best short-term scheduling. This determines the hours for which the baseload thermal units of the electrical power company should be taken either off-line or on-line. The traditional thermal scheduling algorithms, augmented Lagrangian relaxation, branch and bound, successive dynamic programming or heuristic method (genetic algorithms and neural networks), can be used for minimizing the cost of operating the thermal units with a given renewable-battery system. Marwali et al.[2] has recently utilized the successive dynamic programming to find the minimum cost trajectory for battery and the augmented Langrangian to find thermal unit commitment. In a case study of a 300 MW thermal-pv-battery power plant, the authors have arrived at the total production costs shown in Table 13-1, where the battery hybrid system saves $54,000 per day compared to the thermal only.

13.5 Utility Resource Planning Tool

The wind and photovoltaic power, in spite of their environmental, financial, and fuel diversity benefits, are not presently included in the utility resource planning analysis because of the lack of the familiarity and analytical tools for nondispatchable sources of power. The wind and pv powers are treated as nondispatchable for not being available on demand. The Massachusetts Institute of Technology's Energy Laboratory has developed an analytical tool to analyze the impact of nondispatchable renewables on the New England's power systems operation. Cardell and Connors[3] have applied this tool for analyzing two hypothetical wind farms totaling 1,500 MW capacity for two sites, one in Maine and the other in Massachusetts. The average capacity factor at these two sites is estimated to be 0.25. This is good, although some sites in California have achieved the capacity factor of 0.33 or higher. The MIT study shows that the wind energy resource in New England is comparable to that in California. The second stage of their analysis developed the product cost model, demonstrating the emission and fuel cost risk mitigation benefits of the utility resource portfolios incorporating the wind power.

References

1. Stevenson, W. D. 1962. "Elements of Power System Analysis," New York, McGraw Hill Book Company, 1962.
2. Marwali, M. K. C., Halil, M., Shahidehpour, S. M., and Abdul-Rahman, K. H. 1997. "Short-term generation scheduling in photovoltaic-utility grid with battery storage," *IEEE paper No. PE-184-PWRS-16-09*, 1997.
3. Cardell, J. B. and Connors, S. R. 1997. "Wind power in New England, modeling and analyses of nondispatchable renewable energy technologies," *IEEE Paper No. PE-888-PWRS-2-06*, 1997.

14

Electrical Performance

14.1 Voltage Current and Power Relations

The power systems worldwide are 60 Hz or 50 Hz AC three-phase systems. The three coils (phases) of the generator are connected in Y or Δ as shown in Figure 14-1. In the balanced three-phase operation, the line-to-line voltage, the line current and the three phase power in terms of the phase voltage and current are given by the following equations, with notations marked in Figure 14-1.

In the Y connected system:

$$
\begin{aligned}
V_L &= \sqrt{3}\, V_{ph} \\
I_L &= I_{ph} \\
P_{3\text{-}ph} &= \sqrt{3}\, V_L I_L \text{ pf}
\end{aligned}
\qquad (14\text{-}1)
$$

Where pf = power factor

In the Δ connected system:

$$
\begin{aligned}
V_L &= V_{ph} \\
I_L &= \sqrt{3}\, I_{ph} \\
P_{3\text{-}ph} &= \sqrt{3}\, V_L I_L \text{ pf}
\end{aligned}
\qquad (14\text{-}2)
$$

For steady state or dynamic performance studies, the system components are modeled so as to represent the entire system. The power generator, the rectifier, the inverter, and the battery models were discussed in the earlier chapters. The components are accurately modeled to represent the conditions under which the performance is to be determined. This chapter concerns itself with the system level performance.

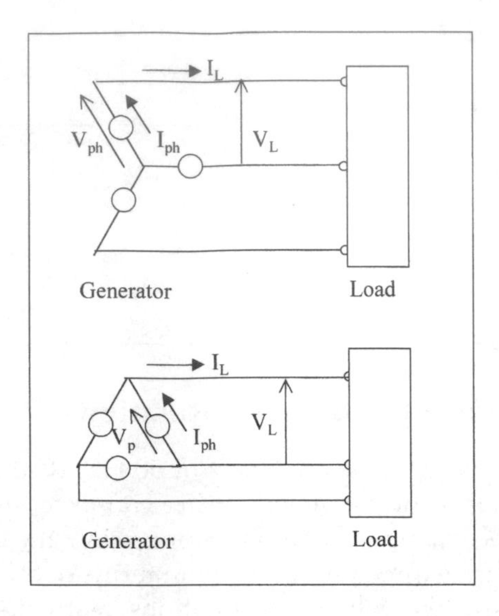

FIGURE 14-1
Three-phase AC systems is connected in Y or Δ.

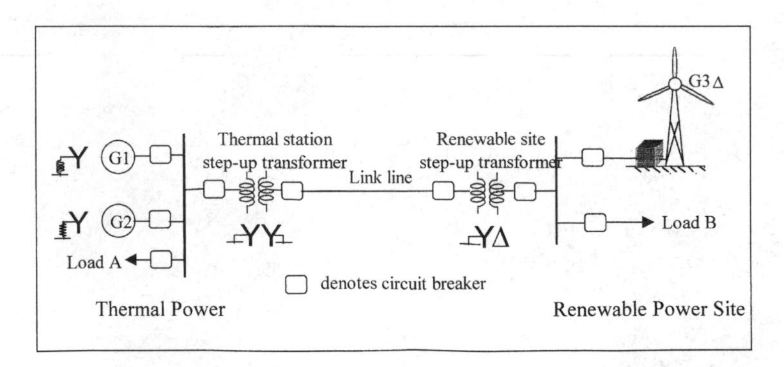

FIGURE 14-2
One-line schematic diagram of grid-connected wind farm.

One-line diagram is widely used to represent the three phases of the system. Figure 14-2 is an example of a one-line diagram of the grid-connected system. On the left hand side are two Y-connected synchronous generators, one grounded through a reactor and one grounded through a resistor, supplying power to load A. On the right hand side is the wind power site with one Δ-connected induction generator, supplying power to load B, feeding the remaining power to the grid via the step up transformer, the circuit breakers, and the transmission line.

The balanced three-phase system is analyzed as the single-phase system. The neutral wire in the Y connection does not enter the analysis in any way, since it is at zero voltage and carries zero current.

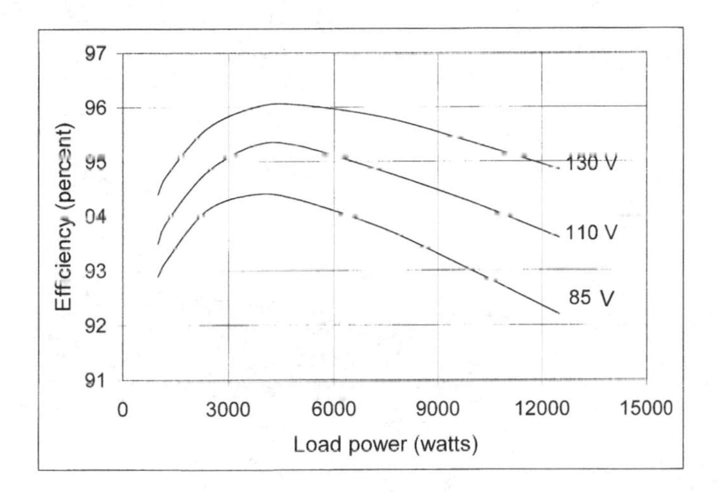

FIGURE 14-3
Power equipment efficiency varies with load with single maximum.

The unbalanced system, the balanced three-phase voltage on unbalanced load and faults need advanced analytical methods, such as the method symmetrical components.

14.2 Component Design for Maximum Efficiency

An important performance criterion of any system is the efficiency, measured as the power output as percentage of the power input. Since the system is as efficient as its components are, designing an efficient system means designing each component to operate at its maximum efficiency.

The electrical and electronic components while transferring power from the input side to the output side lose some power in the form of heat. In practical designs, the maximum efficiency of 90 to 98 percent is typical in large power equipment in hundreds of kW ratings, and 80 to 90 percent in small equipment in tens of kW ratings. The component efficiency, however, varies with load as shown in Figure 14-3. The efficiency increases with load up to a certain point beyond which it decreases. A good design maximizes the efficiency at the load that the equipment supplies most of the time. For example, if the equipment is loaded at 70 percent of its rated capacity most of the time, it is beneficial to have the maximum efficiency at 70 percent load. The method of achieving the maximum efficiency at a desired load level is presented below.

The total loss in any power equipment generally has two components. One remains fixed representing the quiescent no-load power consumption. The fixed loss primarily includes the eddy and hysteresis losses in the magnetic

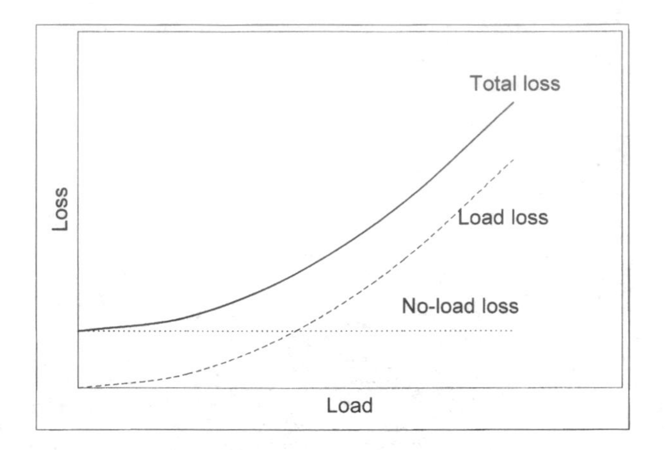

FIGURE 14-4
Loss components varying with load in typical power equipment.

parts. The other component varies with the current squared, representing the I^2R loss in the conductors. For a constant voltage system, the conductor loss varies with the load power squared. The total loss is, therefore, expressed as (Figure 14-4) the following:

$$Loss = L_o + k \cdot P^2 \tag{14-3}$$

where P is the power delivered to the load (output), L_o is the fixed loss and k is the proportionality constant. The efficiency is given by the following:

$$\eta = \frac{output}{input} = \frac{output}{output + loss} = \frac{P}{P + L_o + kP^2} \tag{14-4}$$

For the efficiency to be maximum at a given load, its derivative with respect to the load power must be zero at that load. That is as follows:

$$\frac{d\eta}{dP} = \frac{P(1 + 2kP) - (P + L_o + kP^2)}{(P + L_o + kP^2)^2} = 0 \tag{14-5}$$

This equation reduces to $L_o = kP^2$. Therefore, the component efficiency is maximum at the load under which the fixed loss is equal to the variable loss. This is an important design rule, which can save significant electrical energy in large power systems.

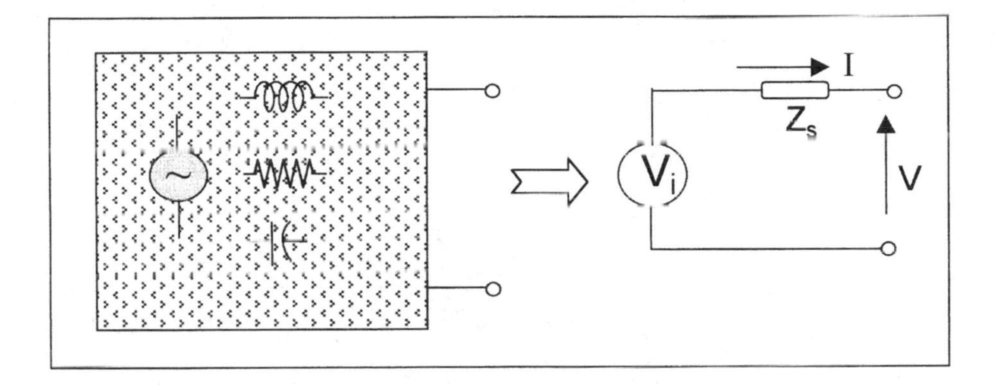

FIGURE 14-5
Thevenin's equivalent circuit model of a complex power system.

14.3 Electrical System Model

The electrical network of any complexity can be reduced to a simple Thevenin's equivalent circuit consisting of a single source voltage V_s and impedance Z_s in series (Figure 14-5). The two parameters are determined as follows:

With the system operating at no load with all other parameters at rated values, the voltage at the open-circuit terminals equals the internal source voltage, since the internal voltage drop is zero. Therefore:

$$V_s = \text{open-circuit voltage of the system.}$$

For determining the source impedance, the terminals are shorted together and the terminal current is measured. Since the internal voltage is now consumed in driving the current through the only source impedance, then:

$$Z_s = \text{open-circuit voltage/short-circuit current} \qquad (14\text{-}6)$$

If Z_s is to be determined by an actual test, the general practice is to short the terminals with the open-circuit voltage only several percentages of the rated voltage. The low level short-circuit current is measured and the full short-circuit current is calculated by scaling to the full-rated voltage. Any nonlinearality, if present, is accounted for.

The equivalent circuit is developed on a per phase basis and in percent or perunit bases. The source voltage, current, and the impedance are expressed in units of their respective base values at the terminal. The base values are defined as follows:

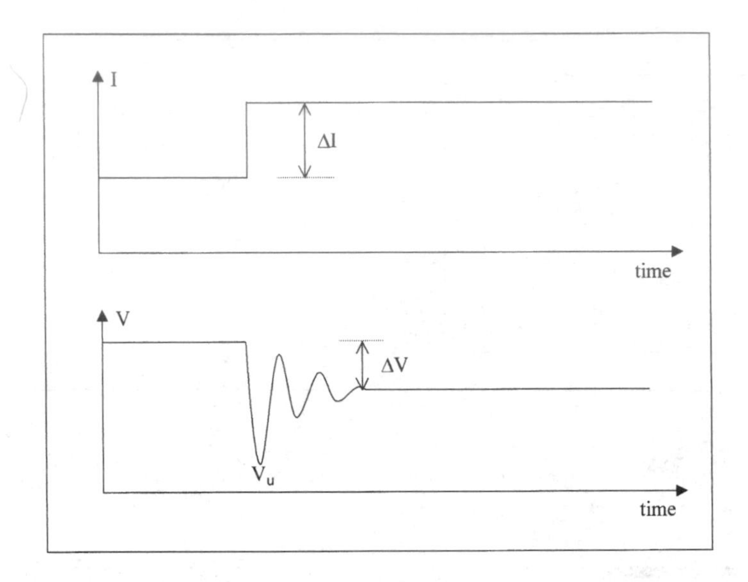

FIGURE 14-6
Transient response of the system voltage under sudden load step.

V_{base} = Rated output voltage
I_{base} = Rated output current, and
Z_{base} = Rated output voltage/Rated output current

14.4 Static Bus Impedance and Voltage Regulation

If the equivalent circuit model of Figure 14-5 is derived under the steady state static condition, the Z_s is called the static bus impedance. The steady state voltage rise on removal of the full load rated current is then $\Delta V = I_{base} \cdot Z_s$, and the voltage regulation is as follows:

$$Voltage\ Regulation = \frac{\Delta V}{V_{base}} \cdot 100 \text{ percent} \tag{14-7}$$

Under a load step transient, partial or full, as in the case of a loss of load due to accidental opening of the load side breaker, the voltage oscillates until the transient settles to the new steady state value. If the load current rises in step as shown in Figure 14-6, the voltage goes through oscillation before settling down to a lower steady state value. The steady state change in the bus voltage is then given by the following:

$$\Delta V = \Delta I \cdot Z_S$$

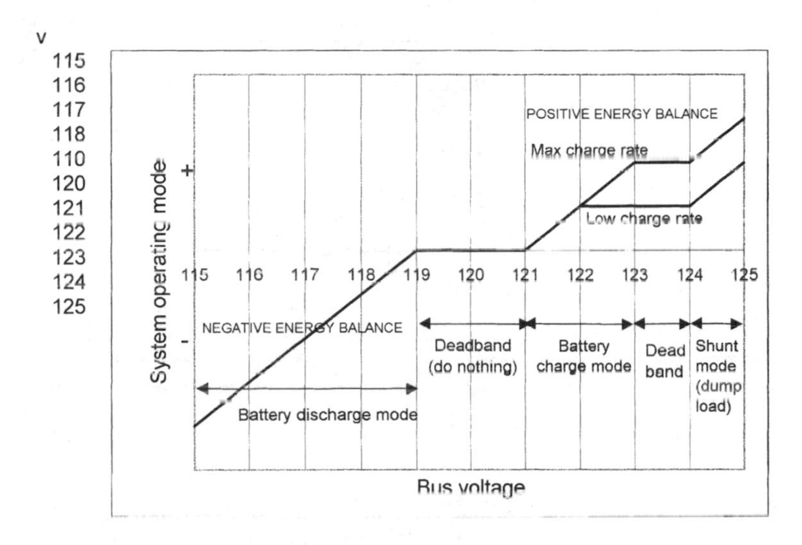

FIGURE 14-7
Deadbands in the feedback voltage control system avoid the system flutter.

The feedback voltage control system responds to bring the bus voltage deviation back to the rated value. However, in order not to flutter the system more than necessary, the control system is designed with suitable deadbands. For example, Figure 14-7 shows a 120 voltage photovoltaic system with battery with two deadbands in its control system.

14.5 Dynamic Bus Impedance and Ripple

If the circuit model of Figure 14-5 is derived under the dynamic condition, that is for an incremental load, the source impedance is called the dynamic bus impedance, and is denoted by Z_d. It can be either calculated or measured as follows. With the bus in operational mode supplying the rated load, inject a small high frequency AC current I_h into the bus using an independent grounded current source (Figure 14-8). The high frequency voltage perturbation in the bus voltage is measured and denoted by V_h. The dynamic bus impedance at that frequency is then:

$$Z_d = \frac{V_h}{I_h} \tag{14-8}$$

The ripple is the term used to describe periodic glitches in the current or the voltage, generally of high frequency. Ripples are commonly found in systems with power electronics components, such as rectifiers, inverters, battery chargers, or other switching circuits. The ripples are caused by the

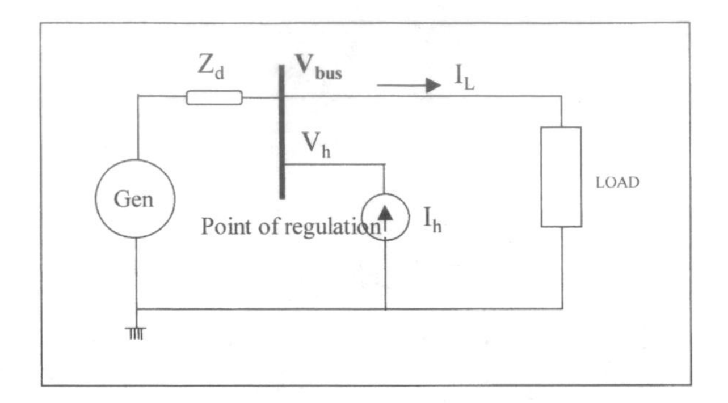

FIGURE 14-8
Harmonic and dynamic source impedance test measurement set-up.

transistors switching on and off. The ripple frequencies are integer multiples of the switching frequency. The ripples are periodic but not sinewave and are superimposed on the fundamental wave.

The ripple voltage induced on the bus due to ripple current is given by the following:

$$V_{ripple} = I_{ripple} \cdot Z_d \tag{14-9}$$

The ripple is minimized by the capacitor connected to the bus or preferably at the load terminals of the component causing ripples. The ripple current is then supplied or absorbed by the capacitor, rather than by the bus, thus, improving the quality of power.

14.6 Harmonics

The harmonics is the term used to describe the higher frequency sinewave currents or voltages superimposed on the fundamental sinewave. Phase-controlled power switching is one source of harmonics. The harmonics are also generated by magnetic saturation in power equipment. With no saturation present in the magnetic circuit, the generator and the transformer behave linearly, but not so with saturation. The saturated magnetic circuit requires non-sinewave magnetizing current.

The usual method of analyzing the system with harmonics is to determine the performance of the system for each harmonics separately and then to superimpose the results. The system is represented by the equivalent circuit for each harmonic.

The fundamental equivalent circuit of the electrical generator is represented by the d-axis and q-axis.[1-2] In the nth harmonic equivalent circuit, the harmonic inductance L_n, being for high frequency, is the average of the subtransient inductance in the d and q axes, that is as follows:

$$L_n = \frac{\left(L_d'' + L_q''\right)}{2} \tag{14-10}$$

and the reactance for the harmonic of order n is given by the following:

$$X_n = 2\pi f_n \cdot L_n = 2\pi n f L_n \tag{14-11}$$

where f_n – n^{th} harmonic frequency
f = fundamental frequency.

In single-phase or three-phase AC currents having positive and negative portions of the cycle symmetrical, the odd number of harmonics are absent. That is, $I_n = 0$ for n – 2, 4, 6, 8, and so on. In three-phase load circuits fed by transformers having the primary windings connected in delta, all triple harmonics are absent in the line currents, that is $I_n = 0$ for n = 3, 9, 15, and so on.

In the inverter circuit having m-pulse full bridge circuit, the harmonics present are of the order n = mk ± 1, where k = 1, 2, 3, 4, and so on. For example, the harmonics present in a 6-pulse inverter are 5, 7, 11, 13, 17, 19. On the other hand, the harmonics present in 12-pulse inverter are 11, 13, 23, 25. The magnitude and phase of the harmonic currents are found to be inversely proportional to the harmonic order n, that is as follows:

$$I_n = \frac{I_1}{n} \tag{14-12}$$

where I_1 is the fundamental current. This formula gives approximate harmonic contents in 6 and 12-pulse inverters as given in the first two columns of Table 14-1, which clearly shows the benefits of using 12-pulse converters.

TABLE 14-1

Harmonic Contents of the 6-pulse and 12-pulse Converters

Harmonic Order n	6-Pulse Converter Eq. 14-12	12-Pulse Converter Eq.14-12	3-pulse and 6-pulse Converters (IEEE Standard 519)
5	20	—	17.5
7	14.5	—	11.1
11	9.1	9.1	4.5
13	7.7	7.7	2.9
17	5.9	5.9	1.5
19	5.3	5.3	1.0

The actually measured harmonic currents are lower than those given by the approximate Equation 14-12. The IEEE Standard-519 gives the current harmonic spectrum in typical 6-pulse converters as listed in the last column of Table 14-1.

The harmonic currents induce harmonic voltage on the bus. The harmonic voltage of order n is given by $V_n = I_n \cdot Z_n$, where Z_n is the nth harmonic impedance. The harmonic impedance can be derived in a manner similar to the dynamic impedance, in that a harmonic current I_h is injected to or drawn from the bus and the resulting harmonic voltage V_n is measured. In a rectifier circuit drawing harmonic currents I_n provides a simple circuit which works as the harmonic current load. If all harmonic currents are measured, then the harmonic impedance of order n is given by the following:

$$Z_n = V_n / I_n, \text{ where } n = mk \pm 1, \text{ and } k = 1, 2, 3, \text{ and so on.} \quad (14\text{-}13)$$

14.7 Quality of Power

The requirement of the quality of power at the grid interface is a part of the power purchase contracts between the utility and the renewable power plant. The rectifier and inverter are the main components contributing to the power quality concerns. The grid-connected power systems, therefore, need converters which are designed to produce high quality, low distortion AC power acceptable for purchase by the utility company. The power quality concerns become more pronounced when the renewable power system is connected to small capacity grids using long low voltage link.

Until recently, there was no generally acceptable definition of the quality of power. However, the International Electrotechnical Commission and the North American Reliability Council have developed working definitions, measurements and design standards.

Broadly, the power quality has three major components for measurements:

- the total harmonic distortion generated by the power electronic equipment, such as rectifier and inverter.
- the transient voltage sags caused by system disturbances and faults.
- periodic voltage flickers.

14.7.1 Harmonic Distortion

Any non-sinusoidal alternating voltage V(t) can be decomposed in the following Fourier series:

$$V(t) = V_1 \sin \omega t + \sum_{n=2}^{\infty} V_n \sin(n\omega t + \alpha_n) \quad (14\text{-}14)$$

The first component on the right hand side of the above equation is the fundamental component, whereas all other higher frequency terms (n = 2,3…∞) are the harmonics.

The Total Harmonic Distortion Factor is defined as follows:

$$THD = \frac{\left[V_2^2 + V_3^2 + \ldots\ldots V_n^2\right]^{\frac{1}{2}}}{V_1} \tag{14-15}$$

The THD is useful in comparing the quality of AC power at various locations of the same power system, or of two or more power systems. In a pure sine wave AC source, THD = 0. The greater the value of THD, the more distorted the sinewave, resulting in more I^2R loss for the same useful power delivered. This way the quality of power and the efficiency are related.

As seen earlier, the harmonic distortion on the bus voltage caused by harmonic current I_n drawn by any nonlinear load is given by $V_n = I_n Z_n$. It is this distortion in the bus voltage that causes the harmonic current to flow even in pure linear resistive load, called the victim load. If the renewable power plant is relatively small, the nonlinear electronic loads may cause significant distortion on the bus voltage, which then supplies distorted current to the linear loads. The harmonics must be filtered out before feeding power to the grid. For a grid interface, having the THD less than 3 percent is generally acceptable. The IEEE Standard-519 limits the THD for the utility grade power to less than 5 percent.

It can be seen that harmonics do not contribute to delivering useful power, but produce I^2R heating. Such heating in generators, motors, and transformers is more difficult to dissipate due to their confined designs, as opposed to open conductors. The 1996 National Electrical Code® requires all distribution transformers to state their k-ratings on the permanent nameplate. This is useful in sizing the transformer for use in a system having a large THD. The k-rated transformer does not eliminate line harmonics. The k-rating merely represents the transformer's ability to tolerate the distortion. The unity k rating means the transformer can handle the rated load drawing pure sinewave current. The transformer supplying only electronic load may require high k-rating of 15 to 20.

A recent study funded by the Electrical Power Research Institute reports the impact of two pv solar parks on the power quality of the grid-connected distribution system[3]. The harmonic current and voltage waveforms were monitored under connection/disconnection tests over a nine month period ending March 1996. The current injected by the pv park had a total distortion below the 12 percent limit set by the IEEE-519-1992 standard. However, even the individual harmonic components between 18 and 48, except the 34th, exceeded the IEEE-519 standard. The total voltage distortion, however, was minimal.

A rough measure of quality of power is the ratio of the peak to rms voltage measured by the true rms voltmeter. In a pure sine wave, this ratio is $\sqrt{2}$ = 1.414. Most acceptable bus voltages will have this ratio in the 1.3 to 1.5 range,

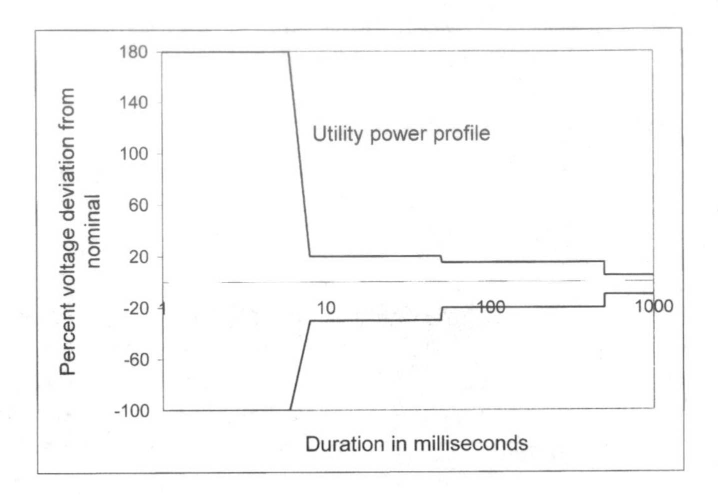

FIGURE 14-9
Allowable voltage deviation in utility-grade power versus time duration of the deviation. (Adapted from the American National Standards Institute.)

which can be used as a quick approximate check on the quality of power at any location in the system.

14.7.2 Voltage Transients and Sags

The bus voltage can deviate from the nominally-rated value due to many reasons. The deviation that can be tolerated depends on its magnitude and the time duration. Small deviations can be tolerated for a longer time than large deviations. The tolerance band is generally defined by voltage versus time (v-t) limits. Computers and business equipment using microelectronic circuits are more susceptible to the voltage transients than the rugged power equipment such as motors and transformers. The power industry has developed an array of protective equipment. Even then, some standard of power quality must be maintained at the system level. For example, the system voltage must be maintained within the v-t envelope shown in Figure 14-9, where the solid line is that specified by the American National Standard Institute (ANSI). The right hand side of the band comes primarily from the steady state performance limitations of motors and transformer-like loads, the middle portion comes from visible lighting flicker annoyance considerations, and the left hand side of the band comes from the electronic load susceptibility considerations. The left hand side curve allows larger deviations in the microsecond range based on the volt-second capability of the power supply magnetics. The ANSI requires the steady state voltage of the utility source to be within 5 percent, and short-time frequency deviations less than 0.1 Hz.

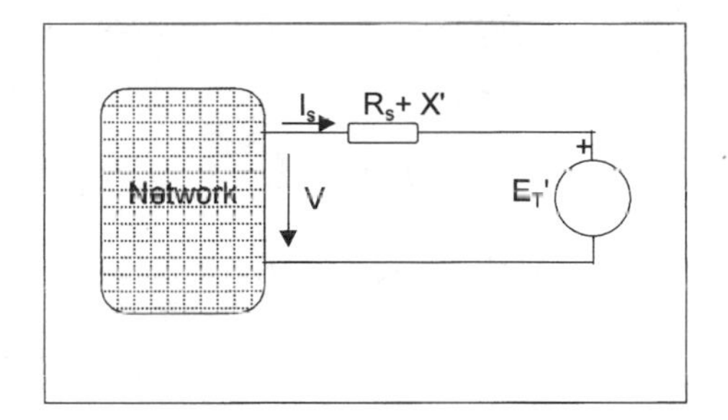

FIGURE 14-10
Thevenin's equivalent circuit model of induction generator for voltage flicker study.

14.7.3 Voltage Flickers

The turbine speed fluctuation under fluctuating wind, causes slow voltage flickers and current variations that are large enough to be detected as flickers in fluorescent lights. The relation between the fluctuation of mechanical power, the rotor speed, the voltage, and the current is analyzed by using the dynamic d and q-axis model of the induction generator or the Thevenin's equivalent circuit model shown in Figure 14-10. If we let:

R_1, R_2 = resistance of stator and rotor conductors, receptively.
x_1, x_2 = leakage reactance of stator and rotor windings, receptively.
x_m = the magnetizing reactance.
x = the open-circuit reactance.
x' = the transient reactance.
τ_o = the rotor open-circuit time constant = $(x_2+x_m)/(2\pi f R_2)$.
s = the rotor slip.
f = frequency.
E_T = the machine voltage behind the transient reactance.
V = the terminal voltage.
I_s = the stator current.

then, the value of E_T is obtained by integrating the following equation:

$$\frac{dE_T}{dt} = -j2\pi fsE_T - \frac{1}{\tau_o}\left[E_T - j(x - x') \cdot I_S\right] \tag{14-16}$$

And the stator current is as follows:

$$I_s = \frac{V - E_T}{R_1 + jX'} \tag{14-17}$$

The mechanical equation, taking into account the rotor inertia, is as follows:

$$2H\frac{ds}{dt} = P_e - \frac{P_m}{1-s} \tag{4-18}$$

where P_e = electrical power delivered by the generator
P_m = mechanical power of the wind turbine
H = rotor inertia constant in seconds = $1/2\ J\ \omega^2/P_{rated}$
J = moment of inertia of the rotor, and
ω = $2\ \pi\ f$

The electrical power is the real part of the product of the E_T and I_s^*:

$$P_e = \text{Real part of}\left[E_T \cdot I_S^*\right] \tag{14-19}$$

where I_s^* is the complex conjugate of the stator current.

The mechanical power fluctuations can be expressed by a sinewave superimposed on the steady value of P_{mo}:

$$P_m = P_{mo} + \Delta P_m \sin \omega_1 t \tag{14-20}$$

where ω_1 = rotor speed fluctuation corresponding to the wind power fluctuation.

Solving these equations by iterative process on the computer, Feijoo and Cidras[4] showed that fluctuations of a few hertz can cause noticeable voltage and current fluctuations. Several hertz fluctuations are too small to be detected at the machine terminals. The high frequency fluctuations, in effect, are filtered out by the wind turbine inertia, which is usually large. That leaves only a band of the fluctuations that could be detected at the generator terminals.

The flickers caused by the wind fluctuations may be of a concern in low voltage transmission lines connecting to the grid. The voltage drop related to the power swing is small in high voltage lines because of small current fluctuation for a given wind fluctuation.

14.8 Renewable Capacity Limit

A recent survey made by Gardner[5] and Risø National Laboratory[6] in the European renewable power industry indicates that the grid interface issue is one of the economic factors limiting full exploitation of the available wind resources. The regions of high wind power potentials have weak existing electrical grids. In many developing countries such as India, China, and

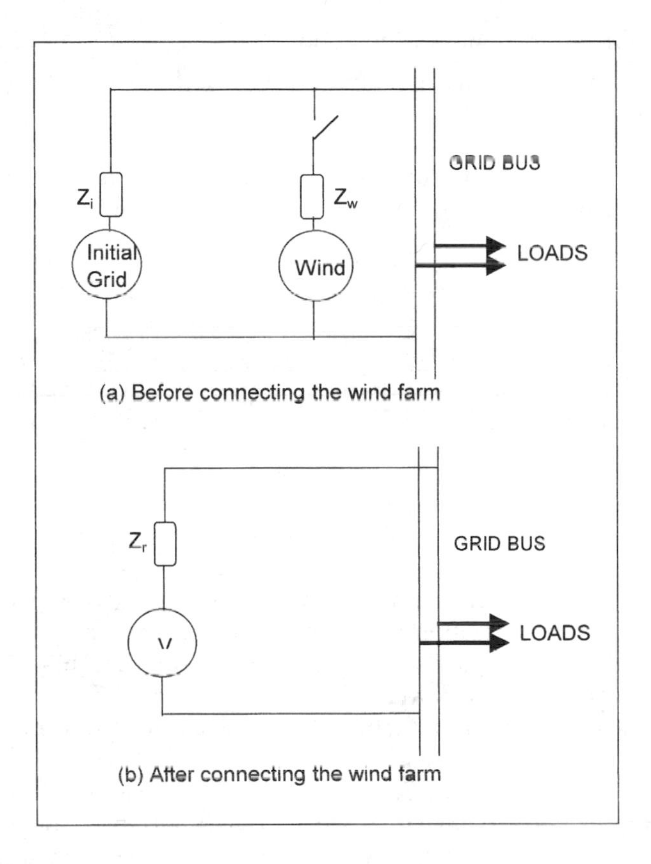

FIGURE 14-11
Thevenin's equivalent circuit model of the grid and wind farm for evaluation of the grid interface performance for evaluating the system stiffness at the interface.

Mexico, engineers are faced with locating sites that are also compatible for interfacing with the grid from the power-quality point of view. The basic consideration in such decisions is the source impedance before and after making the connection. Another way of looking at this issue is the available short-circuit MVA at the point of the proposed interconnection, also known as the system stiffness or the fault level.

14.8.1 System Stiffness

One way of evaluating the system stiffness after making the interconnection is by using the Thevenin's equivalent circuit of the grid and the renewable plant separately as shown in Figure 14-11.

If we let:

V = network voltage at the point of the proposed interconnection.

Z_i = source impedance of the initial grid before the interconnection.

Z_w = source impedance of the wind farm.

Z_l = impedance of the interconnecting line from the wind farm to the grid.

then, after connecting the proposed wind capacity with the grid, the combined equivalent Thevenin's network would be as shown in (b), where:

$$\frac{1}{Z_t} = \frac{1}{Z_i} + \frac{1}{Z_w + Z_l} \tag{14-21}$$

where Z_t = total combined source impedance of the two systems.

The combined short circuit MVA at the point of the interconnection is then:

$$MVA_{SC} = \frac{1}{Z_t} \cdot \frac{1}{10^6} \tag{14-22}$$

The higher the short-circuit MVA, the stiffer the network. Certain minimum grid stiffness in relation to the renewable power capacity is required to maintain the power quality of the resulting network. This consideration limits the total wind capacity that can be added at a site under consideration. The wind capacity exceeding that limit may be a difficult proposal to sell to the company in charge of the grid.

Not only the magnitude of Z_t is important, the resistance R and the reactance X components of Z_t have their individual importance. The fundamental circuit theory dictates that the real and reactive power in any electrical network must be separately maintained in balance. Therefore, the real and reactive components of the wind generator impedance would impact the network, more so in a weaker grid. The R/X ratio is often found to be 0.5, which is generally satisfactory for fixed-speed wind generators. Lower R/X ratio may pose another limitation on designing the system.

The fault current decays exponentially as $e^{-R/X}$. Low R/X ratio causes the fault currents to linger on, making the fault protection relaying more difficult. The voltage regulation (variation from zero to full load on the wind farm) is yet another design consideration that is impacted by the R/X ratio. The estimated voltage regulation of the wind farm must be compared with the contractual limit with the utility. In doing so, the actual continuous maximum load the wind farm can deliver must be taken into account, not the nominal rating which can be much higher. The acceptable voltage regulation in industrial countries is typically 5 percent, and 7 percent in developing countries.

Starting the wind turbine as the induction motor causes an inrush of current from the grid, causing a sudden voltage drop for a few seconds. This can be intolerable by the grid line. Most countries limit this transient voltage dip to 2 to 5 percent, the higher value prevailing in the developing countries.

Fortunately, for wind farms with many machines, the machines can be started in sequence to minimize this effect. For very weak system, however, this issue can limit the number of machines that can be connected with the grid lines. The voltage flicker severity increases as the square root of the number of machines. The flicker caused by starting one machine varies inversely with the fault level at the point of the grid connection, hence can be an issue on the weaker grids.

Harmonics are generated by the power electronics employed for the soft start of large wind turbines and for the speed control during energy producing operation. The former can be generally ignored due to the short duration and use of the sequential starts.

The operating harmonics, however, need to be filtered out. Total harmonic content of the wind farm is empirically found to depend on the square root of the number of machines. Since the high voltage grid side has lower current, designing the pulse-width-modulated (PWM) converters on the grid side can be advantageous. The utility companies around the world are not consistent in the way they limit the harmonics. Some limit the harmonics in terms of the absolute current in amperes, while others set limits proportional to the grid's short-circuit MVA at the point of interconnection.

Utilities find it convenient to meet the power-quality requirements by limiting the total renewable power rating less than a few percent of the short-circuit MVA of the grid at the proposed interface. The limit is generally 2 percent in developed countries and 5 percent in developing countries. This requirement can be more restrictive than the overall power quality requirement imposed in accordance with the national standards.

14.8.2 Interfacing Standards

The maximum capacity the renewable power plant can install is primarily determined from the electrical system and power quality considerations presented in this chapter. A rough rule of thumb to control the power quality has been to keep the renewable power plant capacity in MW less than the grid line voltage in kV if the grid is stiff (large). On a weak grid, however, only 10 or 20 percent of this capacity may be allowed. The regulation generally imposed on the renewable power farms these days are in terms of the short-circuit capacity at the proposed interface site.

We recall that the major power quality issues discussed in this chapter are as follows:

- the acceptable voltage range on the distribution system.
- the step load voltage.
- the steady state voltage regulations.
- flickers caused by the wind fluctuations.
- the harmonics.

TABLE 14-2

Acceptable Voltage Variation at Distribution Points (Low Voltage Consumers May See Wider Variations)

Country	Acceptable range
U.S.A.	±5 percent
France	±5 percent
U.K.	±6 percent
Spain	±7 percent

TABLE 14-3

Allowable Step-Change in Voltages a Customer can Cause by Step Loading or Unloading

Country	Allowable range
France	±5 percent
U.K.	±3 percent
Germany	±2 percent
Spain	±2 percent for wind generators ±5 percent for embedded generators

TABLE 14-4

Renewable Power Generation Limit in Percent of the Grid Short Circuit Capacity at the Point of Interface

Country	Allowable limit
Germany	2 percent
Spain	5 percent
Other countries	Evolving

The acceptable voltage variations in the grid voltage at the distribution point in four countries are given in Table 14-2. The limit on the step-change a customer can cause on loading or unloading is listed in Table 14-3. It is complex to determine the maximum renewable capacity that can be allowed at a given site which will meet all these complex requirements. It may be even more difficult to demonstrate the compliance. Under the situation, some countries impose limits in percent of the grid short-circuit capacity. For example, such limits in Germany and Spain are listed in Table 14-4, but similar limits are evolving in other countries.

The power quality standards applicable specifically to grid-connected wind farms are being drafted by the International Electrotechnical Commission. It is expected that such standards would be more realistic than the rigid criteria based on the ratio of the wind turbine capacity to the short-circuit MVA of the grid. The ultimate results would be consistent and predictable

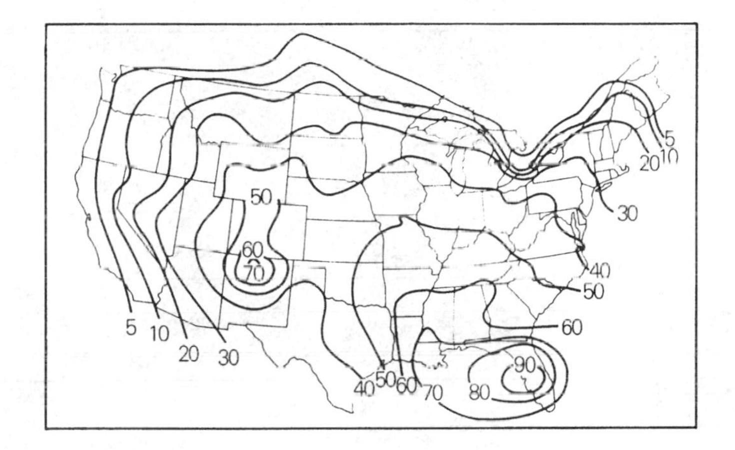

FIGURE 4-12
Thunderstorm frequency in the United States and Southern Canada. The contour lines indicate the average number of days per year with thunderstorm occurrence.

understanding on the power quality requirements and less cost-related in designing wind farms to comply with the power quality issues.

Again, in order to control the quality of power, a rough rule of thumb has been to keep the renewable power plant capacity in MW less than the grid line voltage in kV if the grid is stiff (large). On a weak grid, however, only 10 or 20 percent of this capacity may be allowed.

14.9 Lightning Protection

The risk of mechanical and thermal damage to the blades and the electrical systems due to lightning is minimized by a coordinated protection scheme using lightning arresters and spark gaps[6-7] in accordance with the International Standards. The IEC Standard -1024-1 covers the requirements for protection of structures against lightning. The electrical generator and transformers are designed with certain minimum Basic Insulation Level (BIL) consistent with the lightning risk in the area. The risk is proportional to the number of thunderstorms per year in the area. Figure 14-12 depicts the number of thunderstorms per year in the United States and southern Canada.

Tall towers are more vulnerable to the lightning risk. Among the valuable new experience in the wind power industry is that the offshore wind power towers experience a higher than average incidence of lightning strikes. Based on such experience, the system manufacturer has addressed the design problems. Figure 14-13 is one such solution developed by Vestas Wind Systems of Denmark. The lightning current is dispersed to the ground though a series of spark gaps and equipotential bonding at joints. The transformer is protected by placing it inside the tower.

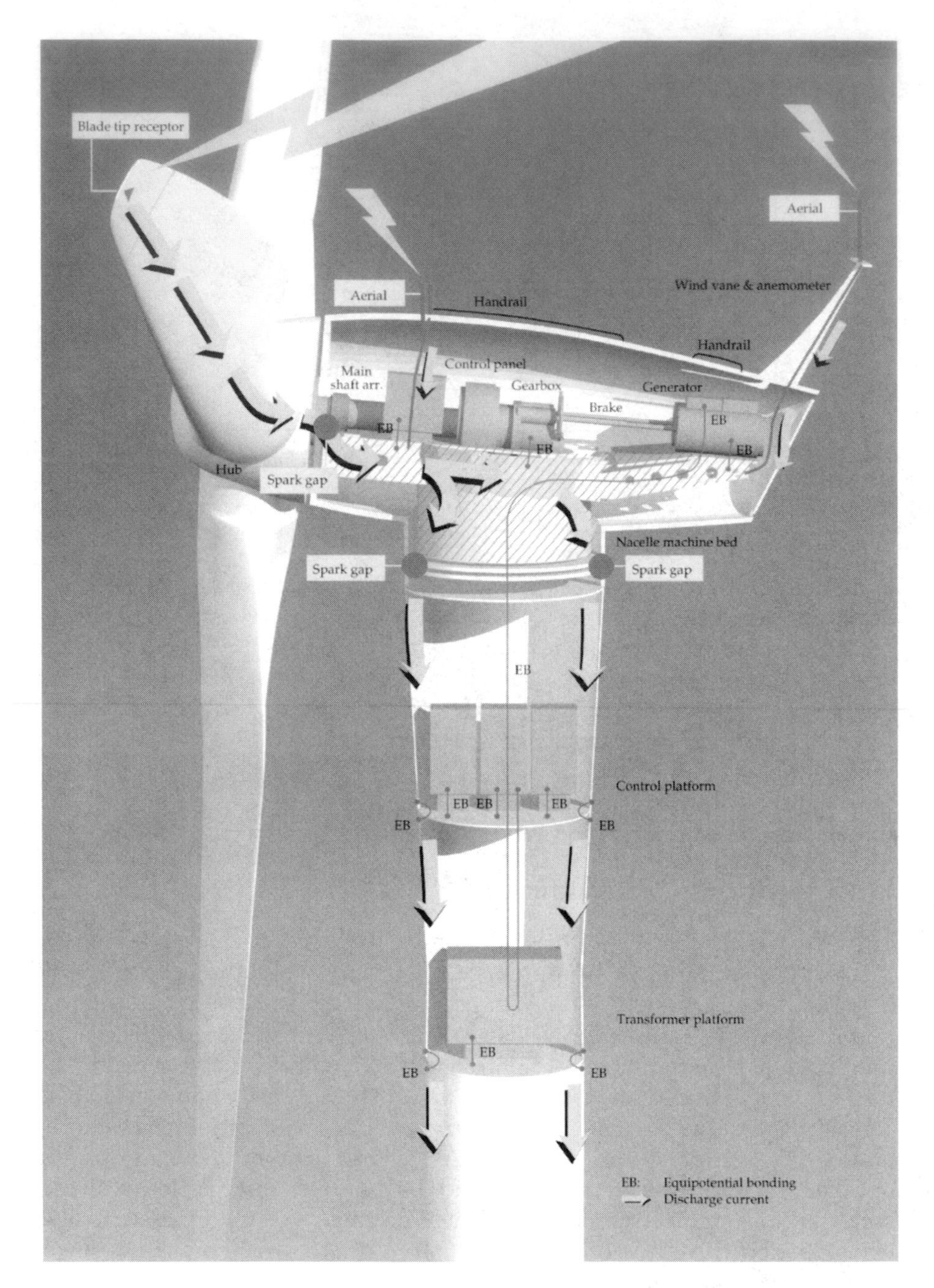

FIGURE 14-13
Lightning protection of the wind tower. (Source: Nordtank Energy Group/NEG Micon, Denmark. With permission.)

14.10 National Electrical Code® on Renewable Power Systems

The 1996 revision of the National Electrical Code[8] has established electrical system requirements on the wind and photovoltaic power systems. The photovoltaics are covered in Article 690. The Article 705 covers all renewable and cogeneration power systems, including the wind, which is reproduced in Appendix 1 of this book.

References

1. Adkins, B. 1964. "The General Theory of Electrical Machines," London, Chapman and Hall, 1964.
2. Kron, G. 1967. "Equivalent Circuits of Electrical Machines," New York, Dover Publications, 1967.
3. Oliva, A. R., Balda, J. C., McNabb, D. W., and Richardson, R. D. 1997. "Power quality monitoring of a pv generator," *IEEE paper No. PE-507-EC0-07*, 1997.
4. Feijoo, A and Cidras, J. 1997. "Analyses of mechanical power fluctuations in asynchronous WECs," *IEEE paper No. PE-030-EC-0-1 0*, 1997.
5. Gardner, P 1997. "Wind farms and weak networks," *Wind Directions, Magazine of the European Wind Energy Association*, London, July 1997.
6. Risø National Laboratory. 1996. "Power quality and grid connection of wind turbines" — *Summary report R-853*, Denmark, October 1996.
7. Lewis, W. W. 1965. "The Protection of Transmission Systems Against Lightning," New York, Dover Publications, 1965.
8. National Fire Protection Association. 1996. National Electrical Code Handbook, 7th Ed., Quincy, MA, 1996.

15

Plant Economy

The economic viability of a proposed plant is influenced by several factors that contribute to the expected profitability. Since the profitability inadvertently varies with variations in the contributing factors, the sensitivity of the expected profitability is analyzed with variances around their expected values. The sensitivity analysis raises the confidence level of potential investors. This is important for both the wind and pv systems, but more so for wind systems whose profitability is extremely sensitive to the wind speed variations.

The primary factors contributing to the economic viability of the plants are discussed in this chapter. The results are displayed in easy-to-use charts for screening the profitability of a proposed plant.

15.1 Energy Delivery Factor

The key economic performance measure of a power plant is the electrical energy it delivers over the year. Not all power produced is delivered to the paying customers. A fraction of it is used internally to power the control equipment, meeting the power equipment losses and for the housekeeping functions such as lighting. In a typical wind farm or pv park, about 90 percent of the power produced is delivered to the customers, and the remaining is self-consumed for the plant operation.

The quantity of energy delivered depends on the peak power capacity of the site and how fully that capacity is utilized over every hour of the year. The normalized measure of the power plant performance is the Energy Delivery Factor (EDF). It is defined as the ratio of the electrical energy delivered to the customers to the energy that can be delivered by the plant if it could be operated at the fully installed capacity during all 8,760 hours of the year, that is as follows:

$$\textit{Average annual EDF} = \frac{\textit{kWh delivered over the year}}{\textit{Installed capacity} \cdot \textit{Number of hours in the year}} \qquad (15\text{-}1)$$

Since the load power varies over the time, the EDF takes the integral form:

$$Average\ annual\ EDF = \frac{\int^{year} P_o \cdot dt}{P_m \cdot 8760} \tag{15-2}$$

where P_m = Plant capacity (the maximum power the plant can deliver)
P_o = Power delivered to the customers at any time t

The EDF is usually determined by bookkeeping the sum of the energy delivered over a continuous series of small discrete time intervals, that is as follows:

$$Average\ annual\ EDF = \frac{\sum^{year} P_{avg} \cdot (\Delta t)}{P_m \cdot 8760} \qquad 15\text{-}3)$$

where P_{avg} = average power delivered over the small time interval Δt.

The EDF is a figure of merit that measures how hard the plant is utilized to deliver the maximum possible energy. Not only does it include the energy conversion efficiencies of various components, it also accounts for the reliability, maintainability, and availability of the overall plant over the entire year.

The EDF is useful in comparing the economic utilization of one site over the other, or the annual performance of a given site. Wind plants operate with the annual average EDF around 30 percent, with some plants reporting EDF as high as 40 percent. This compares with 40 to 80 percent for the conventional plants. The base load plants operate at the higher end of the range.

The wind farm energy delivery factor varies with season and that must be taken into account. For example, the quarterly average EDF in England and Wales from the beginning of 1992 to the end of 1996 are shown in Figure 15-1. The operating data show rather wide variations, ranging from 15 to 45 percent. The EDF is high in the first quarter and low in the last quarter of every year. Even with the same kWh produced per year, the average price per kWh the plant may fetch could be lower if the seasonal variations are wide.

15.2 Initial Capital Cost

The capital cost depends on the size, site and the technology of the plant. Typical ranges for various types of large power plants are given in Table 15-1.

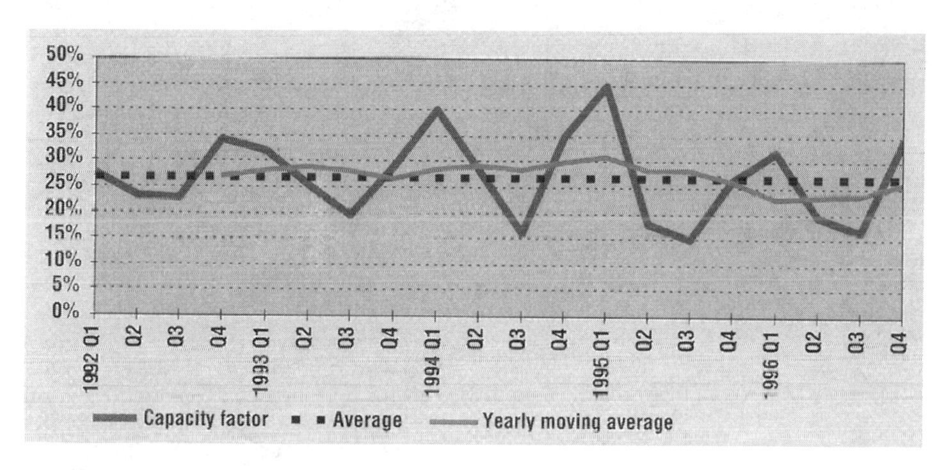

FIGURE 15-1
Capacity factors of operating wind farms in England and Wales. (Source: U.K. Department of Trade and Industry, Renewable Energy Technology Status Report, August 1997. With permission.)

TABLE 15-1
Capital Cost for Various Power Technologies

Plant Technology	Capital Cost, $/kW
Wind Turbines	500–800
Solar Photovoltaics	2000–4000
Solar thermal	3000–5000
(Solar II type)	(land ~15 acres/MW)
Coal Thermal (Steam-Turbine)	400–600
	(land ~ 10 acres/MW)
Combined Cycle Gas-Turbine	800–1200

The costs of technologically matured coal and gas turbine plants are rising with inflation, whereas the wind and pv plant costs are falling with new developments coming in the market. Therefore, the renewable power plant cost must be estimated with their component costs current at the time of procurement, which may be lower than those at the time of planning.

The percent breakdown of the component costs in the total initial capital cost of a typical wind farm is shown in Table 15-2. The single largest cost item is the turbine blades and rotor assembly, as expected.

15.3 Availability and Maintenance

The failure rates and effects determine the maintenance cost and availability of the plant to produce power. The past data on the operating experience is

TABLE 15-2

Wind Power System Component Cost Contribution in Total Capital Cost

Cost Item	Percent Contribution
Rotor assembly	25
Nacelle structure and auxiliary equipment	15
Electrical power equipment	15
Tower and foundation	10
Site preparation and roads	10
Ground equipment stations	8
Maintenance equipment and initial spares	5
Electrical interconnections	4
Other nonrecurring costs	3
Financing and legal	5
TOTAL	100

(Land cost is not included)

used for learning lessons and making improvements. In the wind power industry, the failures, their causes and effects are recorded and periodically published by ISET, the solar energy research unit of the University of Kassel in Germany. Figure 15-2 summarizes the ISET database.[2] The statistics show that 67 percent of the time during the reported failure period, the plant is nonoperational. Among the repairs needed to bring the plant back in operation, 20 percent are in the electrical power equipment and 19 percent are in the electronic controls. The major causes of the failure have been 28 percent in the control systems and 24 percent in the component defects. The failure rates are declining from this level with the design improvements made since then.

The overall availability of the plant is defined as the ratio of hours the plant is operating to deliver full power over the total hours in the year. It is impacted by downtimes due to repairs and routine maintenance. Significant improvements in reliability and maintainability have pushed the availability of modern renewable power plants up to 95 percent in recent years. The availability factor, however, is reflected in the energy delivery factor EDF.

15.4 Energy Cost Estimates

The principal decision-making parameter of an electrical power plant is the unit cost of energy (UCE) per kWh delivered to the paying consumers. It takes into account all economy factors discussed above, and is given by the following:

$$UCE = \frac{ICC \cdot (AMR + TIR) + OMC}{EDF \cdot kW \cdot 8760} \tag{15-4}$$

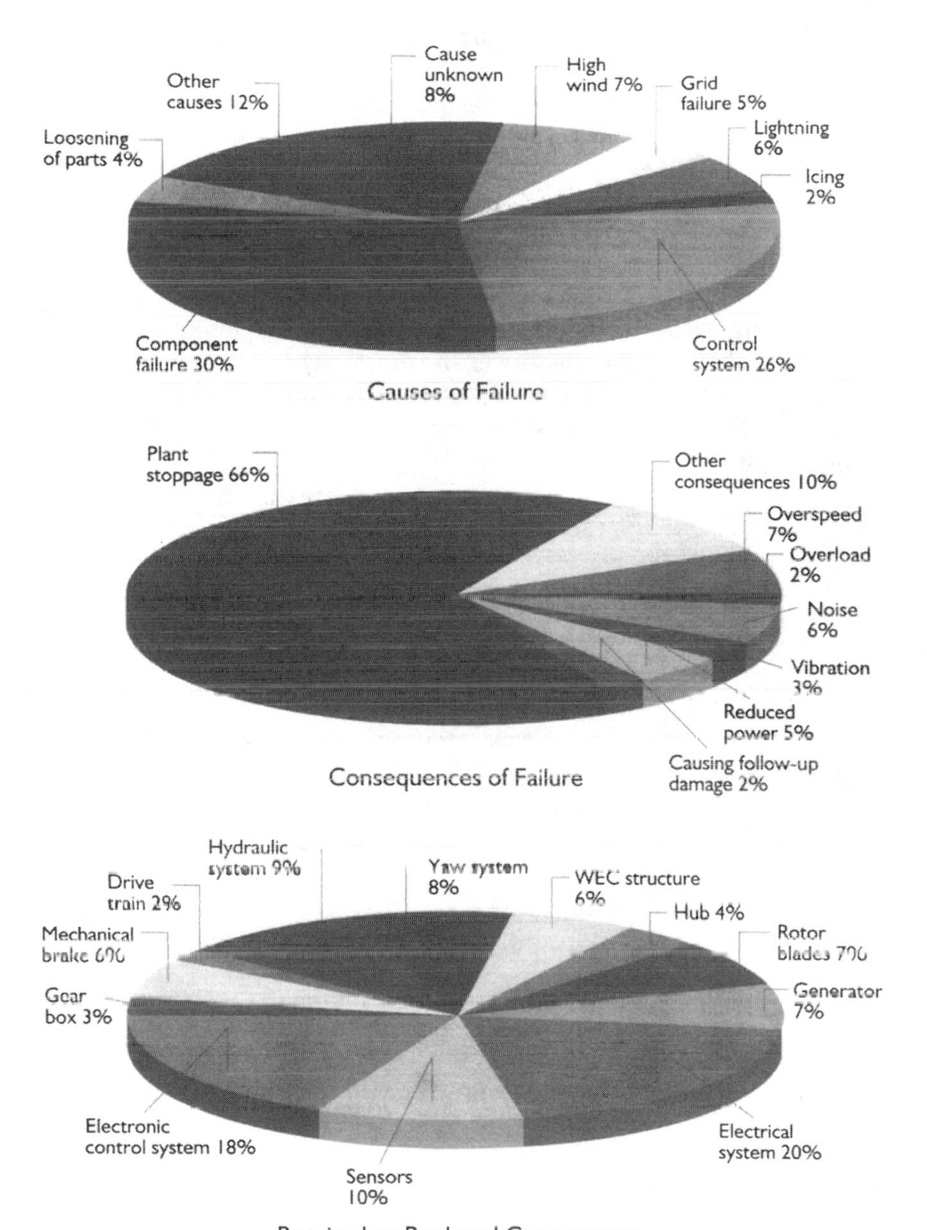

FIGURE 15-2
Wind power plant failure, cause, effect, and repair statistics. (Source: Institute of Solar Energy and Technology, University of Kassel, Germany, With permission.)

where ICC = Initial capital cost, including land cost and startup cost up to the time the first unit of energy is sold

AMR = Amortization rate per year as a fraction of the ICC

TIR = Tax and insurance rate per year as a fraction of the ICC
OMC = Operating and maintenance cost per year
EDF = Energy delivery factor over one year, as defined earlier
kW = Kilowatt electrical power capacity installed

The amortization rate AMR reflects the cost of money, generally taken as 0.10 to 0.20 depending on the prevailing mortgage interest rates available for such a project. The energy delivery factor EDF accounts for the variations in the impinging energy (sun intensity or wind speed) at the proposed site, and all downtimes, full or partial. It is also a strong function of the reliability and maintainability of the plant over the entire year.

The present costs of various energy sources are summarized in Table 15-3. It shows that the wind power now compares well with the conventional coal thermal power.

15.5 Sensitivity Analysis

Accurately estimating the cost of electricity produced is not sufficient. The project planners must also carry out the sensitivity analysis, in that the energy cost is estimated with a series of input parameters deviating on both sides of the expected values. The sensitivity of the energy cost on the following two factors is discussed in this section:

- variation in the wind speed.
- variation in solar irradiation.

15.5.1 Effect of Wind Speed Variations

Since the energy output of the wind plant varies with the cube of the wind speed, a several percent change in the speed can have significant impact on

TABLE 15-3

Energy Generation Cost with Various Power Technologies

Plant Technology	Generation Cost (Fuel + Capital) U.S. cents/kWh
Wind Turbines	5–7
Solar Photovoltaics	15–25
Solar thermal (Solar II type)	8–10
Coal Thermal (Steam-Turbine)	3–4
Combined Cycle Gas-Turbine	5–7

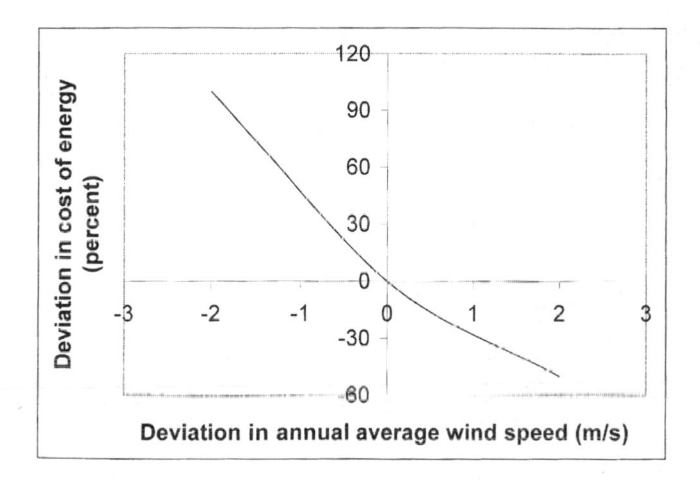

FIGURE 15-3
Sensitivity analysis of wind electricity cost with variation in wind speed around the expected annual mean of 10 meter/second.

the plant economy. For example, if the annual average wind speed is 10 m/s, the unit cost of energy is calculated for wind speeds ranging from 8 to 12 m/s. The plant site is considered economically viable only if the plant can be profitable at the low end of the probable range. Figure 15-3 illustrates the result of a sensitivity study. It shows that the cost of electricity can be as low as one-half or as high as twice the expected value if the wind speed turns out to be 2 m/s higher or lower than the expected value of 8 m/s.

15.5.2 Effect of Tower Height

Tower height varies from approximately four rotor diameters for turbines in a few kW ratings to little over one diameter for turbines in several hundred kW ratings. The tower height for large turbines is primarily determined by the tower structure and the foundation design considerations. The tower height of small turbines is determined from the wind speed at the hub. The higher the hub, the higher the wind speed.

As seen in Table 15-2 earlier, the turbine cost constitutes 25 percent of the total wind farm cost. Since the wind speed increases with tower height, we can produce more energy from the same turbine by installing it on a taller tower, thus reducing the cost of energy produced.

We define E_1 as contribution of the turbine cost in the total energy cost per square meter of rotor swept area. The E_1 is therefore measured in $/kWh·m². Since power is proportional to the rotor diameter squared and the wind speed cubed, we write the following:

$$P \propto D^2V^3 \tag{15-5}$$

The energy captured over the life of the turbine is also proportional to the same parameters, i.e., D^2 and V^3. In a matured competitive market, the rotor cost would be proportional to the swept area, hence D^2, although it is far from that stage in the current rapidly evolving market. However, assuming the competitive market E_1 is as follows:

$$E_1 \propto D^2/D^2\ V^3 \propto 1/V^3 \qquad (15\text{-}6)$$

As seen earlier in Equation 4-34, the wind speed varies with tower height in the exponential relation:

$$V \propto H^{\alpha} \qquad (15\text{-}7)$$

where H = tower height at the hub.

Combining Equations 15-6 and 15-7, we obtain:

$$E_1 \propto \frac{1}{H^{3\alpha}} \qquad (15\text{-}8)$$

The parameter α varies with the terrain type. Table 4-3 in Chapter 4 gives this parameter value of 0.10 over the ocean, 0.40 over urban areas with tall buildings and the often used average of 0.15 (~1/7) in areas with foot high grass on flat ground. Even higher value of 0.43 was estimated for a New England site at Stratton, Mt. Vermont.[1] With such variations in α, the wind speed can vary with H over a wide range from $H^{0.3}$ to $H^{1.3}$.

Equation 15-8 clearly indicates that the turbine contribution in cost of energy per unit swept area decreases at least with the square root of the tower height. With the average value of α of 1/7, the turbine cost contribution is approximately:

$$\textit{Turbine cost per kWh per } m^2,\ E_1 = \frac{\textit{Constant}}{\sqrt{\textit{Tower Height}}} \qquad (15\text{-}9)$$

In rough terrain, the tower height can be extremely beneficial as shown by the following equation with $\alpha = 0.4$:

$$\textit{Turbine cost per kWh per } m^2,\ E_1 = \frac{\textit{Constant}}{H^{1.2}} \qquad (15\text{-}10)$$

Equations 15-9 and 15-10 indicate that there is not much benefit in increasing the tower height of the offshore installations. However, on rough terrain, increasing the tower height from 30 to 60 meters would decrease the contribution of the turbine cost per kWh/m^2 by 56 percent, a significant reduction.

TABLE 15-4

Average Electricity Prices for U.S. Residential Customers in High Cost Areas in 1997

Utility area	Cents/kWh
Long Island Lighting, New York	17
Consolidated Edison, New York	16
Public Service of New Hampshire	15
New York State Electric and Gas	14
Commonwealth Energy Systems, MA	14
Central main Power (Main)	13
PECO Energy, Philadelphia, PA	13
Public Service Electric (New Jersey)	13

(Average price in the U.K. ≈ 8 pence/kWh ≈ 13 cents/kWh)

Therefore, determining the α parameter accurately for the specific site is important in the estimate of the plant economy. An uneconomical site using a short tower or conservative estimate of α can turn out to be profitable with a tall tower or the site specific estimate of α.

15.6 Profitability Index

As with the conventional projects, the profitability is measured by the profitability index, defined as the following:

$$PI = \frac{\text{Present worth of future revenues} - \text{Initial project cost}}{\text{Initial project cost}} \tag{15-11}$$

By definition, the PI of zero gives the break-even point.

The profitability obviously depends on the price at which the plant can sell the energy it produces. In turn, it depends on the prevailing market price the utilities are charging to the area customers. Regions with high energy cost could be more profitable if the capital cost isn't high in the same proportion. Table 15-4 lists the 1997 average electricity prices in certain regions of the U.S.A. The average price in the United Kingdom is about 8 pence/kWh (about 13 cents/kWh).

Inputs to the renewable power plant profitability analysis include the following:

- anticipated energy impinging the site, that is the wind speed at hub height for the wind farm or the insolation rate for the pv park.
- expected initial capital cost of installing the farm.
- cost of capital, usually the interest rate on the loans.

- expected economic life of the plant.
- operating and maintenance cost.
- average selling price of the energy generated from the farm.

A detailed multivariable profitability analysis with the above parameters is always required before making financial investments. Potential investors can make initial profitability assessment using screening charts. The following sections present easy-to-use profitability charts for initial screening of the wind and pv power plant sites.

15.6.1 Wind Farm Screening Chart

Figure 15-4 is a wind farm profitability screening chart based on cost of capital (discount rate) of 8 percent, project life of 15 years, initial cost of 1,100 European currency units per kW capacity, and the operating and maintenance cost of 3 percent of the initial project cost.[3] The chart takes into account the inflation at a constant rate, i.e., the costs and revenues rising at the same rate. The taxes on sales, if any, must be deducted from the average selling price before it is entered in reading the chart. An example for using the chart follows:

At a site with 7.5 m/s wind speed at hub height, the profitability index would be zero (a break-even point) if the energy can be sold at 0.069 ECU/kWh. If the energy price is 0.085 ECU/kWh, the profitability index would be 0.30, generally an attractive value for private investors. At a site with only 7 m/s wind speed, the same profitability can be achieved if the energy can be sold at 0.097 ECU/kWh.

15.6.2 pv Park Screening Chart

Similar screening charts for pv stand-alone or grid-connected systems are shown in Figures 15-6 and 15-7.[4] For these charts, the yearly energy delivery E_y in kWh/year is defined as the following:

$$E_y = K_p \cdot H_y \cdot P_{pk} \quad \text{kWh/year} \tag{15-12}$$

where K_p = performance ratio of the system
H_y = yearly solar irradiation in the plane of the modules, kWh/m²/year
P_{pk} = installed peak power under the standard test conditions of 1 kW/m² solar irradiation, the cell junction operating temperature of 25°C and the air mass of 1.5.

The pv energy conversion efficiency of the modules is included in P_{pk}. The typical values of K_p for well designed systems are 0.7 to 0.8 for gird-connected

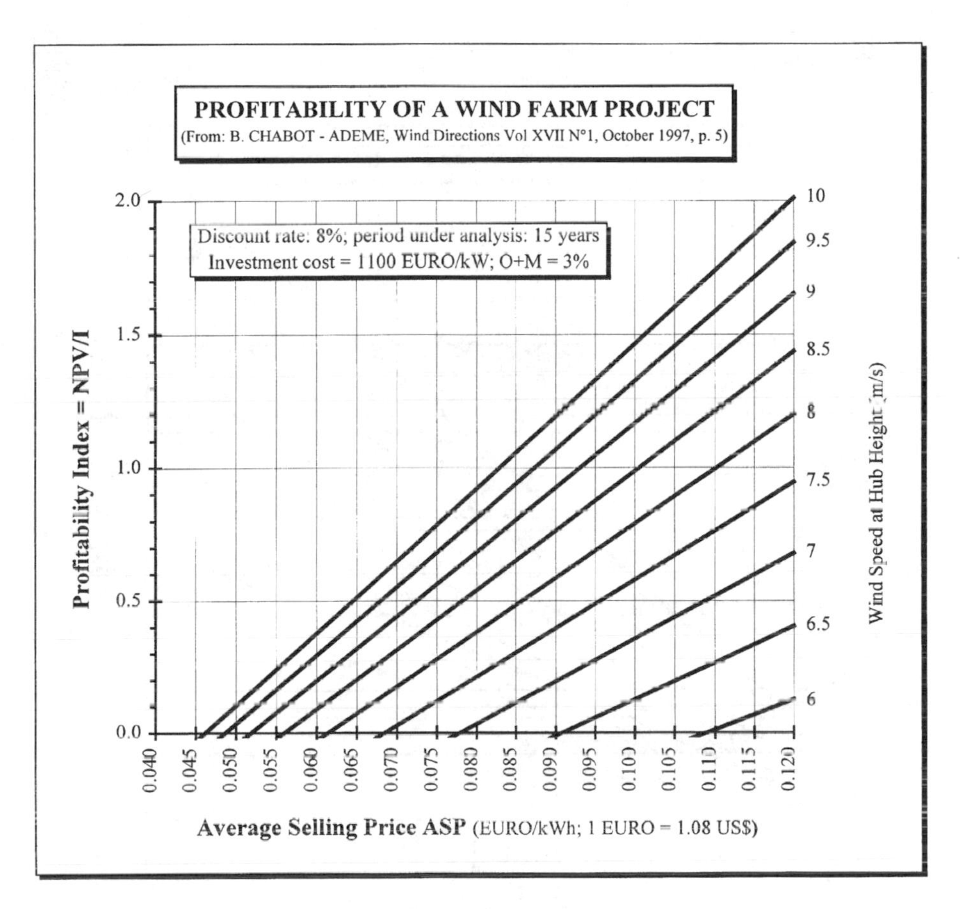

FIGURE 15.4
Profitability chart for wind farms. (Source: B. Chabot – ADME, France, and Wind Directions, Magazine of the European Wind Energy Association, London, October 1997. With permission.)

systems without batteries and delivering AC power using inverters, and 0.5 to 0.6 for stand-alone systems using batteries and delivering DC power.

The input to the figures are the initial capital cost of the pv system in dollars per peak watt capacity I_{up}, the cost of capital, the life of the plant, the operating and maintenance cost K_{om}, the yearly solar irradiation H_y, and the performance parameter K_p as defined above. With these inputs, the chart gives the overall discounted cost (ODC) of electricity delivered. The life of the plant in years, n, can be the desired payback period to recover the initial investment, or the full economic life of the plant. In the latter case, the ODC is known as the life cycle cost (LCC) of the energy delivered. The chart is prepared with typical values for large pv farms, such as K_p of 0.75, 8 percent cost of capital, 20 years life and 2 percent operating and maintenance cost.

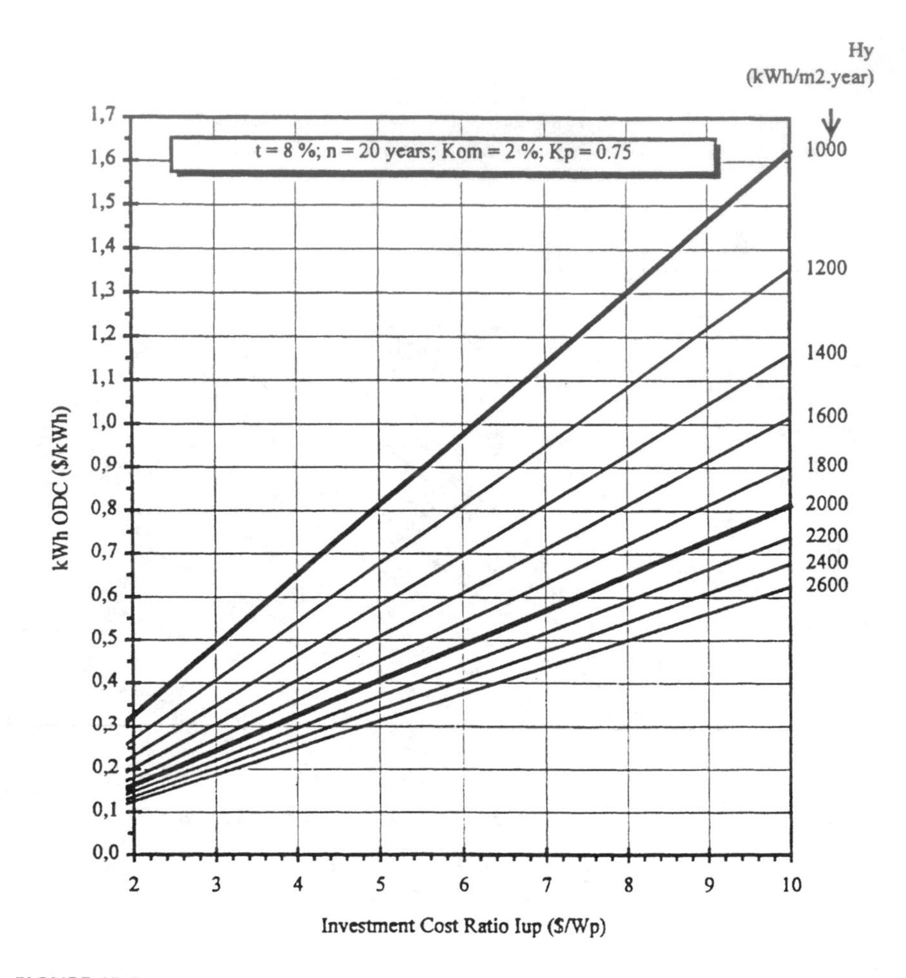

FIGURE 15-5
Profitability chart for photovoltaic power. (Source: B. Chabot – ADME, France and Progress in Photovoltaic Research and Applications, Volume 6, p. 62, 1998. John Wiley & Sons Limited, Bristol, U.K.. With permission.)

An example of using Figure 15-5 follows. At a solar site with H_y of 2,000 kWh/m^2/yr and the capital cost of 5 $/Wp, we see that the energy cost would be $0.40 per kWh delivered. For the cost of energy delivered to be below 0.20 $/kWh, the capital cost of the plant must be $2.50 per Wp. With new pv technologies being developed and made commercially available at lower costs, the capital cost below 3 $/$W_p$ is realizable.

Being costlier than the wind power systems, the pv systems at present are more likely to be installed in remote areas where higher cost can be justified and partially offset by low interest loans. Figure 15-6 is cast in terms of the profitability index defined earlier. The parameters for this chart are the capital cost of 4 $/$W_p$, 3 percent cost of capital, 30 years economic life, 1 percent operating and maintenance cost and the performance parameter

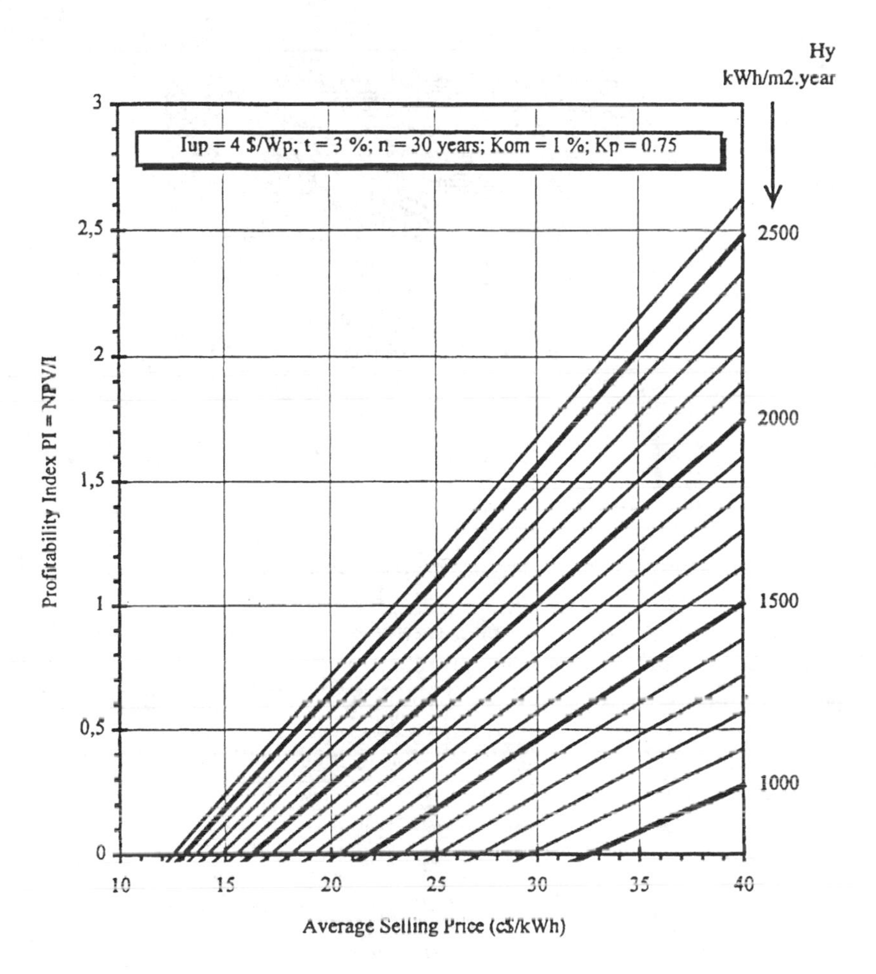

FIGURE 15-6
Overall discounted cost of pv energy. (Source: B. Chabot – ADME, France and Progress in Photovoltaic Research and Applications, Volume 6, p. 58, 1998. John Wiley & Sons Limited, Bristol, U.K.. With permission.)

of 0.75 percent. For a solar irradiation of 2,000 kWh/m²/year and an average selling price of 20 cents/kWh, the profitability index would be 0.25, a reasonable number for private investors.

15.6.3 Stand-Alone pv Versus Grid Line

In remote areas, where the grid power is not available and the wind is not an option, the customer who wants electricity has two options:

- pay the utility company for extending the line.
- install a pv power system at the site.

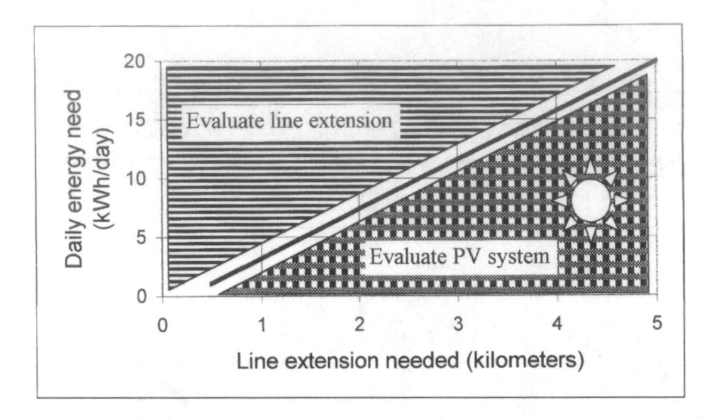

FIGURE 15-7
Stand-alone pv plant or line extension costs for various levels of energy need based on average 1997 costs of Idaho Power Company.

The cost of extending the line may be prohibitive if the stretch is long. For a given distance of the line extension, the line cost is fixed. The economy of the electrical energy, therefore, clearly depends on the energy demand in kWh per day. Figure 15-7 is a screening chart for making economic choice between the line extension and a pv plant at site. It is developed using the 1997 average utility charges for line extension and pv modules price. As an example of using the chart, if the customer needs 10 kWh/day and lives three miles away from the nearest power line, the stand-alone pv at the site would be economical. If the distance were one mile instead, the line extension would be economical. The break-even distance for this customer is about 1.7 miles. For another customer who needs 20 kWh/day, the break-even distance is three miles.

In getting the line extension, rural utilities and their customers face conflicting interest. The utility is interested in controlling the cost, and the customer is interested in reliable power. In selected low population density areas of the U.S.A., some utility companies offer their customers a choice of installing a pv system as an alternative to costly emergency service and long stretches of the line dedicated to serve one or a few customers.

This is how some U.S. utility companies in rural areas are responding to potential pv customers:

- customer interested in electric power contacts the utility company.
- determination is made whether the point-of-service will be economically served by extending a line or by installing a pv system at the site.
- utility company develops the pv design.
- the system is installed by an independent contractor.

- the total capital cost is paid by the utility company.
- the utility company is responsible for maintenance of the system.
- customer is billed a fixed monthly rate, rather than the amount based on the energy consumed.

This approach spreads out the burden of the initial capital cost on the customer.

15.7 Hybrid Economics

A detailed analytical tool to evaluate the economics of hybrid systems has been developed by the National Renewable Energy Laboratory. The newer version, Hybrid2[5] allows the user to determine basic economic figures for a particular simulation run. The economics module uses information from the performance simulation run and economic data supplied by the user. It computes parameters as the payback period, internal rate of return, cash flow, and equipment replacement expenses. The user has a wide versatility in determining the expense of the project and what inputs to include. Parameters such as grid extension, importation tariffs, systems administration costs, and taxes can be included in the analysis. The user may conduct comparisons between differing hybrid possibilities and power solutions, and determine approximate costs. Figure 15-8 is the computer screen displaying the choices available to the user. As indicated by the menu buttons, it includes wind, pv, diesel, and battery.

Hybrid2 allows user to conduct parametric analyses on certain cost parameters, such as fuel cost, discount rate, and inflation rate to help determine how the value of certain parameters can affect the viability of the project.

As for the project funding, there have been strong financial incentives for the renewable power in many countries around the world. However, the incentives have been declining because the renewables are becoming economically competitive on their own merits. More projects are being funded strictly on the commercial basis. Enviro Tech Investment Funds in the U.S.A. is a venture-capital fund supported by Edison Electric Institute (an association of investor-owned electric utility companies providing 75 percent of the nation's electricity). The World Bank now includes the wind and pv in its landing portfolios. The first such loan was made to India in 1993. Several units in the World Bank group are jointly developing stand-alone projects in developing countries.

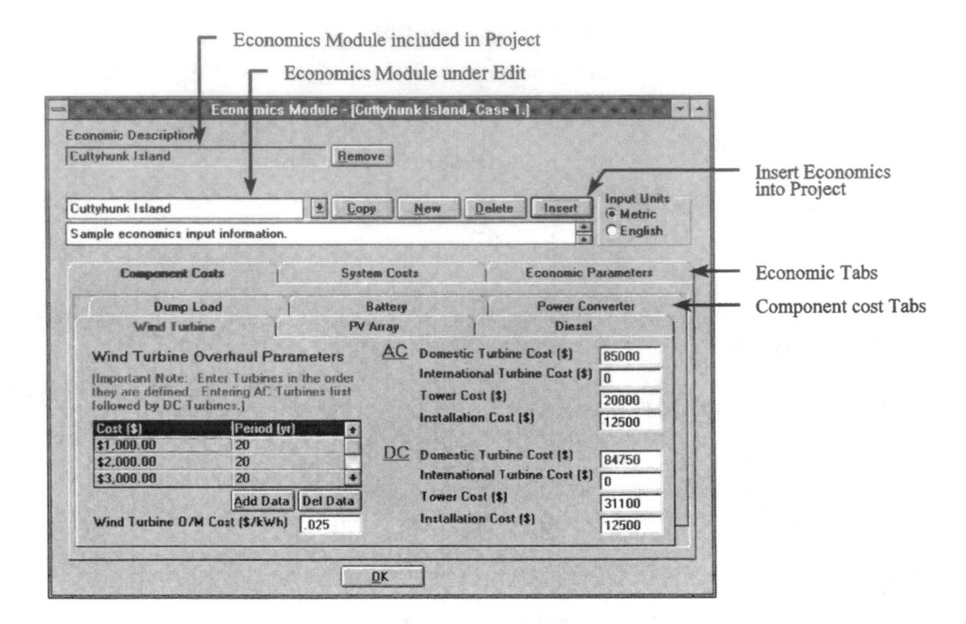

FIGURE 15-8
Hybrid2 economic simulation model. (Source: Ian Baring-Gould, National Renewable Energy Laboratory, Golden, Colorado. With permission.)

References

1. Cardell, J. B. and Connors, S. R. 1997. "Wind power in New England, modeling and analyses of nondispatchable renewable energy technologies," *IEEE Paper No. Power Engineering-888-PWRS-2-06,* 1997.
2. Institute of Solar Energy and Technology. 1997. "Annual Report," Germany, University of Kassel, 1997.
3. Chabot, B. 1997. "L'analyse economique de l'energie ecolienne," *Revue de l'Energie,* ADME, France, No. 485, Feb. 1997.
4. Chabot, B. 1998. "From Costs to Prices: economic analyses of photovoltaic energy and services," *Progress in Photovoltaic Research and Applications,* Vol. 6, p. 55-68, 1998.
5. Baring-Gould, E. I. 1996. "Hybrid2, The hybrid system simulation model user manual," *NREL Report No. TP-440-21272,* June 1996.

16

The Future

The presented trends and future forecasts of the electricity demand, and how photovoltaic and wind power may grow in percentage of the total demand are discussed in this chapter. The historical data on the market developments in similar basic-need industries and the Fisher-Pry market growth model indicate that the pv and wind power may reach their full potential in the year 2065. The probable impact of the U.S. utility restructuring on the renewables is reviewed.

16.1 World Electricity Demand to 2015

According to the U.S. Department of Energy, electric power is expected to be the fastest growing source of energy to the end users throughout the world over the next two decades. The electricity demand is projected to grow to 19 trillion kWh by the year 2015 (Figure 16-1) at the annual growth rate of 2.6 percent.[1] The electricity demand in the industrial and developing countries projected to 2015 is shown in Figure 16-2, indicating a fast growth rate in the developing countries. Table 16-1 shows the per capita consumption by region and country. The Organization for Economic Cooperation and Development (OECD) countries constitutes 20 percent of the world's population, but consumes over 60 percent of the world's electricity. The non-OECD Asia is expected to grow faster than the Eastern European and Former Soviet Union (EE/FSU) countries. This is a result of the relatively high economic growth rate projected for the non-OECD Asia. Although the United States is the largest consumer of electricity in the world, it has the lowest projected growth rate of 1.3 percent versus the world average of 2.6 percent. Mexico has the highest projected growth rate of 4.7 percent per year to the year 2015.

The developing Asia is projected to experience the strongest growth than any other region in the world at a growth rate of 5 percent per year (Figure 16-3). As the electricity demand grows, the coal will remain the primary fuel for power generation, especially in China and India. The share of nuclear generation worldwide has reached the peak, and is expected to decline in the future. The coal, the natural gas, and the renewable, all are

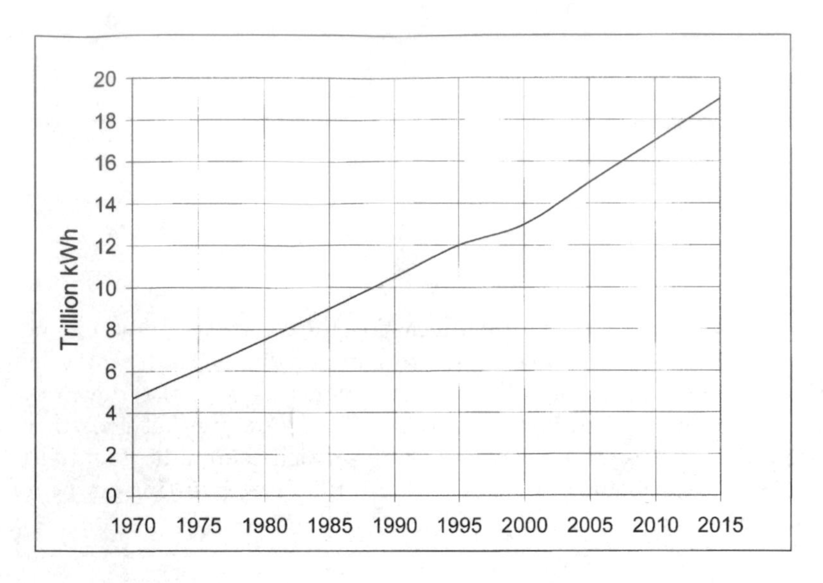

FIGURE 16-1
World electricity consumption 1970-2015. (Source: U.S. Department of Energy, International Energy Outlook 1997 with Projections to 2015, DOE Office of the Integrated Analysis and Forecasting, Report No. DE-97005344, April 1997.)

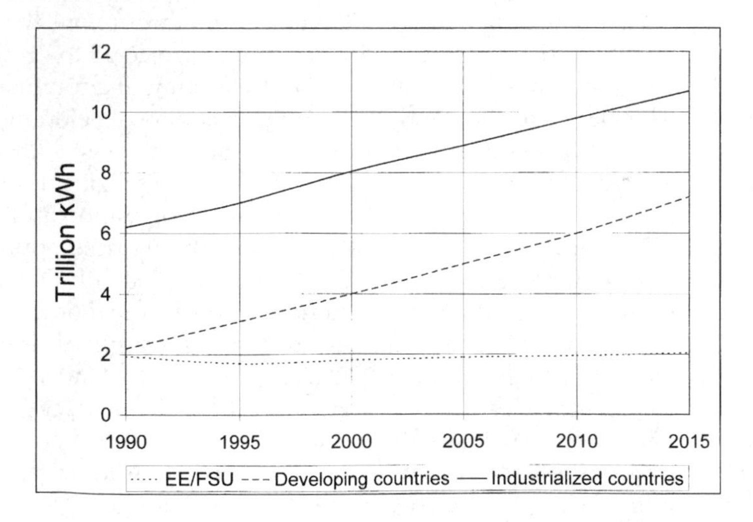

FIGURE 16-2
Electricity consumption by industrialized and developing countries 1990-2015. (Source: U.S. Department of Energy, International Energy Outlook 1997 with Projections to 2015, DOE Office of the Integrated Analysis and Forecasting, Report No. DE-97005344, April 1997.)

expected to grow to replace the retiring nuclear power plants. However, the renewables are expected to take a greater percentage of the growth. For example, in February 1997, the Swedish government announced that it

TABLE 16-1

Electricity Consumption Per Capita by Region, 1980, 1993 and 2015 in MWh Per Person

Region/Country	1980	1993	2015 (projected)
OECD North America			
United Sates	9.2	11.1	12.6
Canada	12.6	15.4	19.2
Mexico	0.9	1.2	2.5
Other Regions			
OECD Europe	4.0	5.0	7.8
EE/FSU	4.0	4.0	5.1
Japan	4.4	6.3	10.3
Non-OECD Asia	0.3	0.5	1.2
Middle East	0.9	1.3	1.6
Africa	0.4	0.4	0.5
Central and South America	0.9	1.3	1.7

(Source: U.S. Department of Energy, International Energy Outlook 1997 with Projections to 2015, DOE Office of the Integrated Analysis and Forecasting, Report No. DE-97005344, April 1997.)

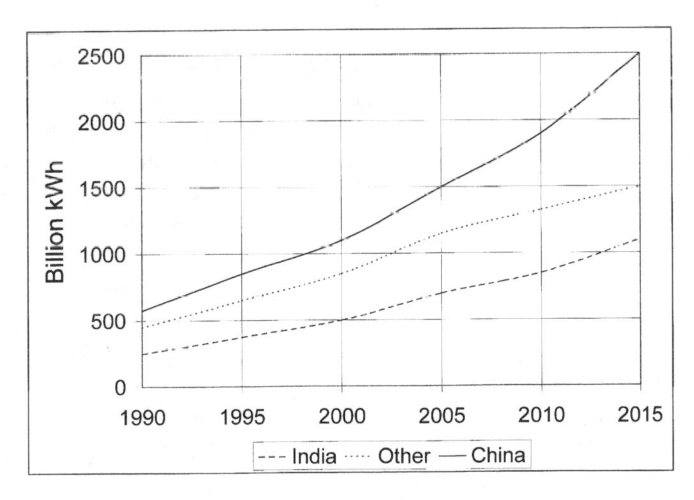

FIGURE 16-3

Electricity consumption in developing Asia 1990-2015. (Source: U.S. Department of Energy, International Energy Outlook 1997 with Projections to 2015, DOE Office of the Integrated Analysis and Forecasting, Report No. DE-97005344, April 1997.)

would begin shutting down the country's large nuclear power industry, whose plants supply 50 percent of Sweden's electricity. This announcement was in response to a referendum vote to end nuclear power in Sweden. To

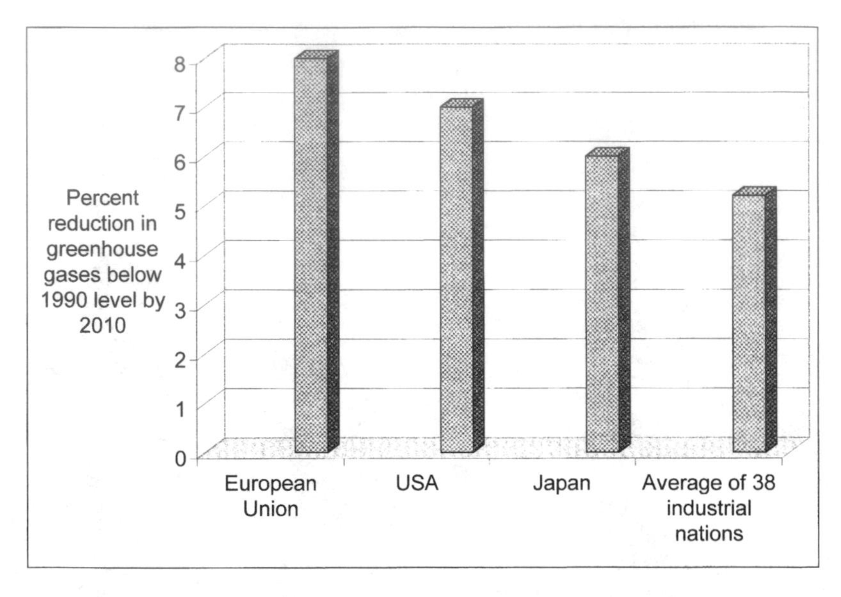

FIGURE 16-4
Cutting emissions under Kyoto treaty. (Source: United Nations.)

compensate, the government is to step up intervention in energy market and force a switch to greater use of nonfossil alternatives. Other industrial countries have serious commitments to reduce pollution of the environment. According to the Environmental Protection Agency of the U.S.A., the electric power industry is the nation's largest polluter. It is responsible for emitting 66 percent of SO_2, 29 percent of NOx, 36 percent of manmade CO_2, and 21 percent of mercury. The 160 countries, gathered in Kyoto, Japan, in December 1997 under the U.N. Convention on Climate Change, agreed to work on reducing emissions of carbon dioxide and other greenhouse gases as per the national and regional goals depicted in Figure 16-4. The U.S. has promised a 7 percent cut by the year 2008-2012 measured against the base year of 1990. Because the U.S. economy would grow by then since 1990, the real cut required would be over 30 percent from the current level of emission. This is a significant challenge and would certainly require an aggressive policy on the renewable energy sources.

The developing countries have additional reason for promoting wind and pv programs. With significant rural population, they would benefit from the distributed power generation near the loads.

16.2 Wind Future

The U.S. Department of Energy reports that wind is experiencing the strongest growth among the renewable energy sources with the falling generation

TABLE 16-2

Wind Power Technology, Past, Present and Future

Technology Status	1980	1997	After 2000
Cost per kWh*	$0.35–$0.40	$0.05–$0.07	<$0.04
Capital cost per kW	$2,000–3,000	$500–800	<$500
Operating life	5–7 years	20 years	30 years
Capacity factor (average)	15 percent	25–30 percent	>30 percent
Availability	50–65 percent	95 percent	>95 percent
Size range	50–150 kW	300–1000 kW	500–2,000 kW

* For wind sites with average annual wind speed of 7 meters/second (15.6 miles per hour) at 30 meters (100 feet) hub height.

cost. The wind power cost has dropped to 5 cents per kWh, and the DOE projections suggest that further reduction to about 2 cents is possible by 2010 with favorable financing. The American Wind Energy Association forecasts that the five top-growth markets for wind energy through 2005 will be the U.S.A., India, China, Germany, and Spain, with capacity additions of between 1,275 and 2,730 MW projected in each country. Table 2-2 in Chapter 2 listed the cumulative installed wind-generating capacity in selected countries. In 1995, India and Germany together accounted for two-thirds of the entire world's added wind capacity.

The renewable power sources are clean, abundant, and do not need to be imported. However, they must be economical on their own merit. The new developments are meeting this challenge on the both fronts, the initial capital cost and the cost per unit of electricity generated.

Since the early 1980s, wind technology capital costs have declined by 80 percent, operation and maintenance costs have dropped by 80 percent and availability factor of grid-connected plants has risen to 95 percent (Table 16-2). For the wind plans at present, the capital cost has dropped below $800 per kW and the electricity cost to about 6 cents per kWh. The goal of current research programs is to bring the wind power cost below 4 cents per kWh by the year 2000. This is highly competitive with the cost of conventional power plant technologies. According to the National Renewable Energy Laboratory, several research partners are negotiating with U.S. electrical utilities to install additional 4,200 MW of wind capacity at a capital investment of about $2 billion throughout the nation during the next several years. This amounts to the capital cost of $476 per kW, which is comparable with the conventional power plant costs. According to the Electric Power Research Institute, the continuing technology developments and production economy would make the wind the least-cost energy within a decade.

The industry experts make this forecast based on the following ongoing research programs:

- more efficient airfoil and blade design and manufacturing.
- better understanding on the structure and foundation loads under turbulence, operating fatigue loads and their effect on life.

- computer prototyping by accurate system modeling and simulation.
- integrated electrical generators and power electronics to eliminate the mechanical gearbox.
- efficient low cost energy storage at large scale.
- better wind speed characterization, particularly within a large wind farm.

Successful design, development, and demonstration based on the results of these research programs are expected to increase the share of wind power in the U.S.A. from a fraction of 1 percent to more than 10 percent over the next two decades.

The offshore wind potential is much greater than onshore due to good wind speed and the large area available for commercial installations. It is limited only by practical working depths and other maritime activities in the area of interest. Fishing, shipping routes, and military test grounds are some of the activities that may conflict with the wind farm. Water depth of 30 meters is practical. Taking only the water depth as a constraint, the accessible U.K. offshore wind resource is estimated to be 380 TWh per year. Taking other constraints also into account, this estimate is reduced to 120 TWh per year.

A study commissioned by the European Union has suggested that the offshore wind farms in the coastal regions of Germany, Holland, and Denmark could meet all electricity demand in those countries. PowerGen Corporation of Britain plans to build 37 MW (25 towers, each rated 1.5 MW) of offshore wind capacity near Yarmouth, Norfolk, by 1999. Germany plans to install a large wind farm in the Baltic Sea, near Schleswig-Holstein.

Denmark's ambitious offshore plan announced in the government's Energy-21 report published in 1997 aims to have at least 750 MW offshore capacity by 2005, 2,300 MW by 2015 and 4,000 MW by the year 2030. This will amount to about 50 percent of that country's electricity coming from wind. This would push the percentage limit to a practical maximum for any intermittent source of power. The first group of Denmark's offshore sites is shown in Figures 16-5 and 16-6.

The offshore wind speed is generally higher, typically 8 to 10 m/s. However, due to lack of long term data, these estimates must account for the inherent variability in the estimate and the associated sensitivity to the cost of generated electricity.

The onshore wind technology is applicable to the offshore installations. The major difference, however, is the hostile environment and the associated increase in the installation cost. The electrical loss in transmitting power to the shore needs to be accounted for. Overall, it is estimated that the offshore wind power plant cost can be at least 30 percent higher than the comparable inland plant.

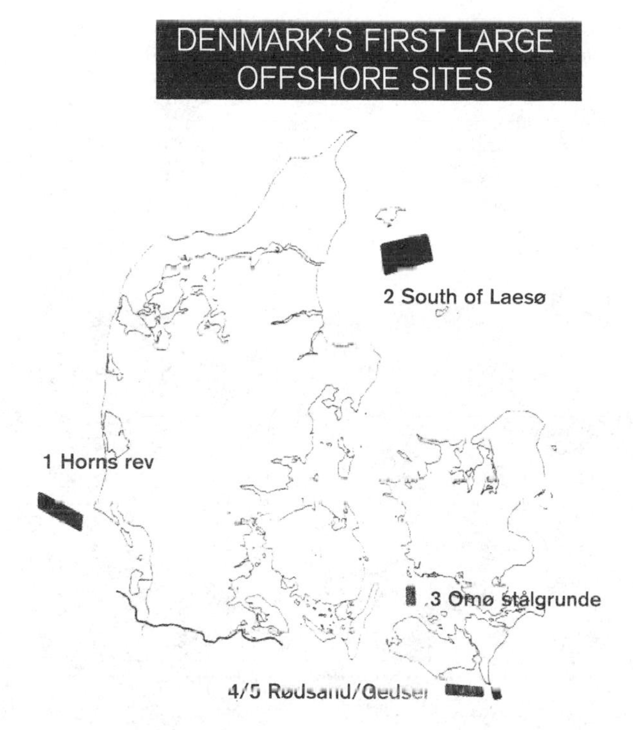

FIGURE 16-5
Denmark's first large offshore wind power sites. (Source: Wind Directions, Magazine of the European Wind Energy Association, January 1998, London. With permission.)

As of December 1997, Europe's wind capacity by country was shown in Figure 2-10. Germany and Denmark lead Europe in wind power. Both have achieved phenomenal wind energy growth through a guaranteed tariff based on the domestic electricity prices, perhaps a blueprint for the rest of the world to follow. Germany has a 35-fold increase between 1990 and 1996, a 70 percent annual rate of growth. With 2,000 MW installed, Germany is now the world leader. The former global leader, the U.S.A., has seen only a small increase during this period, from 1,500 MW in 1990 to 1,650 MW in 1996.

In 1997, the European Wind Energy Association adopted an ambitious target of 40,000 MW of wind capacity in Europe by the year 2010 and 100,000 MW by 2020 (Table 16-3). Each European country would be obligated to meet its committed share of renewable electricity, including the wind energy obligation, towards the overall target of 12 percent of the primary energy by wind in Europe by the year 2010.

In selected countries, the present wind capacity and future targets are listed in Table 16-4. The present and future targets for meeting the total electricity demand by wind capacity are listed in Table 16-5. Based on these targets, it is reasonable to expect that the wind may contribute 10 to 25 percent of the total electricity demand in some countries by 2010.

FIGURE 16-6
Vindeby offshore wind farm in Denmark. (Source: Vestas Wind Systems, Denmark. With permission.)

TABLE 16-3
European Wind Energy Association Target for the Future

Year	Installed wind capacity target	Annual growth rate
1997 (actual)	4,425 MW	Reference
2000	8,000 MW	21.8 percent
2010	40,000 MW	17.5 percent
2020	100,000 MW	10 percent

(Source: Wind Directions, Magazine of the European Wind Energy Association, London. With permission.)

The environmental benefit is generally the primary contributing factor to the development of the wind and pv systems. Wind plants, however, introduce one area of concern on birds and their habitat. The Avian bird issue in

TABLE 16-4

Present Wind Capacity and Future Targets in Selected Countries

Country	Capacity
Germany	1,764 MW was operational as of June 1997
Netherlands	2,750 MW target by 2020
Denmark off-shore target	750 MW target by 2005
	2,300 MW target by 2015
	4,000 MW target by 2030
United Kingdom	160 MW was operational as of January 1996
Ireland	Government target 500 MW by 2010
	IWEA target 1,000 MW by 2010
Italy	100 MW was operational as of November 1997
USA	1,650 MW was operations as of December 1996
	6,000 MW target by 2006
	12,000 MW target by 2015
India	1,000 MW was operational as of December 1997

TABLE 16-5

Present and Future Target for Percent of Total Electricity Demand Met by Wind Capacity

Country	Wind as Percent of Total Electricity Demand
Denmark	Was 6 percent as of June 1997
	Target 50 percent by 2030 (aggressive)
Germany	Was 10 percent as of June 1997
	Target 20-25 percent by 2010
Ireland	Government target 5 percent by 2010
	IWEA target 10 percent by 2010
USA	Was <1 percent in 1997
	EPRI estimate 10 percent by 2020

(Source: Wind Directions, Magazine of the European Wind Energy Association, London, With permission.)

California is under research funded by the DOE and the wind industry to define the magnitude of the problem and ways to prevent the bird kills. The noise factor and the impact on rural landscape are other concerns, which may impact the wind market development. These factors have begun to contribute and make the planning process difficult in some countries. The sheer speed at which new development has taken place has far outpaced the capacities and capabilities of the decision makers. The regulations are inadequate to cover the installations of wind turbine in rural areas. The closer people watch the wind turbine rotating, and the smaller the local benefits in terms of jobs and tax revenue becomes, the stronger the opposition. Some governments have started publishing clear guidelines in order to help local authorities in the planning procedures.

16.3 pv Future

On the other hand, the cost of solar pv electricity is still high in the neighborhood of 15 to 20 cents per kWh. With the consumer cost of utility power ranging from 10 to 15 cents per kWh nationwide, pv cannot economically compete directly with the utility power as yet, except in remote markets where the utility power is not available and the transmission line cost would be prohibitive. Many developing countries have large areas falling in this category. With the ongoing research in pv technologies around the world, the pv energy cost is expected to fall to 12 to 15 cents per kWh or less in the next several years as the learning curve and the economy of scale come into play. The research programs jointly funded by DOE and NREL have the goal of bringing down the pv electricity cost below 12 cents per kWh by the year 2000.

The pv cell manufacturing process is energy intensive. Newer processes are being developed and continuously implemented to lower the energy use per square centimeter of the finished cell. In the current manufacturing process, the front and the back of the cells are diffused separately, taking one to three hours of manufacturing time. Researchers at the Georgia Institute of Technology are considering rapid-thermal-processing technology. This process simultaneously forms the front and the back of the cell in several minutes, and the metal contacts are applied by screen printing. It not only cuts the processing time, it also lowers the temperature and energy needed for fabrication.

Total worldwide production of pv modules has reached 100 MW, and is projected to grow to almost 200 MW by 2000 (Figure 16-7). Approximately ⅔ of the current production in the U.S.A. is by two companies, Siemens Solar Industries in California and Solarex in Maryland.

Among the new advances underway in the pv technology, the thin film is expected to lead the group. The amorphous silicon particularly holds the most near-term promise of quickly penetrating the market due to its low material and manufacturing costs. In this technology, 2 μm thick amorphous silicon vapor is deposited on glass or stainless steel rolls, typically 2,000 feet long and 13 inches wide. Compared to the thick crystalline silicon cells, amorphous silicon technology uses less than 1 percent material. The sheet manufacturing, instead of single cells from ingots, offers a low-cost base. The disadvantage in single junction amorphous silicon is that the cell efficiency degrades about 20 percent in the first several months under the sunlight before stabilizing. At present, the stabilized efficiency is half and the retail price is also half, keeping the price per watt the same. However, this is the market-based price, and not the technology-based price. The crystalline-silicon technology has been with us for several decades and has entered a plateau of a slow learning curve. On the other hand, the amorphous-silicon technology is new and, hence, the steep learning curve is expected. The

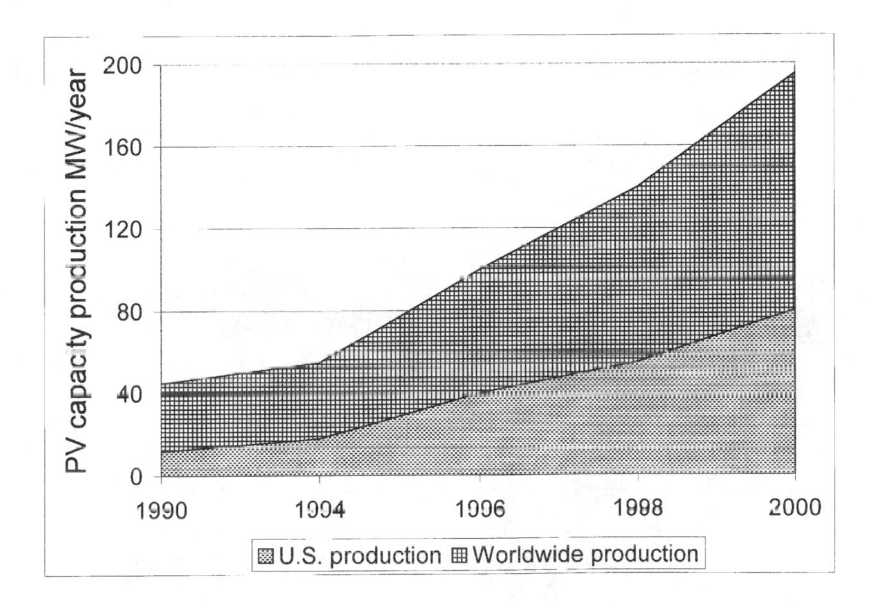

FIGURE 16-7
PV annual capacity and cumulative installed capacity worldwide. (Source: PV News, an industry publication.)

expected result is a rapid decline in the amorphous silicon price per watt by increasing the efficiency and decreasing the manufacturing cost. On this premise, two large amorphous-silicon-panel manufacturing plants started in the U.S.A. in 1996. The technology and competition are expected to take hold on the market and the price, as they always have.

In the U.S.A., the pv and the wind power today are not competing on a uniform playing field, as their costs do not fully reflect benefits of the fuel diversity and pollution free operation. For example, if the renewables get due credits for pollution elimination of 600 tons of CO_2 per million kWh electricity consumed, it would get a further boost in the incentives presently offered by the United States Government. For the American pv and wind industries, there is an additional competition in the international market. Other governments support their companies with well-funded research, low-cost loans, and favorable tax rate tariffs not generally available to their U.S. counterparts.

16.4 Declining Production Costs

The economy of scale is expected to continue in contributing to the declining prices. Future wind plants will undoubtedly be larger than those installed in the past, and the cost per square meter of the blade swept area decline

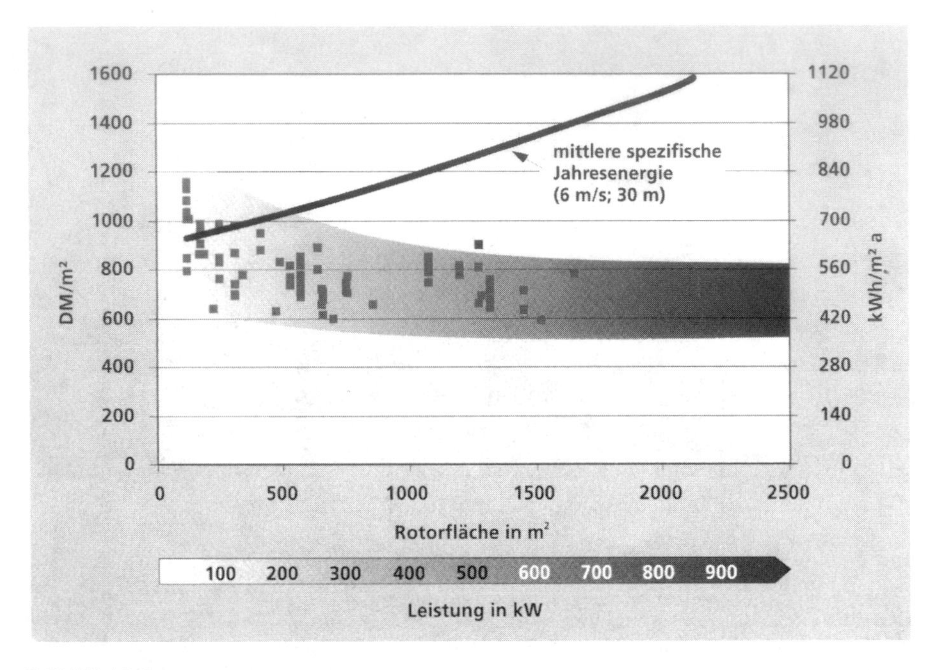

FIGURE 16-8
Economy of scale trends for capital cost in Deutsche mark/m2 and annual energy produced in kWh/m2 versus blade swept area. (Source: Institute of Solar Energy and Technology, University of Kassel, Germany. With permission.)

with size. Figure 16-8 shows the prevailing costs of wind turbines of various sizes in Germany. The price band is DM/m^2 of the blade area, falling from 800 to 1200 DM/m^2 for small turbines to 500 to 800 DM/m2 for large turbines of 50 meters diameter. The line showing the annual energy potential per square meter rises from 630 kWh/m^2 in small turbines to 1,120 kWh/m^2 in large turbines of 46 meters diameter.

With the technology and the scale of economy combined, the manufacturing cost of new technologies has historically shown declining patterns. The growth of new product eventually brings with it a stream of competitors and the leaning curve benefits. A learning curve hypothesis has been commonly used to model such cost declines. The cost is modeled as an exponentially decreasing function of the cumulative number of units produced up to that time. A standard form for a cost decline is constant doubling, where the cost is discounted by a fraction λ when the cumulative production doubles. For renewable electricity, the production units are MW of capacity produced and kWh of electricity generated. If we want to monitor only one production unit, the kWh generated includes both the MW capacity installed and the length of the operating experience, thus, making it an inclusive unit of production.

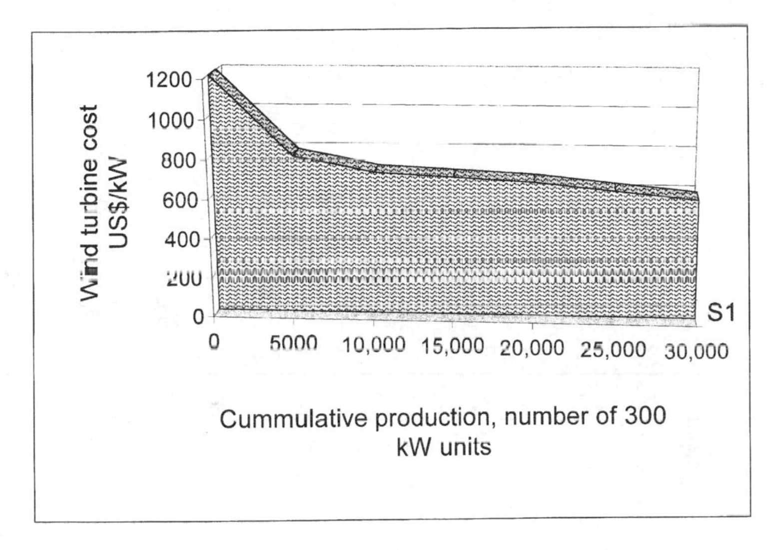

FIGURE 16-9
Learning curve of the wind turbine costs versus number of 300 kW units produced. (Source: American Wind Energy Association, Washington, D.C. With permission.)

For new technologies, the price keeps declining during the early phase of a new product at a rate which depends on the nature of the technology and the market. The pattern varies widely as seen in new technologies such as the computer, telephone, and airline industries. For wind turbines, the cost per kW capacity has declined with the number of units produced, as seen in Figure 16- 9. The declining electricity price was seen earlier in Figure 2-2.

In general, the future cost of a new product can be expressed as follows:

$$C(t) = C_o \lambda^{\log_2(N_t/N_o)}, \quad 0 < \lambda < 1 \qquad (16\text{-}1)$$

where $C(t)$ = cost at time t
C_o = cost at the reference time
N_t = cumulative production at time t
N_o = cumulative production at reference time

Estimating the parameter λ based on the historical data is complex, and the estimate may be debated among the experts. However, the expression in itself has shown to be valid time and again for new technologies in the past. It may be used to forecast the future trends in the capital cost decline in any new technology, including the wind and photovoltaic power.

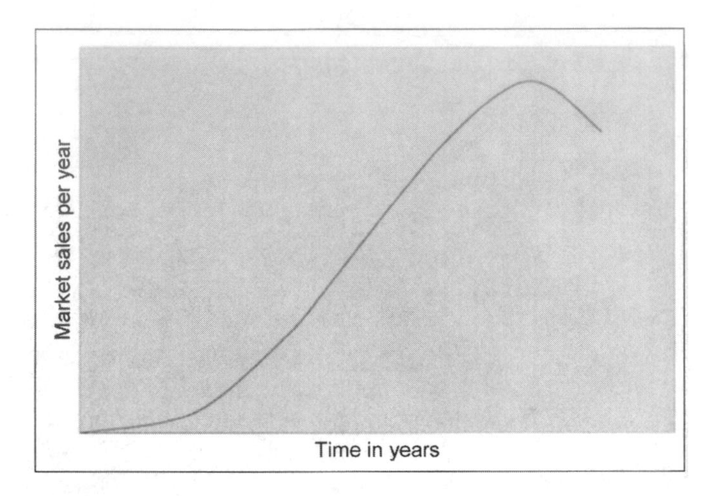

FIGURE 16-10
S-shape growth and maturity of new products.

16.5 Market Penetration

With all market forces combined and working freely, the penetration of a new technology over a period of time takes the form of an s-shaped curve shown in Figure 16-10. It is characterized by a slow initial rise, followed by a period of more rapid growth, tapering off to a saturation plateau and finally declining to make room for a newer technology. The s-shape hypothesis is strongly supported by empirical evidences.

One model available for predicting the market penetration of a new product is the diffusion model proposed by Bass.[2] However, it can be argued that the renewable energy is not a really new product. It merely substitutes an existing product on a one-to-one basis. The Bass model, therefore, may not be an appropriate model for the renewable power. Since electricity is a basic need of the society, the penetration of renewable power technologies is better compared with similar substitutions in the past, such as in the steel making industry as shown in Figure 16-11. The solid lines are the actual penetration rates seen in those industries.

With ongoing developments in the wind and pv power technology and with adjustments in social attitudes, the market penetration rate may follow the line parallel to the historical experience on similar products depicted in Figure 16-11. It can be analytically represented by the Fisher and Pry substitution model.[3] In this model, the rate at which the wind and pv energy may penetrate the market can be expressed as a fraction of the total MWh energy consumed every year.

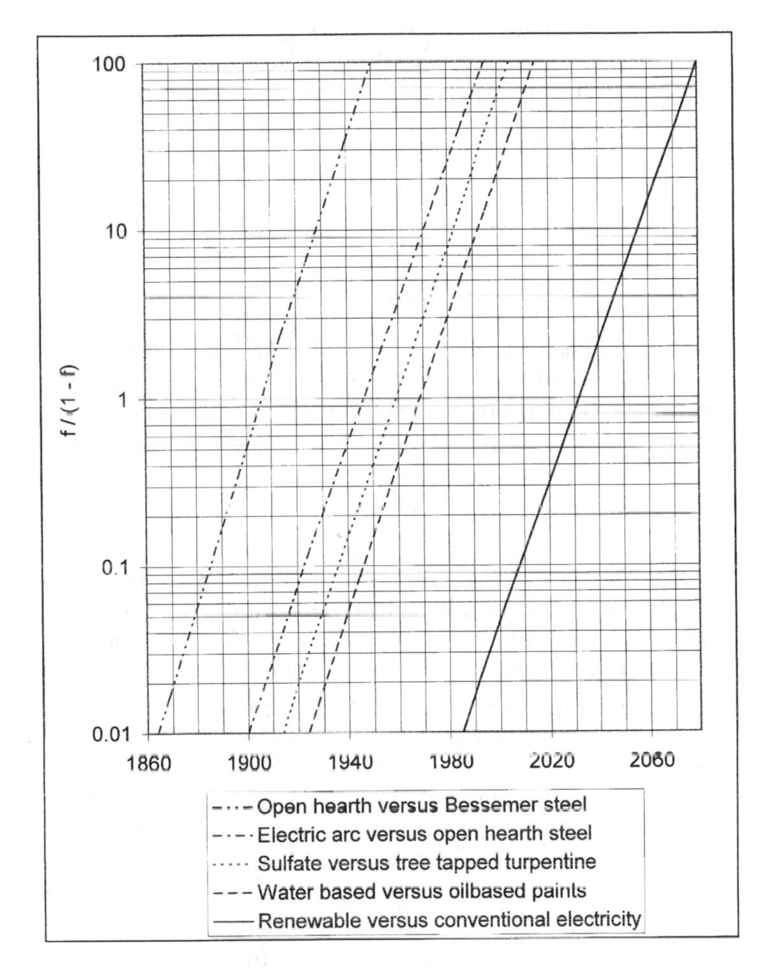

FIGURE 16-11
Market penetration of new product substituting existing products. Wind and pv shown in parallel with historical data.

If we let:

f = fraction of the market captured by the renewables at time t, and
t_o = time when f equals ½

then f can be expressed as follows:

$$f = \frac{1}{1 + e^{\left[-b\left(t-t_o\right)\right]}}, \quad or$$

$$\frac{f}{1-f} = e^{-b\left(t-t_o\right)} \tag{16-2}$$

Here b is the growth constant, characterizing the growth to the potential associated with a particular technology. The equation gives log-linear straight lines shown in Figure 16-11.

We take that the renewable power was successfully demonstrated to be commercially viable in the 1970s and made a commercial start in 1985. This is the time when a group of wind farms in California were installed and operated for profit by private investors. Based on this premise and based on the Fisher-Pry model just described, we draw the dotted line in Figure 16-11, beginning in 1985 and then running parallel to the historical lines followed by similar substitution products in the past. That is the rate of market penetration we can expect the wind and pv to follow in the future.

At the rate projected by the dotted line, the full 100 percent potential of the wind and pv will perhaps be realized around 2065. The wind and pv reaching the 100 percent potential in the year 2065 does not mean that they will completely replace the thermal and other power. It merely means it will reach its fully attainable potential. Experts would argue on the upper limit of this potential for a given country. However, the wind and solar, being intermittent sources of energy, cannot be the baseload provider. They may augment the baseload plants, thermal, or other types, which can dispatch energy on demand. Such reasoning puts an upper limit on the pv and wind power to well below 50 percent, perhaps around 25 percent. According to the Royal Institute of the International Affairs study published in the U.K. in May 1997, all renewable sources could provide between 25 to 50 percent of European electricity by the year 2030. The contributions of the wind and pv in the total electricity demand, however, will largely depend on the operating experience with the grid-connected plants and energy storage technologies developed during the next few decades that can remedy the nondispatchable nature of the pv and wind energy. New energy storage products are being developed that incorporate batteries, flywheel, fuel cells, and superconducting coils into system solutions for the power industry such as load leveling and power quality. Electric utility restructuring and the subsequently increased competitive market are creating new and higher-value market for energy storage. The renewable power is the new market opportunity for large-scale energy storage products.

The sophisticated prediction models have been developed by the DOE, NREL, and others based on mathematical considerations and regression analyses through applicable data. However, there is a limit to how such models can be confidently used to commit huge sums of capital investments. Small errors in data can cause large errors in projections, particularly decades out in the future. The time-proven data on how people make investment decisions under long-term uncertainties can also help. The experience indicates that investors commit funds only if the payback period is less than five years, the shorter it is, the quicker the investment penetrates. The upper limit of the payback period that starts having a significant market is four to five years, and large market penetration requires a payback period less than three years.

16.6 Effect of Utility Restructuring

Although the U.S. electricity market was $210 billion in 1997, the electric utilities have grown into regulated monopolies, with no competition at the consumer level. Both the generation and the transmission of electricity remained with one company. Even when electricity could be produced by alternate sources, the transmission access to the loads was denied or the owners set the price so high that the access was effectively denied to protect their own position in the generation market.

The Public Utility Regulatory Policy Act of 1978 (PURPA) required utilities to purchase electricity from power generators using renewable energy sources or co-generation. The PURPA also required the utilities to pay the "qualifying facilities (QFs)" at rates based on their "avoided cost". The United States Congress defined the avoided cost as the incremental cost of electrical energy the utility would incur if it generated the energy itself or purchased from another source. This definition is vague at least in one sense. That is, does the incremental cost include the share of the capital cost needed to generate the incremental unit of energy? For example, if the plant cost is taken as sunk cost, the avoided cost would primarily mean the fuel cost, which for thermal power plants may be 2 cents per kWh. On the other hand, if a new plant has to be built to meet the added demand, which a renewable power plant can supply, the avoided cost would include the capital cost of about 5 to 6 cents per kWh, making the total avoided cost of 7 to 8 cents per kWh. Congress left the exact definition of the avoided cost and the implementation of PURPA to the states. However, with the regulatory advantage of PURPA, renewable and cogeneration power plants started coming up across the nation.

Another issue debated in the renewable power industry is definition of the "avoided cost" which are paid by the utility companies. The issue is complex as illustrated by the following example. For a utility company, say the sunk cost of generation is 10 cents per kWh. This includes 2 cents for the fuel cost, and 8 cents for the capital and administrative cost of generation and transmission. If that utility has an excess capacity, its incremental avoided cost is only 2 cents, the cost of adding a customer or saving by deleting a customer. And this is what it would pay to the renewable power generator. On the other hand, it charges 12 cents/kWh to its average consumer. If the cost of self-generation using a wind power plant is 5 cents per kWh, the customer would save 7 cents by self-generating, but the society at large would lose 3 cents per kWh (utility would save 2 cents in fuel cost but the self-generating customer would spend 5 cents). However, in situations where the energy demand is growing but the generation and transmission capacities cannot keep up, the avoided cost includes the capital cost also. For example, Pacific Gas and Electric Company in California found that by purchasing power to meet loads at the end of heavily loaded lines, it could avoid having to construct new generation and transmission capacity. The "avoided cost" in this situation is

10 cents per kWh, and the society at large would gain 5 cents (utility will save 10 cents but the self-generator would spend 5 cents). This example highlights the basic difficulty in accounting the benefits of renewable power. With such common situations, the impact of the electric utility restructuring on the renewable would depend on how fairly it is treated in defining some of the costs of both the conventional and renewable electricity.

16.6.1 Energy Policy Act of 1992

By passing the Energy Policy Act of 1992, Congress encouraged even greater wholesale competition by reducing the market barriers for independent generators interested in selling electrical power. This act permits wholesale customers to have a choice of generators and obliges utilities to wheel power across their transmission lines at the same cost that they would charge to all others, including themselves. For effective implementation of the act, the utilities are required to break up the generation, transmission, and distribution businesses in separate companies. One will own the generating plants and the other will own the wires. In the past, electricity has been sold as a delivered product. The new restructuring merely unbundles the product price and the delivery charge, with the wires owned and operated separately as common carriers obligated to charge the same rate to all customers.

The new law has separated the power industry in two parts: (1) the generation and/or sale of electricity by licensed power supplier, and (2) the delivery of electricity to the consumers by regulated companies.

In the new retail competition, all customers will choose their energy suppliers, each supplier having direct open access to the transmission and distribution wires. The concept is similar to what happened to the telephone industry more than a decade ago. The idea is as old as the American free enterprise, that is the open unrestricted competition at the consumer level. The United Kingdom, Chile, Norway, and parts of Australia have systems similar to that presently taking effect in the U.S.A.

This is a major restructuring of the electrical utility industry in the U.S.A. since the beginning of the industry some 100 years ago. The uncertainty and skepticism around the EPAct of 1992 is, therefore, causing significant confusion and uncertainties of new investments in this industry.

Like PURPA, the implementation of the EPAct of 1992 is also left to individual states. The states with relatively higher energy cost (12 to 16 cents per kWh), such as California, Illinois, Massachussetts, Maine, New Hampshire, and Pennsylvania, have active pilot programs underway. In 1996, NH started a pilot project in which 16,500 customers were chosen randomly and given to choose their electricity producer. This 2-year experiment will lead to full implementation by the end of 1998. In late 1997, Pennsylvania became the largest state in the U.S.A. to implement an extensive program of competitive electric power sources. The state, home of 12 million people, has licensed 47 companies to sell power to public. Presently, there are four major established conventional utility companies (PECO Energy, PP&L Resources, GPU Energy, and Allegheny Power). Approximately one million customers volunteered for the pilot program, in which they were guaranteed savings. Full scale competition will begin in 1999, when the rate cuts will be at least 8 percent. The rate reduction is expected to reach 15 percent in the years to follow.

In Massachusetts, the most immediate effect of the restructuring is an automatic 10 percent savings effective March 1998, and 15 to 18 percent later. New Jersey plans a pilot program in 1999 and full implantation in 2003. Fact-finding and investigations are underway in mid range cost states, but no significant activity is reported in states having lower energy cost (6 to 8 cents per kWh).

The 1992 EPAct is expected to greatly benefit the consumer. When fully implemented, the end users are expected to save 10 to 20 percent in the electricity price.[4] This will be achieved partly by promoting competition among utility and nonutility power generators, and partly by the most efficient use of the generation and transmission assets by generating and wheeling the power as dictated by the regional economies. This becomes rather evident when viewed from an operational research point of view. In finding an optimized solution with constraints, the farther the constraint boundaries are, the more economical is the solution. No constraint at all results in the most economical solution.

With the expected decline in the electricity price, some of the high-cost power plants would become uneconomical compared to new power plants, including the renewables. One of the debated issues in the conventional power industry at present is who would pay the "stranded cost" of these uneconomical plants which cannot be recovered in the new competitive market. Examples of stranded costs are the cost of nuclear plants which is not recovered as yet, and the cost of locked-in power contracts with the independent generators at a higher market price. According to the Competitive Enterprises Institute, the stranded cost in the U.S. power industry is expected to add up to $200 billion.

16.6.2 Impact on Renewable Power Producers

Industry leaders expect the generating business, conventional or renewable, to become more profitable in the long run under the EPAct of 1992. The reasoning is that the generation business will be stripped of regulated prices and opened to competition among electricity producers and resellers. The transmission and distribution business, on the other hand, would still be regulated. The American experience indicates that the free business has historically made more profits than the regulated business. Such is the experience in the U.K. and Chile, where the electrical power industry has been structured similar to the EPAct of 1992. The generating companies there make good profits as compared to the transmission and distribution companies.

Until the renewable electricity cost matches with the conventional one, the renewable market penetration may be slow. On the other hand, under the restructured electricity market, it is possible that the environmentally conscious consumer may choose green power even with higher prices. Surveys strongly indicate that they would.[5] In 1995, several utility companies in the U.S.A. surveyed 300 customers of each. Sixty-four percent said the renewable power is very important and 36 percent said they would pay more for electricity from those sources. Twenty-four percent said they would pay at least 10 percent more. Several U.S. utilities are developing green marketing programs under which they would allow their customers to volunteer to choose to pay premium for renewable power. A pilot program, called ClearChoice was initiated in the city of San Angelo, Texas, beginning in October 1997.

As for the renewable electricity producers, they are likely to benefit as much as other producers of electricity, as they can now freely sell power to the end users through truly open access to the transmission lines (Figure 16-12).

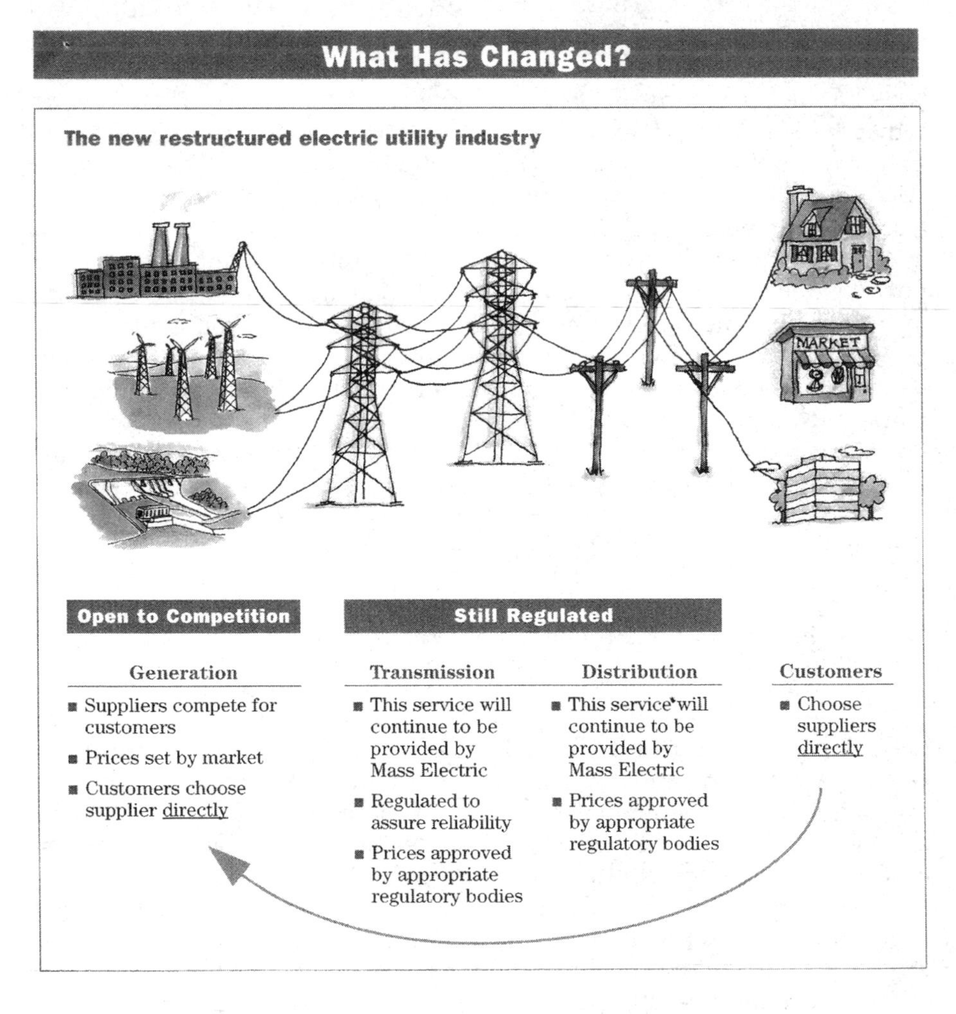

FIGURE 16-12
New restructured electric utility industry in U.S.A. (Courtesy: Massachusetts Electric, Boston, MA.)

Another benefit is that the renewable power price would be falling as the technology advances, whereas the price of the conventional power would rise with inflation, making the renewable even more advantageous in the future.

References

1. U.S. Department of Energy. 1997. "Office of the Integrated Analysis and Forecasting, International Energy Outlook 1997 with Projections to 2015," *Report No. DE-97005344,* April 1997.
2. Bass, F. M. 1969. "A new product growth model," *Management Science,* Vol. 15, p. 215-227, 1969.
3. Fisher, J. C. and Pry, R. H. 1971. "A simple substitution model of technological change," *Technology Forecasting and Social Change,* Vol. 3, p.79-88, 1971.
4. Backus, G. and Baylis, S. 1996. "Dynamics of U.S. electrical utility deregulation," *DOE Office of Utility Technology, Report No. DE-9600052,* December 1996.
5. Puttgen, H. B., Truly, R., Hyde, D. G., Walker, R., Terrado, E., Cohen, G., and Hurwitch, J. W. 1998. "Realization of alternative energy generation and storage," *IEEE Power Engineering Review,* May 1998, p. 5-18.

Further Reading

1. Boer, K. W. 1996. "Advances in Solar Energy," *Annual Review of R&D*, Boulder, CO, American Solar Energy Society, 1996.
2. Eggleston, D. M. and Stoddard, F. S. 1987. "Wind Turbine Engineering Designs," Van Nostrand Reinhold Company, 1987.
3. Freris, L. L. 1990. "Wind Energy Conversion Systems," London, Prentice Hall, 1990.
4. Frost, W. 1978. "Engineering handbook for use in wind turbine generator development," *NASA Technical Report No. 1359*, 1978.
5. Gipe, P. 1995. "Wind Energy Comes of Age," New York, John Wiley & Sons, 1995.
6. Johnson, G. L. 1985. "Wind Energy Systems," Prentice Hall, 1985.
7. Komp, R. J. 1995. "Practical Photovoltaics," *Aatec Publications*, Ann Arbor, Michigan, 1995.
8. Lof, G. 1993. "Active Solar Systems," Cambridge, MA, MIT Press, 1993.
9. Park, J. 1981. "The Wind Power," Palo Alto, CA, Cheshire Books, 1981.
10. Spera, D. A. 1994. "Wind Turbine Technology," New York, American Society of Mechanical Engineers, 1994.

APPENDIX 1

National Electrical Code® – Article 705

Portions reprinted with permission from NFPA 70-1996, the *National Electrical Code® Handbook,* 7th Edition, Copyright© 1996, National Fire Protection Association, Quincy, MA 02269. This reprinted material is not the complete and official position of the National Fire Protection Association on the referenced subject which is represented only by the standard in its entirety.

Article 705 Interconnected Electric Power Production Sources

Contents

705-1. Scope. This article covers installation of one or more electric power production sources operating in parallel with a primary source(s) of electricity.

(FPN): Examples of the types of primary sources are a utility supply, on-site electric power source(s), or other sources.

This article sets forth basic safety requirements for the installation of generators and other types of power production sources that are interconnected and operate in parallel. Power sources include any system that produces electric power. They include not only electric utility sources, but also on-premises sources, ranging from rotating generators (see Article 445) to solar photovoltaic systems (see Article 690) to fuel cells.

The purpose of this article is to address the basic safety requirements, specifically related to parallel operation, for the generators and other power sources, the power system that interconnects the power sources, and equipment that is connected to these systems. Proper application of these systems requires a thorough review of the entire power system.

705-2. Definition. For purposes of this article, the following definition applies:

Interactive System: An electric power production system that is operating in parallel with and capable of delivering energy to an electric primary source supply system.

705-3. Other Articles. Interconnected electric power production sources shall comply with this article and also the applicable requirements of the following articles:

	Article
Generators	445
Solar Photovoltaic Systems	690
Emergency Systems	700
Legally Required Standby Systems	701
Optional Standby Systems	702

705-10. Directory. A permanent plaque or directory, denoting all electrical power sources on or in the premises, shall be installed at each service equipment location and at locations of all electric power production sources capable of being interconnected.

Exception: Installations with large numbers of power production sources shall be permitted to be designated by groups.

705-12. Point of Connection. The outputs of electric power production systems shall be interconnected at the premises service disconnecting means. See Section 230-82, Exception No. 6.

Exception No. 1: The outputs shall be permitted to be interconnected at a point or points elsewhere on the premises where the system qualifies as an integrated electric system and incorporates protective equipment in accordance with all applicable sections of Article 685.

Exception No. 2: The outputs shall be permitted to be interconnected at a point or points elsewhere on the premises where all of the following conditions are met:

a. The aggregate of nonutility sources of electricity has a capacity in excess of 100 kW, or the service is above 1000 volts;
b. The conditions of maintenance and supervision ensure that qualified persons will service and operate the system; and
c. Safeguards and protective equipment are established and maintained.

The point of interconnection is required to be at the premises service disconnecting means. This requirement may be difficult to meet for systems 100 kW or smaller. It is intended to prevent the indiscriminate interconnection of small generators or other sources of power without proper protection against fire and electric shock. See Figure 705-1. It is important to utilize disconnect devices (switches, etc.) suitable for the purpose.

The requirement specifying "at the premises service disconnecting means" permits connection ahead of the disconnect or on the load side. This is to accommodate the safe work practices of many utilities that provide a readily accessible disconnect for dispersed generation. It is a contract matter between the utility and the customer and does not adversely affect the safety of the premises wiring; therefore, it is not within the scope of the *Code*.

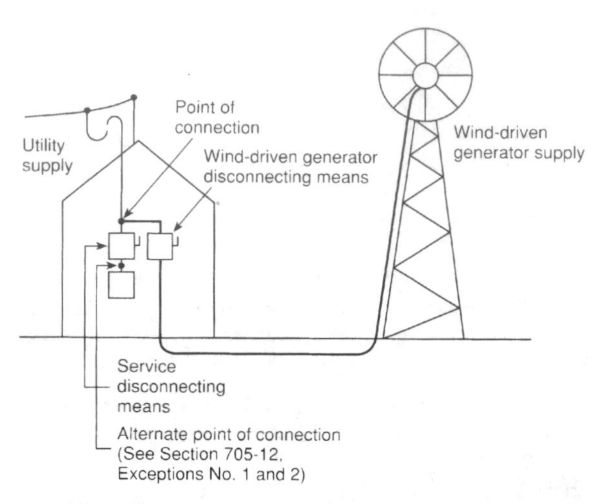

Figure 705-1. The point of interconnection is required to be at the premises service disconnecting means.

The two exceptions recognize that generators and other power sources can be safely connected elsewhere on the premises system. These locations include where the premises has an integrated electrical system as set forth in Article 685, where the total generator capacity on premises is greater than 100 kW, or where the service is greater than 1000 volts.

Experience has proved that a higher level of design input and more responsible installation are used for larger systems than are commonly used for small systems (100-kW or less, 600-volt or less services).

705-14. Output Characteristics. The output of a generator or other electric power production source operating in parallel with an electric supply system shall be compatible with the voltage, wave shape, and frequency of the system to which it is connected.

(FPN): The term compatible does not necessarily mean matching the primary source wave shape.

The level of output voltage and frequency must be controlled to permit real power and reactive power to flow in the intended amount and proper direction. Control of the driver speed causes real power (kW) to flow from a rotating generator. Control of voltage causes reactive power (kVAr) to flow to or from a synchronous generator. The parallel operation of generators is a complex balance of several variables. These are design parameters and are beyond the scope of the *Code*. A considerable amount of data is available for equipment application and design.

The output characteristics of a rotating generator are significantly different from those of a solid-state power source. Their compatibility with other sources and with different types of loads will be limited in different ways.

Where either the power source or the loads have solid-state equipment, such as inverters, uninterruptible power supplies (UPS), or solid-state variable speed drives, harmonic currents will flow in the system. See Figures 705-2, 705-3, and 705-4. These multiples of the basic supply frequency (usually 60 Hz) cause additional heating, which may require derating of generators, transformers, cables, and motors. Special generator voltage control systems are required to avoid erratic operation or destruction of control devices. Circuit breakers may require derating where the higher harmonics become significant.

Significant magnitudes of harmonics may be inadvertently matched to system resonance and result in opening of capacitor fuses, overheating of circuits, and erratic operation of controls. The usual solution is to detune the system by rearrangement or installation of reactors, or both.

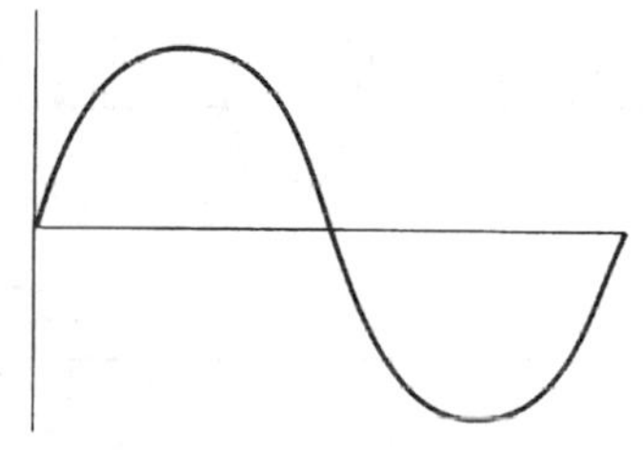

Figure 705-2. *Typical output wave shape of rotating generator and system wave shape normally encountered with motor, lighting, and heating loads.*

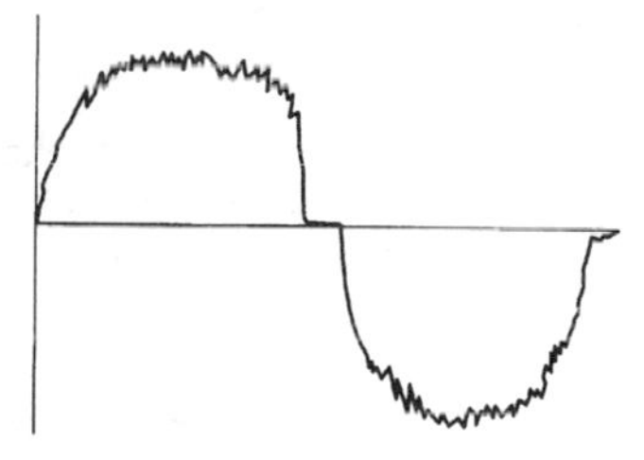

Figure 705-3. *Typical output wave shape with inverter source. Motors and transformers will be driven by harmonic-rich voltage and may require derating.*

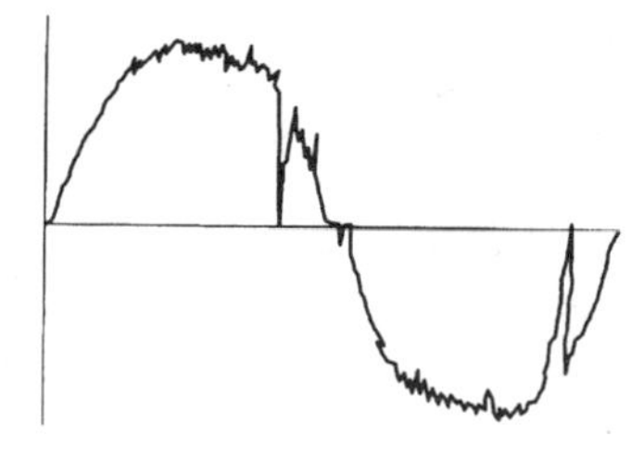

Figure 705-4. *Wave shape typical of system with variable speed drive, rectifier, elevator, and uninterruptible power supply loads. Source generator may require derating, and special voltage control may be needed.*

705-16. Interrupting and Withstand Rating. Consideration shall be given to the contribution of fault currents from all interconnected power sources for the interrupting and withstand ratings of equipment on interactive systems.

705-20. Disconnecting Means, Sources. Means shall be provided to disconnect all ungrounded conductors of an electric power production source(s) from all other conductors. See Article 230.

705-21. Disconnecting Means, Equipment. Means shall be provided to disconnect equipment, such as inverters or transformers, associated with a power production source, from all ungrounded conductors of all sources of supply.

Exception: Equipment intended to be operated and maintained as an integral part of a power production source exceeding 1000 volts.

An example of equipment covered under the exception is a generator designed for 4160 volts and connected to a 13,800-volt system through a transformer. The transformer and generator are to be operated as a unit. A disconnecting means is not required between the generator and transformer. However, a disconnecting means is required between the transformer and the point of connection to the power system.

705-22. Disconnect Device. The disconnecting means for ungrounded conductors shall consist of a manually or power operable switch(es) or circuit breaker(s):

(1) Located where accessible;

(2) Externally operable without exposing the operator to contact with live parts and, if power operable, of a type that can be opened by hand in the event of a power supply failure;

(3) Plainly indicating whether in the open or closed position; and

(4) Having ratings not less than the load to be carried and the fault current to be interrupted.

For disconnect equipment energized from both sides, a marking shall be provided to indicate that all contacts of the disconnect equipment may be energized.

(FPN No. 1): In parallel generation systems, some equipment, including knife blade switches and fuses, are likely to be energized from both directions. See Section 240-40.

(FPN No. 2): Interconnection to an off-premises primary source could require a visibly verifiable disconnecting device.

The requirements for disconnects in Section 705-22 are probably the most significant and important requirements in Article 705. A disconnecting means is to serve each generating source. This device or another will be the service-entrance disconnect. Still another may be applied to separate the generating systems.

The basic requirement in Section 705-22 recognizes the success of applying switches as well as molded-case circuit breakers in this service. Most safe work practices on the premises utilize these disconnect devices.

The disconnect at the service entrance is required for disconnecting the premises wiring system from the utility. The utility safe work practices may also utilize this disconnect device. Utility work practices may require a visibly verifiable disconnect device. For this reason, some utility contracts require that a visible break be provided. The second Fine Print Note is intended to bring attention to this common utility requirement.

705-30. Overcurrent Protection. Conductors shall be protected in accordance with Article 240. Equipment overcurrent protection shall be in accordance with the articles referenced in Article 240. Equipment and conductors connected to more than one electrical source shall have a sufficient number of overcurrent devices so located as to provide protection from all sources.

(1) Generators shall be protected in accordance with Section 445-4.

(2) Solar photovoltaic systems shall be protected in accordance with Article 690.

(3) Overcurrent protection for a transformer with a source(s) on each side shall be provided in accordance with Section 450-3 by considering first one side of the transformer, then the other side of the transformer, as the primary.

705-32. Ground-Fault Protection. Where ground-fault protection is used, the output of an interactive system shall be connected to the supply side of the ground-fault protection.

Exception: Connection shall be permitted to be made to the load side of ground-fault protection provided that there is ground-fault protection for equipment from all ground fault current sources.

The larger systems have become very sophisticated and involve many special relays. The more common relays are under- and over-voltage, under- and over-frequency, voltage-restrained overcurrent, anti-motoring, loss-of-field, over-temperature, and shutdown for derangement of the mechanical driver. Small-generator installations cannot justify the cost of these switchgear-type relays. Small-generator protection is comprised of more common devices with fewer features. Application guides are available in technical literature and manufacturers' data.

It is important to note that the requirements in Article 705 are only for protection against conditions that may occur because generating sources are being operated in parallel. It is essential that the installation also meet the requirements of the referenced articles to provide the basic protection and safeguards for all equipment, whether or not it is involved with more than one source.

705-40. Loss of Primary Source. Upon loss of primary source, an electric power production source shall be automatically disconnected from all ungrounded conductors of the primary source and shall not be reconnected until the primary source is restored.

(FPN No. 1): Risks to personnel and equipment associated with the primary source could occur if an interactive electric power production source can operate as an island. Special detection methods can be required to determine that a primary source supply system outage has occurred, and whether there should be automatic disconnection. When the primary source supply system is restored, special detection methods can be required to limit exposure of power production sources to out-of-phase reconnection.

(FPN No. 2): Induction-generating equipment on systems with significant capacitance can become self-excited upon loss of primary source and experience severe over-voltage as a result.

When two interconnected power systems separate, they will drift out of synchronism. When the utility separates, there is a risk of damage to the system if restoration of the utility occurs out of phase. If the timing of the reconnection is random, there will be violent electromechanical stresses. These can destroy mechanical components such as gears, couplings, and shafts, and can displace coils. It is essential to disconnect the premises or generator from the primary source. Many technical guides are available from which to select appropriate protective systems and equipment for large systems. A limited choice of low-cost devices is still available for application to small systems.

Induction generators are commonly used today. They have characteristics quite different from synchronous machines. They are more rugged because of the construction of the rotor. They are less expensive because of their basic design, availability, and type of starting and control equipment. Theoretically, an induction machine can continue to run on an isolated system if a large capacitor bank provides excitation. In reality, an induction machine will lose stability and be shut down quickly by one of the protective devices.

705-42. Unbalanced Interconnections. A 3-phase electric power production source shall be automatically disconnected from all ungrounded conductors of the interconnected systems when one of the phases of that source opens.

Exception: An electric power production source providing power for an emergency or legally required standby system.

705-43. Synchronous Generators. Synchronous generators in a parallel system shall be provided with the necessary equipment to establish and maintain a synchronous condition.

705-50. Grounding. Interconnected electric power production sources shall be grounded in accordance with Article 250.

Exception: For direct-current systems connected through an inverter directly to a grounded service, other methods that accomplish equivalent system protection and that utilize equipment listed and identified for the use shall be permitted.

Appendix 2

Sources of Further Information on Renewable Energy

U.S. National Renewable Energy Laboratory (NREL)

Technical Inquiry Service
1617 Cole Boulevard
Golden, CO 80401
Phone: (303) 275-4099
Fax: (303) 275-4091

NREL offers a number of outreach services, ranging from consumer information clearinghouses to electronic bulletin boards, to its various audiences. The following is a list of these informative resources for wind, solar, and all other renewable energy.

Electronic Information Server

This service provides general and technical information about NREL and may be accessed through Gopher at gopher.nrel.gov or through the Web at http://www.nrel.gov or via modem at (303) 275-INFO.

Document Distribution Service

The DDS is the central repository for technical documents published by NREL.

Phone: (303) 275-4363
Fax: (303) 275-4053
E-mail: evanss@tcplink.nrel.gov

Energy Efficiency and Renewable Energy Clearinghouse

EREC provides information and technical assistance on subjects related to energy efficiency and renewable energy technologies.

P.O. Box 3048
Merrifield, VA 22116
Phone: (800) 363-3732
TDD: (900) 273-2957
BBS: (800) 273-2955
E-mail: doe.erec@nciinc.com

Energy Efficiency and Renewable Energy Network

This electronic service on the Internet serves as a gateway and repository of multimedia information on energy efficiency and renewable energy technologies.

Phone: (900) 363-3732
Web: http://www.eren.doe.gov

Renewable Electric Plant Information System

REPIS, which is supported by DOE's Office of Utility Technologies, is an electronic database of core information on grid-connected renewable electric plants in the United States. RFPIS contains information on more than 103,000 megawatts of installed renewable capacity; approximately 89 percent of which is hydroelectric.

Phone: (303) 275-4643
Fax: (303) 275-4611
E-mail: sinclaik@tcplink.nrel.gov

Renewable Resource Data Center

RREDC is the on-line information distribution service for the Resource Assessment Program at NREL. Sponsored by DOE, RREDC provides information about a variety of renewable resources in the United States. The information includes publications, data, maps, algorithms, and links to other sources of renewable energy data.

Phone: (303) 275-4638
Web: http://rredc.nrel.gov

Technical Inquiry Service

The TIS informs the scientific, industrial, and business communities about NREL's R&D activities and those of NREL subcontractors.

Phone: (303) 275-4065
Fax: (303) 275-4091
E-mail: rubin@tcplink.nrel.gov

Technology and Business Ventures Office

Access to collaborative and sponsored research, cost-shared subcontracts, NREL facilities, visiting researcher program, and cooperative research and development agreements is available to individuals and businesses through this office.

Phone: (303) 275-3008
Fax: (303) 275-3040
E-mail: technology_transfer@nrel.gov
Web: http://www.nrel.gov

Visitors Center

The NREL visitor center provides an interactive environment in which you can explore how solar, wind, and other types of renewable resources can provide energy to your industry, farm, community, and home.

Phone: (303) 384-6566
E-mail: griegog@tcplink.nrel.gov

Solar Energy Information Sources

American Solar Energy Society
2400 Central Avenue
Boulder, CO 80301, USA
Phone: (303) 443-3130
Fax: (303) 443-3212

British Photovoltaic Association
The Warren, Bramshill Road
Eversley, Hamshire RG27 OPR
United Kingdom
Phone: +44 (0) 118 932 4418
Fax: +44 (0) 118 973 0820

Solar Trade Association Ltd
Pengillan, Lerryn, Lostwithiel
Cornwall, United Kingdom

Manufacturers of Solar Cells and Modules in the U.S.A.

Advanced Photovoltaic Systems Inc.
PO Box 7093
Princeton, NJ 08543

Applied Solar Energy Corporation
15751 East Don Julian Road
City of Industry, CA 91746

ASE Americas, Inc.
4 Suburban Park Drive
Billerica, MA 01821

AstroPower, Inc.
Solar Park
Newark, DE 19716-2000

ENTECH
PO Box 612
DFW Airport, TX 75261

EPV Energy Photovoltaics
PO Box 7456
Princeton, NJ 08543

Evergreen Solar
49 Jones Road
Waltham, MA 02154

Iowa Thin Film Technologies
ISU Research Park #607
Ames, IA 50010

SunPower Corporation
430 India Way
Sunnyvale, CA 94086

SunWatt Corporation
RFD Box 751
Addison, ME 04606

Kyocera America, Inc.
8611 Balboa Avenue
San Diego, CA 92123

A.Y. McDonald Mfg. Co.
4800 Chavenelle Road-Box 508
Dubuque, IA 52004-0508

Midway Labs, Inc.
2255 East 75th Street
Chicago, IL 60649

Sanyo Energy Corporation
2001 Sanyo Avenue
San Diego, CA 92073

Siemens Solar Industries
4650 Adohr Lane
Camarillo, CA 93011

Solarex Corporation
630 Solarex Court
Frederick, MD 21701

Solec International, Inc.
12533 Chadron Avenue
Hawthorne, CA 90250

Spire Corporation
One Patriots Park
Bedford, MA 01730

Tideland Signal Co.
PO Box 52430
Houston, TX 77052

United Solar Systems Corporation
9235 Brown Deer Road
San Diego, CA 92121-2268

Utility Power Group
9410 DeSoto Avenue, Unit G
Chatsworth, CA 91311

Wind Energy Information Sources

As wind energy technology gains more widespread acceptance as an economically and technically viable resource, an increasing number of people representing electric utilities, the news media, the research community, political and regulatory leaders, and local communities are searching for up-to-date, accurate information on this clean, renewable energy resource. The following sources can provide a range of general and technical information about wind.

American Wind Energy Association (AWEA)
122 C Street, NW, Fourth Floor
Washington, D.C. 20001
Phone: (202) 383-2500
Publications: (202) 383-2520
Fax: (202) 383-2505

Energy Efficiency & Renewable Energy Network
Phone: (800) 363-3732

This electronic service on the internet serves as a gateway and repository of multimedia information on energy efficiency and renewable energy technologies.

National Climatic Data Center
151 Patton Avenue, Room 120
Asheville, NC 28801-2733
Phone: (704) 271-4800
Fax: (704) 271-4876

National Wind Technology Center
1617 Cole Boulevard
Golden, CO 80401
Phone: (303) 384-6900
Fax: (303) 384-6901

National Technical Information Service
U.S. Department of Commerce
5285 Port Royal Road
Springfield, VA 22161
Phone: (703) 487-4650

Sandia National Laboratories
P.O. Box 5800-MS 0899
Albuquerque, NM 87185
General Phone: (505) 845-0111
Technical Info.: (505) 845 8287

National Institute of Science & Technology
Energy Inventions & Innovations
U.S. Department of Commerce
Room Al15, Building 411
Gaithersburg, MD 20899-0001
Phone: (301) 975-5500

University Wind Energy Programs in the U.S.A.

A number of universities offer courses in renewable energy technology through environmental science or engineering departments. In addition to classes, they work with industry partners to conduct valuable research. Institutions offering specific wind energy curriculum and research programs include the following:

Alternative Energy Institute
West Texas A&M University
P.O. Box 248
Canyon, TX 79016
Phone: (806) 656-2295

University of Colorado
Electrical & Computer Engineering Department
Campus Box 425
Boulder, CO 80309
Phone: (303) 492-7010

Montana State University
Chemical Engineering Department
302 Cableigh Hall
Bozeman, MT 59717
Phone: (406) 994-4543

New Mexico State University
Southwest Technology Development Institute
Dept. 3 SOLAR
P.O. Box 30001
Las Cruces, New Mexico 88003-9001
Phone: (505) 646-1846

Ohio State University
Aero and Astronautical Research
2300 West Case Road
Columbus, OH 43220
Phone: (614) 292-5491

Oregon State University
Mechanical Engineering Department
Corvallis, OR 97331
Phone: (503) 737-2218

Southwestern Technical College
401 West Street
Jackson, MN 56143
Phone: (507) 847-3320 xi7i

Tennessee State University
3500 John A. Merritt Boulevard
Nashville, TN 37209-1561
Phone: (615) 320-3268

Tennessee Technological University
Department of Mechanical Engineering
Cookeville, TN 38505-5014
Phone: (615) 372-3273

University of Arizona
AME Department
Tucson, AZ 85721
Phone: (602) 621-2235

University of Massachusetts
Renewable Energy Research Laboratory
College of Engineering
E Lab Building
Amherst, MA 01003
Phone: (413) 545-4359

University of Northern Iowa
Industrial Technology Center
Cedar Falls, IA 50614-0178
Phone: (319) 273-2756

University of Texas, El Paso
Mechanical Engineering Department
500 W. University Avenue
El Paso, TX 79968-0521
Phone: (915) 747-5450

University of Utah
Mechanical Engineering Department
Salt Lake City, UT 84112
Phone: (801) 581-4145

Periodicals on Wind Energy

Wind Energy Weekly
American Wind Energy Association
122 C Street. NW, Fourth Floor
Washington, D.C. 20001
Phone: (202) 383-2500

Published 50 times a year, Wind Energy Weekly provides up-to-date information on utility requests for proposals, federal & state regulatory activity, project development, international business opportunities, and energy and environmental policy.

Home Power Magazine
P.O. Box 520
Ashland, OR 97520
Phone: (916) 475-3179

This bimonthly magazine provides information about residential wind systems as well as information on other renewable technologies and products for residential use.

Independent Energy
620 Central Avenue, North
Milaca, MN 56353
Phone: (612) 983-6892

Published 10 times a year, this magazine provides information for the independent power industry and includes periodic updates on the wind industry.

Wind Energy Technology: Generating Power From the Wind
National Technical Information Service
U.S. Department of Commerce
5285 Port Royal Road
Springfield, VA 22161

This bimonthly publication provides current worldwide information on wind turbine design, performance, wind resource identification, legal/institutional implications, and environmental aspects of wind power.

Windpower Monthly
P.O. Box 496007, Suite 217
Redding, CA 96099
Phone: (415) 775-1985

This Danish magazine provides up-to-date information on all aspects of the international wind industry.

WindDirections
European Wind Energy Association
26 Spring Street
London W2 1JA
United Kingdom
Phone: +44 (0) 171 402 7122
Fax: +44 (0) 171 402 7125
E-mail: ewea@ewea.org
Web: http://www.ewea.org

WindDirections contains current technical and commercial information on wind power plants and technologies throughout the world.

International Wind Energy Associations

British Wind Energy Association
Lincoln's Inn House
42 Kingsway
London WC2B 6EX
UNITED KINGDOM
Phone: +44 171 404 3433
Fax: +44 171 404 3432

Canadian Wind Energy Association
2415 Holly Lane, Suite 250
Ottawa, Ontario KIV 7P2
CANADA
Phone: (613) 736-9077

Danish Wind Power Association
(Wind Turbine Owners' Organization)
Phone: +45 53 83 13 22
Fax: +45 53 83 12 02

European Wind Energy Association
26 Spring street
London W2 1JA
UNITED KINGDOM
Phone: +44 (0) 171 402 7122
Fax: +44 (0) 171 402 7125
E-mail: ewea@ewea.org
Web: http://www.ewea.org

Finnish Wind Power Association
Phone: +358 0 456 6560
Fax: +358 0 456 6538

FME Groep Windenergie
(Dutch wind energy trade association)
Phone: +31 79 353 1100
Fax: +31 79 353 1365

French Wind Energy Association
63 Collet de Darbousson
Valbonne 06560
FRANCE
Phone: +33 93 65 0970

German Wind Energy Association
Dorfangerweg 15
Unterfoehring 85774
GERMANY
Phone: +49 899506411

Gujarat Energy Development Agency
Suraj Plaza II, 2nd Floor, Sayaji Gunj
Vadodara – 390 005, INDIA
Fax: +91-265-363120

Hellenic Wind Energy Association (Greece)
Phone: +30 1 603 9900
Fax: +30 1 603 9905

Indian Renewable Energy Development Agency
Core-4 A, East Court, 1st Floor
Habitat Center Complex, Lodi Road
New Delhi-110 003, INDIA
Fax: +91-11-460-2855

Indian Ministry of Non-Conventional Energy Sources
Block 14, CGO Complex
Lodi Road
New Delhi – 110 003, INDIA
Fax: +91-11-436-1298

Institute Catalia d'Energia
Avenue Diagonal, 453 Bis. Atic
08036 Barcelona, SPAIN
Fax: +343-439-2800

Norwegian Wind Energy Association
Phone:+47 66 84 63 69
Fax: +47 66 98 11 80

Romanian Wind Energy Association
Phone: +40 1 620 67 30/260
Fax: +40 1 312 93 15

Risø National Laboratory
PO Box 49
4000 Roskilde, DENMARK
Tel: + 45 46 77 50 35
Fax: + 45 46 77 50 83
Web: www.risoe.dk/amv/index.html

Tata Energy Research Institute
Darbari Seth Block, Habitat Place
New Delhi – 110 003, INDIA
Phone: +91-11-460-1550
Fax: +91-11-462-1770
E-mail: mailbox@teri.res.in
Web: http://www.teriin.org

Vindmolleindustrien
(Association of Danish wind turbines Manufacturers)
Norre Voldegadc 48, OPGB.
Kobenhaven K DK-1358
DENMARK
Phone: +45 33 –779988

Wind Power System Suppliers in the U.S.A.

The following is a partial list of wind turbine manufacturers and developers in the U.S.A. A complete list and company profile can be obtained from:

American Wind Energy Association
122 C St., NW, Fourth Floor
Washington, D.C. 20001
Phone: (202) 383-2520
Fax: (202) 383-2505 fax
E-mail: windmail@awea.org

Atlantic Orient Corp.
P.O. Box 1097
Norwich, VT 05055
Phone: (802) 649-5446
Fax (801) 649-5404

Bay Winds
1533 Kimball Street
Green Bay, WI 54302
Phone: (920) 468-5500

Bergey Windpower Company
2001 Priestley Avenue
Norman, OK 73069
Phone: (405) 364-4212
Fax: (405) 364-2078
E-mail: mbergey@bergey.com

Lake Michigan Wind & Sun Ltd.
East 3971 Bluebird Road
Forestville, WI 54213-9505
Phone: (414) 837-2267
Fax: (414) 837-7523
E-mail: lmwands@itol.com

Southwest Windpower
2131 North First Street
Flagstaff, AZ 86001
Phone: (520) 779-WIND
Fax: (520) 779-1485
Web: http://www.windenergy.com

Wind Turbine Industries Corp.
16801 Industrial Circle, SE
Prior Lake, MN 55372
Phone: (612) 447-6064
Fax: (612) 447-6050

WindTech International, L.L.C.
P.O. Box 27
Bedford, NY 10506
Phone: (914) 232-2354
Fax: (914) 232-2356
E-mail: info@windmillpower.com
Web: http://www.windmillpower.com

World Power Technologies, Inc.
19 N. Lake Avenue
Duluth, MN 55802
Phone: (218) 722-1492
Fax: (218) 722-0791
E-mail: wpt@cp.duluth.mn.us
Web: http://www.webpage.com/wpt/

Utility-scale wind turbines manufacturers and developers

Advanced Wind Turbines
NEG Micon U.S.A., Inc.
Mitsubishi Heavy Industries
NedWind (Netherlands)
Vestas American Wind
Wind Turbine Company
Enron Wind Corporation
Cannon Energy Corporation
Energy Unlimited, Inc.
FORAS Service Company
SeaWest Energy Corporation
Tomen Power Corporation
York Research Corporation

European Wind Energy Association

List of the EWEA Members as of September 1998. Reproduced with permission from EWEA Head Office

European Wind Energy Association
26 Spring Street
London, W2 1JA, UK
Tel: +44 171 402 7122
Fax: +44 171 402 7125
E-mail: ewea@ewea.org
Web: www.ewea.org

Manufacturers and Developers

ABB Motors OY
PL 633
65101 Vaasa
Finland
Tel: + 358 10 22 4000
Fax: + 358 10 22 47372

Aerpac b.v.
PO Box 167
Bedrijvenpark Twente 96
Almelo 7600 A
The Netherlands
Tel: + 31 546 549 549
Fax: + 31 546 549 599
E-mail: aerpac@worldaccess.nl

ATV Entreprise
Actipole St Charles
13710 Fuveau
France
Tel: + 33 4 42 29 14 62
Fax: + 33 4 42 29 14 61

A/S Wincon West Wind
Hedemoelle Erhvervsvej 4
DK-8850 Bjerringbro
Denmark
Tel: + 45 86 68 1700
Fax: +45 86 68 1734

Bonus Energy A/S
Fabriksvej 4
7330 Brande
Denmark
Tel: + 45 97 18 11 22
Fax: + 45 97 18 30 86
E-mail: bonus@bonus.dk

Desarrollos Eolicos SA
Avenida de la Buhaira, 2
41018 Sevilla
Spain
Tel: + 34 95 493 7000
Fax: + 34 95 493 7017

Ecotecnia SCCL
Amistat 23 1st
08005 Barcelona
Spain
Tel: +34 93 225 7600
Fax: +34 93 221 0939
E-mail: ecotecnia@ecotecnia.com
Web: www.ecotecnia.com

Enercon GmbH
Dreekamp 5
D-26605 Aurich
Germany
Tel: +49 421 24 9920
Fax: +49 421 24 9960
E-mail: export@enercon.de
Web: www.enercon.de

KK Electronic A/S
Cypresvej 6
7400 Herning
Denmark
Tel: + 45 97 22 1033
Fax: + 45 97 21 1431
E-mail: main@kk-electronic.dk
Web: www.kk-electronic.dk

Italian Vento Power Corp. (IVPC)
Via Circumvallazione 54/H
83100 Avellino
Italy
Tel: + 39 825 781 473
Fax: + 39 825 781 472
E-mail: 104124.2261@compu-serve.com

L M Glasfiber A/S
Rolles Mollevej 1
6640 Lunderskov
Denmark
Tel: + 45 75 58 51 22
Fax: + 45 75 58 62 02
E-mail: info@lm.dk
Web: www.lm.dk

National Windpower Ltd
Riverside House, Meadowbank
Furlong Road
Bourne End, SL8 5AJ, UK
Tel: + 44 1628 532300
Fax: + 44 1628 531993
Web: www.enterprise.net/manxwind.nwp.htm

NedWind bv
Postbus 118
3910 AC Rhenen
The Netherlands
Tel: 31 317 619 004
Fax: 31 317 612 129
E-mail: wind.turbines@nedwind.nl
Web: www.nedwind.nl

NEG Micon A/S
Alsvej 21
DK-8900 Randers
Denmark
Tel: + 45 87 10 5000
Fax: + 45 87 10 5001
E-mail: mail@neg-micon.dk

NEG Micon (UK) Ltd
Taywood House, 345 Ruislip Road
Southall, Middlesex, UB1 2QX
UK
Tel: + 44 181 575 9428
Fax: + 44 181 575 8318

Nordex Balke-Durr GmbH
Svindbaek
7323 Give
Denmark
Tel: + 45 75 73 44 00
Fax: + 45 75 73 41 47
E-mail: nordex@nordex.dk
Web: www.nordex.dk

Renewable Energy Systems Ltd
Pilgrims Lodge, Holywell Hill
St Albans, Herts, AL1 1ER, UK
Tel: + 44 1727 797900
Fax: + 44 1727 797929
Web: www.res-ltd.com

Risø National Laboratory
PO Box 49
4000 Roskilde
Denmark
Tel: + 45 46 77 50 35
Fax: + 45 46 77 50 83
Web: www.risoe.dk/amv/index.html

Riva Wind Turbines
Via Emilia Ponente 72
40133 Bologna
Italy
Tel: + 39 51 413 0511
Fax: + 39 51 413 0650

SeaWest Energy Corporation
1455 Frazee Road, Suite 900
CA 9210, San Diego
USA
Tel: +1 619 293 3340
Fax: +1 619 293 3347
E-mail: seawestsd@aol.com

Stork Product Engineering
PO Box 379
1000 AJ Amsterdam
The Netherlands
Tel: + 31 205 563 444
Fax: + 31 205 563 556

Valmet Power Transmission Inc
PO Box 158
40101 Jyvaskyla
Finland
Tel: + 358 14 296 611
Fax: + 358 14 296 868

Vergnet SA
6 Rue Henri Dunant
45410 Ingre
France
Tel: + 33 2 38 227 500
Fax: + 33 2 38 227 522
E-mail: vergnet@wanadoo.fr

Vestas Wind Systems A/S
Smed Hanseen Vej 27
6940 Lem
Denmark
Tel: + 45 97 34 11 88
Fax: + 45 97 34 14 84
E-mail: vestas@vestas.dk
Web: www.vestas.dk

West S.P.A.
Alenia/WEST
Viele Maresciallo Pilsudsky 92
I-00131 Roma
Italy
Tel: + 39 6 80 77 88 33
Fax: + 39 6 80 77 88 48

Western Windpower Ltd
Stroud House, Russell Street
Stroud, Gloucestershire
GL5 3AN, UK
Tel: + 44 1453 579408
Fax: + 44 1453 766770

Zond International Ltd
Prince Consort House
27-29 Albert Embankment
London
SE1 7TJ, UK
Tel: + 44 171 793 2800
Fax: + 44 171 820 3401
E-mail: zond@compuserve.com
Web: www.zond.com

Research & Consultancy

Bond Pearce
Ballard House
West Hoo Road
Plymouth, PL1 3AE, UK
Tel: + 44 1752 266633
Fax: + 44 1752 225350

ECN Renewable Energy
Postbus 1
1755 ZG Petten
The Netherlands
Tel: + 31 224 56 41 84
Fax: + 31 224 56 32 14
Web: www.ecn.nl

Elsamprojekt A/S
Kraftvaerksvej 53
7000 Fredericia
Denmark
Tel: + 45 79 23 3333
Fax: + 45 75 56 4477
Web: www.elsamprojekt.dk

Garrad Hassan & Partners Ltd
The Coach House
Folleigh Lane, Long Ashton
Bristol, BS18 9JB, UK
Tel: +44 1275 394360
Fax: + 44 1275 394361

Banque Paribas
37 Place du Marche, St Honore
75001 Paris
France
Tel: + 33 1 42 98 77 38
Fax: + 33 1 42 98 19 89

BTM Consult A/S
I C Christensens Alle 1
6950 Ringkobing
Denmark
Tel: + 45 97 32 5299
Fax: + 45 97 32 5593
E-mail: btmcwind@post4.tele.dk
Web: www.home4.inet.tele.dk/btmc-wind

China Fulin Windpower Development Corp.
No 1, Dongbinhe Road
Youanmenwai, Fengtai District
100054 Beijing
P.R. China
Tel: + 86 10 635 304 17
Fax: + 86 10 635 304 15

Eole
BP 72
13702 La Ciotat Cedex
France
Tel: + 33 4 42 08 14 66
Fax: + 33 4 42 08 16 56
E-mail: eole@topnet.fr

ESD Ltd
Overmoor Farm
Neston
Wiltshire, SN13 9TZ, UK
Tel: + 44 1225 816804
Fax: + 44 1225 812103
Web: www.esd.co.uk

Espace Eolien Developpement
16 Rue Faidherbe
59000 Lille
France
Tel: + 33 3 20 74 04 00
Fax: + 33 3 20 74 04 07
Web: www.espace-eolien.fr

Grant Rush & Co Ltd
Preston Park Station
Clermont Road
Brighton, BN1 6SG, UK
Tel: + 44 1273 540410
Fax: + 44 1273 504028
E-mail: grant_rush@gb3.global.ib-mail.com

Institutt for Energiteknikk
PO Box 40
2007 Kjeller
Norway
Tel: + 47 63 80 6180
Fax: + 47 63 81 2905
Web: www.ife.no

La Compagnie du Vent
Horizon 21, 650 rue Louis Lepine
Le Millenaire, 34000 Montpellier
France
Tel: + 33 4 99 52 64 70
Fax: + 33 4 99 52 64 71
E-mail: cabinet.germa@wanadoo.fe
Web: perso.wanadoo.fr/cabinet.germa

LMW Renewables BV
PO Box 279
3770 AG Barneveld
The Netherlands
Tel: + 31 342 421 986
Fax: + 31 342 421 759

Mees Pierson N.V.
Camomile Court
23 Camomile Street
London, EC3A 7PP
UK
Tel: + 44 171 444 8712
Fax: + 44 171 444 8810

NRG Systems Inc.
110 Commerce Street
05461 Hinesburg, VT
USA
Tel: +1 802 482 2255
Fax: +1 802 482 2272

Shell International Renewables
Head Biomass & Wind
Shell Centre
London, SE1 7NA, UK
Tel: + 44 171 934 3386
Fax: + 44 171 934 7470

Synergy Power Corporation
20/F Wilson House
19-27 Wyndham Street, Central
Hong Kong
Tel: +852 2846 3168
Fax: +852 2810 0478
E-mail: SynergyPowerCorp@compuserve.com

Teknikgruppen AB
PO Box 21
19121 Sollentuna
Sweden
Tel: + 46 8 444 5121
Fax: + 46 8 444 5129

Tractebel Energy Engineering
Avenue Ariane 7
B-1200 Brussels
Belgium
Tel: +32 2 773 8345
Fax: +32 2 773 9700

Trillium Pakistan (PVT) Ltd
10th Floor, AWT Plaza
5 The Mall
Rawapindi-Cantt.
Pakistan
Tel: + 92 51 56 21 07
Fax: + 92 51 56 80 44
E-mail: Trillium@PakNet1.ptc.pk

Tripod Consult APS
Gladsaxe Mollevej 21
2860 Soborg
Denmark
Tel: + 45 39 66 66 22
Fax: + 45 39 66 66 99
E-mail: tripod@tripod.dk

National Associations

APPA
Paris 205
08008 Barcelona
Spain
Tel: + 34 3 414 22 77
Fax: + 34 3 209 53 07
E-mail: appa@adam.es

Austrian Wind Energy Association
IG Windkraft -Osterreich
Mariahilferstrasse 89/22
1060 Vienna
Austria
Tel: + 43 1 581 70 60
Fax: + 43 1 581 70 61
E-mail: IGW@atmedia.net

British Wind Energy Association
26 Spring Street
London
W2 1JA, UK
Tel: + 44 171 402 7102
Fax: + 44 171 402 7107
E-mail: bwea@gn.apc.org
Web: www.bwea.com

Bundesverband Windenergie
Am Michelshof 8-10
53177 Bonn
Germany
Tel: + 49 228 35 22 76
Fax: + 49 228 35 23 60
E-mail: bwe_bonn@t-online.de
Web: www.wind-energie.de

Danish Wind Turbine Manufacturers Assoc.
Vester Voldgade 106
DK-1552 Copenhagen
Denmark
Tel: + 45 33 73 0330
Fax: + 45 33 73 0333
E-mail: danish@windpower.dk
Web: www.windpower.dk

Danmarks Vindmolleforening
Egensevej 24
Postboks 45
4840 Nr Alstev
Denmark
Tel: + 45 54 43 13 22
Fax: + 45 54 43 12 02
E-mail: info@danmarks-vindmoelle-forening.dk
Web: www.danmarks-vindmoelle-forening.dk

Dutch Wind Energy Bureau
Postbus 10
6800AA Arnhem
The Netherlands
Tel: + 31 26 355 7400
Fax: + 31 26 355 7404

Finnish Wind Power Association
PO Box 846
FIN-00101 Helsinki
Finland
Tel: + 358 9 1929 4160
Fax: + 358 9 1929 4129
Web: www.fmi.fi

Fordergesellschaft Windenergie E.V.
Elbehafen
25541 Brunsbuttel
Germany
Tel: + 49 48 52 83 84 16
Fax: + 49 48 52 83 84 30
E-mail: info@egeb.de

France Energie Eolienne
Institut Aerotechnique
15 rue Marat
78210 Saint-Cyr-l'Ecole
France
Tel: + 33 1 30 45 86 01
Fax: + 33 1 30 58 02 77

Hellenic Wind Energy Association
c/o PPC, DEME
10 Navarinou Str
10680 Athens
Greece
Tel: + 30 1 36 21 465
Fax: + 30 1 36 14 709

Irish Wind Energy Association
Kellystown
Slane, County Meath
Ireland
Tel/fax: + 353 41 267 87

ISES – France
c/o ADEME
500 Route des Lucioles
05560 Valbonne
France
Tel: + 33 4 93 95 79 18
Fax: + 33 4 93 95 79 87

ISES – Italia
P.zza Bologna, 22 A/9
00162 Roma
Italy
Tel: + 39 6 44 24 9241
Fax: + 39 6 44 24 9243
E-mail: info@isesitalia.it
Web: www.isesitalia.it

Japanese Wind Energy Association
c/o Mechanical Engineering Lab.
1-2 Namiki, Tsukuba
305 Ibaraki-Ken
Japan
Tel: + 81 298 58 7014
Fax: + 81 298 58 7275

Netherlands Wind Energy Association
TU Delft, Institute for Wind
Stevinweg 1
2628 CN, Delft
The Netherlands
Tel: + 31 15 27 85 178
Fax: + 31 15 27 85 347

Romanian Wind Energy Association
Power Research Institute Bd
Energeticienilor 8
76619 Bucharest
Romania
Tel/fax: + 40 1 32 14 465
or + 40 1 32 22 790

Turkish Wind Energy Association
EiE Idaresi Genel Mudurlugu
Eskisehir Yolu 7km No. 166
06520 Ankara
Turkey
Tel: + 90 312 287 8440
Fax: + 90 312 287 8431

Acronyms

ASES	American Solar Energy Society
AWEA	American Wind Energy Association
Ah	Ampere·hour of the battery capacity
BIPV	Building Integrated Photovoltaics
C/d	Charge/discharge of the battery
DOD	Depth of Discharge of the battery from its rated capacity
DOE	Department of Energy
ECU	European Currency Unit
EDF	Energy Delivery Factor
EPRI	Electric Power Research Institute
EWEA	European Wind Energy Association
GW	Gigawatt (10^9 watts)
IEA	International Energy Agency
kW	Kilowatt
kWh	Kilowatt·hour
MW	Megawatt
MWh	Megawatt·hour
NEC	National Electric Code
NREL	National Renewable Energy Laboratory
PV	Photovoltaic
QF	Qualifying Facility
SOC	State of Charge of the battery
THD	Total Harmonic Distortion in measuring the quality of power
UCE	Unit Cost of Energy

Conversion of Units

The information contained in the book came from many sources and many countries using different units in their reports. The data are kept in the form they were received by the author rather than converting to a common system of units. The following is the conversion table for the most commonly used units in the book.

To change from	Into	Multiply by
mile per hour	meter/second	0.447
knot	meter/second	0.514
mile	kilometer	1.609
foot	meter	0.3048
inch	centimeter	2.540
pound	kilogram	0.4535
Btu	watt·second	1054.4
Btu	kilowatt·hour	$2.93 \cdot 10^{-4}$
Btu/ft^2	watt·second/m^2	11357
Btu/hr	watts	0.293
Btu/ft^2/hr	kW/meter^2	$3.15 \cdot 10^{-7}$
Btu/h·ft^2·°F	watt/m^2°C	5.678
horsepower	watts	746
gallon of oil (U.S.)	kWh	42
gallon (U.S.)	liters	3.785
barrel of oil	Btu	$6 \cdot 10^6$
barrel	gallons (U.S.)	42

Index

D

E

F

G

N

O

P

Q

R

S

T

U

V

W

Y

Z